AMERICAN MATHEMATICAL SOCIETY TRANSLATIONS

Series 2

Volume 111

9 PAPERS ON ANALYSIS

by

V. M. Adamjan
D. Z. Arov
M. L. Cetlin
I. M. Gel′fan
I. C. Gohberg
V. S. Gurfinkel
M. M. Karčevskiĭ
M. G. Kreĭn
N. Ja. Krupnik
A. D. Ljaško
Petr Mandl
D. E. Men′šov

Coden: AMTRCU

Published by the
AMERICAN MATHEMATICAL SOCIETY
Providence, Rhode Island
1978

Library of Congress Catalog Number A51-5559

International Standard Book Number 0-8218-3061-9

AMS (MOS) subject classifications (1970). Primary 15A57, 28A20, 42A20, 47B10, 47G05, 49–01, 49B10, 49B15, 65N05, 65N10, 65N30, 68A35, 92A05; Secondary 28A20, 45E05, 45L05, 47A50, 49C05, 49D05, 65L10.

Library of Congress Cataloging in Publication Data

Main entry under title:
9 papers on analysis.
(American Mathematical Society translations; ser. 2, v. 111)
Translation from Russian.
Bibliography: p.
CONTENTS: Men'šov, D. E. On limits of indeterminacy of partial sums of universal trigonometric series.—Men'šov, D. E. On limit functions of trigonometric series.—Men'sov, D. E. Universal sequences of functions. [etc.]
1. Mathematical analysis—Addresses, essays, lectures. I. Adamian, V. M. II. Series: American Mathematical Society. Translations; ser. 2, v. 111.
QA3.A572 ser. 2, vol. 111 [QA300] 510'.8s
ISBN 0-8218-3061-9 [515'.08] 78-5442

Printed in the United States of America

TABLE OF CONTENTS

Men'šov, D. E. On limits of indeterminacy of partial sums of universal trigonometric series. [Меньшов, Д. Е. О пределах неопределенности частных сумм универсальных тригонометрических рядов.] Moskov. Gos. Univ. Učen. Zap. 165, Mat. 7 (1954), 3–33 1

Men'šov, D. E. On limit functions of trigonometric series. [Меньшов, Д. Е. О предельных функциях тригонометрического ряда.] Trudy Moskov. Mat. Obšč. 7 (1958), 291–334 35

Men'šov, D. E. Universal sequences of functions. [Меньшов, Д. Е. Об универсальных последовательностях функций.] Mat. Sb. 65 (107) (1964), 272–312 81

Gohberg, I. C. and Krupnik, N. Ja. On the quotient norm of singular integral operators. [Гохберг, И. Ц. и Крупник, Н. Я. О фактор-норме сингулярных интегральных операторов.] Mat. Issled. 4 (1969), No. 3, 136–139 117

Gohberg, I. C. and Krupnik, N. Ja. On composite linear singular integral operators. [Гохберг, И. Ц. и Крупник, Н. Я. О сложных линейных сингулярных интегральных операторах.] Mat. Issled. 4 (1969), no. 4, 20–32 121

Adamjan, V. M., Arov, D. Z. and Kreĭn, M. G. Infinite Hankel block matrices and related extension problems. [Адамян, В. М., Аров, Д. З. и Крейн, М. Г. Бесконечные блочно-ганкелевы матрицы и связанные с ними проблемы продолжения.] Izv. Akad. Nauk Armjanskoĭ SSR 6, No. 2–3 1971, 87–112 133

Ljaško, A. D. and Karčevskiĭ, M. M. Study of difference schemes for nonlinear equations by a variational method. [Ляшко, А. Д и Карчевский, М. М. Исследование разностных схем для нелинейных уравнений с помощью вариационного метода.] Izv. Vysš. Učebn. Zaved. Matematika 1967, no. 3 (58), 59–65 157

Mandl, Petr. Foundations of multistage system optimization. [Mandl, Petr. Základy optimalizace vícestupňových soustav.] Kybernetika (Prague) 8 (1972), 3–52 165

Gel'fand, I. M., Gurfinkel', V. S. and Cetlin, M. L. On control tactics for complex systems and their relation to physiology. [Гельфанд, И. М., Гурфинкель, В. С. и Цетлин, М. Л. О тактиках управления сложными системами в связи с физиологией.] Biologičeskie Aspekti Kibernetiki, Izdat. Akad. Nauk SSSR, Moscow, 1962, 66–73 213

Amer. Math. Soc. Transl.
(2) Vol. 111, 1978

ON LIMITS OF INDETERMINACY OF PARTIAL SUMS OF UNIVERSAL TRIGONOMETRIC SERIES*

D. E. MEN'ŠOV

Introduction

The trigonometric series

$$\frac{a_0}{2}+\sum_{n=1}^{\infty}(a_n\cos nx+b_n\sin nx) \tag{1}$$

is said to be *universal* (see [1]) if to every measurable function $f(x)$ that is defined([1]) almost everywhere on the segment $[-\pi, \pi]$ there corresponds an increasing sequence of natural numbers n_i, $i = 1, 2, \ldots$, such that

$$\lim_{i\to\infty} S_{n_i}(x)=f(x) \tag{2}$$

almost everywhere on $[-\pi, \pi]$, where

$$S_n(x)=\frac{a_0}{2}+\sum_{k=1}^{n}(a_k\cos kx+b_k\sin kx). \tag{3}$$

We now give the definition of a universal trigonometric series with respect to a pair of functions $F_1(x)$ and $F_2(x)$.

Suppose that we are given two measurable functions $F_1(x)$ and $F_2(x)$ that are defined almost everywhere([2]) on $[-\pi, \pi]$ and satisfy the inequality

$$F_1(x)\leqslant F_2(x) \tag{4}$$

almost everywhere on this segment.

DEFINITION A. We shall call the trigonometric series (1) *universal with respect to* $F_1(x)$ *and* $F_2(x)$ if to any measurable function $f(x)$ that is defined almost everywhere on $[-\pi, \pi]$ and satisfies the inequality

$$F_1(x)\leqslant f(x)\leqslant F_2(x) \tag{5}$$

almost everywhere on this segment there corresponds an increasing sequence of natural numbers n_i, $i = 1, 2, \ldots$, such that (2) holds almost everywhere on $[-\pi, \pi]$.

We shall call a trigonometric series that is universal with respect to $F_1(x)$ and $F_2(x)$ a *series of type* $A(F_1, F_2)$. If $F_1(x) = -\infty$ and $F_2(x) = +\infty$ for all x on $[-\pi, \pi]$, it is clear that a series of type $A(F_1, F_2)$ is a universal trigonometric series in the original sense.

AMS (MOS) subject classifications (1970). Primary 42A20.
*Translation of Moskov. Gos. Univ. Učen. Zap. **165**, Mat. 7 (1954), 3–33. MR **16**, 467.

([1]) $f(x)$ may be $+\infty$ or $-\infty$ on a set of positive measure.

([2]) $F_1(x)$ and $F_2(x)$ may assume infinite values on sets of positive measure.

In this paper we shall study the limits of indeterminacy in measure and the limits of indeterminacy in the usual sense of universal trigonometric series. First we shall recall the definition of the limits of indeterminacy in measure.

Let us take a sequence of measurable functions

$$f_1(x),\ f_2(x),\ \ldots,\ f_k(x),\ \ldots, \tag{6}$$

that are defined almost everywhere on some segment $[a, b]$. We shall say that $F(x)$, which is defined almost everywhere(3) on $[a, b]$, is an *upper limit in measure on* $[a, b]$ *of the sequence* (6) if $F(x)$ is measurable and satisfies the following conditions:

a)

$$\lim_{k\to\infty} \operatorname{meas}\{E[f_k(x) > \varphi(x)] \cdot E[\varphi(x) > F(x)]\} = 0 \tag{7}$$

for any measurable function $\varphi(x)$ defined almost everywhere on $[a, b]$;

b)

$$\overline{\lim_{k\to\infty}} \operatorname{meas}\{E[f_k(x) > \chi(x)] \cdot E[F(x) > \chi(x)]\} > 0 \tag{8}$$

for any measurable function $\chi(x)$ that is defined almost everywhere on $[a, b]$ and is such that meas $E[F(x) > \chi(x)] > 0$(4).

We shall say that a function $G(x)$ that is defined almost everywhere on $[a, b]$ is a *lower limit in measure on* $[a, b]$ *of the sequence* (6) if $G(x)$ is measurable and satisfies the following conditions:

α)

$$\lim_{k\to\infty} \operatorname{meas}\{E[f_k(x) < \varphi(x)] \cdot E[\varphi(x) < G(x)]\} = 0 \tag{9}$$

for any measurable function $\varphi(x)$ defined almost everywhere on $[a, b]$;

β)

$$\overline{\lim_{k\to\infty}} \operatorname{meas}\{E[f_k(x) < \chi(x)] \cdot E[G(x) < \chi(x)]\} > 0 \tag{10}$$

for any measurable function $\chi(x)$ that is defined almost everywhere on $[a, b]$ and is such that meas $E[G(x) < \chi(x)] > 0$.

It can be proved that the upper and lower limits in measure are well defined up to a set of measure zero and have many properties that are similar to the corresponding properties of the usual limits superior and inferior(5).

We now give the following definition.

DEFINITION B. Let the functions $F_1(x)$ and $F_2(x)$ satisfy the previous conditions. We shall call the trigonometric series (1) a *series of type* $B(F_1, F_2)$ if to any four measurable functions $\varphi_1(x)$, $\psi_1(x)$, $\psi_2(x)$ and $\varphi_2(x)$ that are defined almost

(3) $f_k(x)$ and $F(x)$ may assume infinite values on a set of positive measure.

(4) See [2], §§2 and 3. If $\varphi_1(x)$ and $\varphi_2(x)$ are any two functions that are defined almost everywhere on $[a, b]$, then we shall as usual denote by $E[\varphi_1(x) > \varphi_2(x)]$ the set of all points on $[a, b]$ for which $\varphi_1(x) > \varphi_2(x)$.

(5) In [2] we assumed that the $f_k(x)$, $k = 1, 2, \ldots$, are finite almost everywhere on $[a, b]$; but the definition of the limits superior and inferior make sense even without these restrictions. The proofs of Theorems A–E in [2], §2, are also valid without these restrictions, and their proofs may remain without change.

everywhere on $[-\pi, \pi]$ and satisfy

$$F_1(x) \leqslant \varphi_1(x) \leqslant \psi_1(x) \leqslant \psi_2(x) \leqslant \varphi_2(x) \leqslant F_2(x) \tag{11}$$

almost everywhere on this segment there corresponds an increasing sequence of natural numbers n_i, $i = 1, 2, \ldots$, such that $\psi_1(x)$ and $\psi_2(x)$ are the limits inferior and superior respectively in measure on $[-\pi, \pi]$ of the sequence of functions $S_{n_i}(x)$, $i = 1, 2, \ldots$, and in addition we have

$$\underline{\lim}_{i\to\infty} S_{n_i}(x) = \varphi_1(x), \quad \overline{\lim}_{i\to\infty} S_{n_i} = \varphi_2(x) \tag{12}$$

almost everywhere on this segment.

If $F_1(x) = -\infty$ and $F_2(x) = +\infty$, then we shall call a trigonometric series of type $B(F_1, F_2)$ a *series of type B.*

We obtain the definition of a series of type B directly if in Definition B we replace the string of inequalities (11) by

$$\varphi_1(x) \leqslant \psi_1(x) \leqslant \psi_2(x) \leqslant \varphi_2(x). \tag{13}$$

We shall prove the following theorem.

THEOREM I. *For any two measurable functions $F_1(x)$ and $F_2(x)$ that are defined almost everywhere on $[-\pi, \pi]$ and satisfy* (4) *almost everywhere on this segment the classes of trigonometric series of type $A(F_1, F_2)$ and of type $B(F_1, F_2)$ coincide.*

In particular, we have the following theorem.

THEOREM II. *The class of universal trigonometric series coincides with the class of series of type B.*

It is clear that any trigonometric series of type $B(F_1, F_2)$ is a series of type $A(F_1, F_2)$. Thus for the proof of Theorem I it is sufficient to prove the following theorem

THEOREM III. *If the functions $F_1(x)$ and $F_2(x)$ satisfy the conditions given above, then any trigonometric series of type $A(F_1, F_2)$ is at the same time a series of type $B(F_1, F_2)$.*

§1. Preliminary lemmas

For the proof of Theorem III we shall need several definitions and lemmas.

We take two sequences of numbers

$$U_1, U_2, \ldots, U_m, \ldots \tag{1.1}$$

and

$$V_1, V_2, \ldots, V_m, \ldots, \tag{1.2}$$

under the assumption that each of the numbers U_m and V_m is either a finite quantity or is $+\infty$ or $-\infty$.

DEFINITION 1. We shall say that the sequences (1.1) and (1.2) are *equiconvergent* if the following conditions hold:

a) $U_m - V_m$ tends to zero as m increases without bound in such a way that it takes values for which both U_m and V_m are finite.

b) U_m and V_m tend to $+\infty$ as m increases without bound in such a way that for each value of m at least one of U_m or V_m is $+\infty$.

c) U_m and V_m tend to $-\infty$ as m increases without bound in such a way that for each value of m at least one of U_m or V_m is $-\infty$.(6)

Let us suppose now that the two sequences of functions

$$f_1(x),\ f_2(x),\ \ldots,\ f_m(x),\ \ldots \tag{1.3}$$

and

$$g_1(x),\ g_2(x),\ \ldots,\ g_m(x),\ \ldots \tag{1.4}$$

are defined almost everywhere(7) on some set E. We shall say that (1.3) and (1.4) are *equiconvergent almost everywhere* on E if they are equiconvergent at each point of this set except possibly for a set of measure zero.(8)

It is easy to prove the following assertions.

LEMMA A. *If the sequences* (1.1) *and* (1.2) *are equiconvergent and if for some nondecreasing sequence of natural numbers* p_k, $k = 1, 2, \ldots$, *that satisfy the condition* $\lim_{k\to\infty} p_k = \infty$ *one of the two subsequences* U_{p_k} *or* V_{p_k}, $k = 1, 2, \ldots$, *converges to a definite limit L, then the other of these sequences converges to the same limit.*(9)

LEMMA B. *If the sequences* (1.1) *and* (1.2) *are equiconvergent, then the sets of their limits points coincide.*

LEMMA C. *If the sequences* (1.1) *and* (1.2) *are equiconvergent, then*

$$\varliminf_{m\to\infty} U_m = \varliminf_{m\to\infty} V_m,$$

$$\varlimsup_{m\to\infty} U_m = \varlimsup_{m\to\infty} V_m.$$

LEMMA D. *Consider a sequence of numbers*

$$W_1,\ W_2,\ \ldots,\ W_m,\ \ldots \tag{1.5}$$

and suppose that the sequences (1.1) *and* (1.5) *are equiconvergent and that the sequences* (1.2) *and* (1.5) *are equiconvergent. Then the sequences* (1.1) *and* (1.2) *are equiconvergent.*

DEFINITION 2. Let the functions $f_m(x)$ and $g_m(x)$, $m = 1, 2, \ldots$, be defined at all points of some set M. We shall say that the sequences (1.3) and (1.4) are *uniformly equiconvergent on M* if for any positive number ϵ it is possible to find a natural number N such that for all $m > N$ and for all $x \in M$ we have the following conditions:

(6)When we say that U_m tends to $+\infty$ or to $-\infty$ as $m \longrightarrow \infty$ we do not exclude the possibility that $U_m = +\infty$ or $V_m = -\infty$ for some values of m.

(7)$f_m(x)$ and $g_m(x)$ may be equal to $+\infty$ or $-\infty$ on sets of positive measure.

(8)The definition of sequences of functions that are equiconvergent almost everywhere on some segment $[a, b]$ was given in [2] (p. 27; transl. p. 216).

(9)The number L may be equal to infinity with a definite sign.

α)
$$|f_m(x)-g_m(x)|<\varepsilon,$$

when both $f_m(x)$ and $g_m(x)$ are finite;

β)
$$f_m(x)>\frac{1}{\varepsilon},\quad g_m(x)>\frac{1}{\varepsilon},$$

when at least one of the functions $f_m(x)$ or $g_m(x)$ is equal to $+\infty$;

γ)
$$f_m(x)<-\frac{1}{\varepsilon},\quad g_m(x)<-\frac{1}{\varepsilon},$$

when at least one of the functions $f_m(x)$ or $g_m(x)$ is equal to $-\infty$.

LEMMA E. *Consider functions $f_m(x)$ and $g_m(x)$, $m = 1, 2, \ldots$, that are defined almost everywhere on some measurable set E and are measurable on this set, and suppose that* (1.3) *and* (1.4) *are equiconvergent almost everywhere on E.*

Then for any positive number σ it is possible to define a measurable set M that satisfies the conditions

$$M\subset E,\quad \operatorname{meas} M>\operatorname{meas} E-\sigma \tag{1.6}$$

and is such that (1.3) *and* (1.4) *are uniformly equiconvergent on M.*([10])

PROOF. Let us denote by E' the set of all points $x\in E$ at each of which all the functions $f_m(x)$ and $g_m(x)$, $m = 1, 2, \ldots$, are defined with (1.3) and (1.4) equiconvergent. Then by the hypotheses of the lemma

$$E'\subset E,\quad \operatorname{meas} E'=\operatorname{meas} E. \tag{1.7}$$

We define functions $\Delta_m(x)$, $m = 1, 2, \ldots$, on E' using the equalities

I)
$$\Delta_m(x)=f_m(x)-g_m(x), \tag{1.8}$$

if

$$-\infty<f_m(x)<+\infty,\quad -\infty<g_m(x)<+\infty. \tag{1.9}$$

II)
$$\Delta_m(x)=\frac{1}{f_m(x)}, \tag{1.10}$$

if

$$0<f_m(x)<+\infty,\quad g_m(x)=+\infty \tag{1.11}$$

or

$$-\infty<f_m(x)<0,\quad g_m(x)=-\infty. \tag{1.12}$$

III)
$$\Delta_m(x)=\frac{1}{g_m(x)}, \tag{1.13}$$

if

$$f_m(x)=+\infty,\quad 0<g_m(x)<+\infty \tag{1.14}$$

([10])Lemma E is of interest only in the case meas $E > 0$, but we do not need to introduce this condition into its statement.

or

$$f_m(x) = -\infty, \quad -\infty < g_m(x) < 0. \tag{1.15}$$

IV)

$$\Delta_m(x) = 0, \tag{1.16}$$

if

$$f_m(x) = +\infty, \quad g_m(x) = +\infty \tag{1.17}$$

or

$$f_m(x) = -\infty, \quad g_m(x) = -\infty. \tag{1.18}$$

V)

$$\Delta_m(x) = m, \tag{1.19}$$

if none of the conditions (1.9), (1.11), (1.12), (1.14), (1.15), (1.17) or (1.18) hold.

In the latter case one of the following four conditions must hold:

$$f_m(x) = -\infty, \quad 0 \leqslant g_m(x) \leqslant +\infty, \tag{1.20}$$

$$-\infty < f_m(x) \leqslant 0, \quad g_m(x) = +\infty, \tag{1.21}$$

$$0 \leqslant f_m(x) \leqslant +\infty, \quad g_m(x) = -\infty \tag{1.22}$$

or

$$f_m(x) = +\infty, \quad -\infty < g_m(x) \leqslant 0. \tag{1.23}$$

By the definition of E', at each of its points (1.3) and (1.4) are equiconvergent, and hence, by Lemma A, at each point $x \in E'$ none of the conditions (1.20), (1.21), (1.22) or (1.23) may hold for infinitely many values of m. It follows from this that for each point x of the set E' it is possible to define a natural number m_x such that for any $m > m_x$ one of the conditions (1.9), (1.11), (1.12), (1.14), (1.15), (1.17) or (1.18) must hold, and consequently one of the equalities (1.8), (1.10), (1.13) or (1.16) holds.

We shall prove that

$$\lim_{m \to \infty} \Delta_m(x) = 0 \tag{1.24}$$

for any point x of E'.

Indeed, as we already know, (1.3) and (1.4) are equiconvergent at each point of E'. In this case, if m increases without bound while taking values for which (1.9) is satisfied, then by (1.8) we have $\Delta_m(x) \to 0$.(11)

Now, if m increases without bound while taking values for which (1.11) or (1.12) holds, then $f_m(x) \to +\infty$ or $f_m(x) \to -\infty$,(12) and consequently, by (1.10), $\Delta_m(x)$ tends to zero. For the same reason $\Delta_m(x) \to 0$ as m increases without bound while taking values for which (1.14) or (1.15) holds. Finally, by (1.16), $\Delta_m(x) \to 0$ as m increases without bound while taking values for which (1.17) or (1.18) holds.

(11)See condition a) in Definition 1 of equiconvergent sequences.

(12)See conditions b) and c) in the definition of equiconvergent sequences.

As we have already seen, for $x \in E'$ and for $m > m_x$, one of (1.9), (1.11), (1.12), (1.14), (1.15), (1.17) or (1.18) must necessarily hold. Consequently, (1.24) holds for every x in E'. In this case we may with the use of (1.7) define for any $\sigma > 0$ a measurable set M that satisfies conditions (1.6) and on which $\Delta_m(x)$ tends uniformly to zero.

We shall prove that (1.3) and (1.4) are uniformly equiconvergent on M. Indeed, let us take an arbitrary positive number ϵ. Since the sequence of functions $\Delta_m(x)$, $m = 1, 2, \ldots$, converges uniformly to zero on M, we may choose a natural number N such that

$$|\Delta_m(x)| < \varepsilon \qquad (m > N, \ x \in M). \tag{1.25}$$

Without loss of generality we may suppose that $\epsilon < 1$, from which it follows that

$$|\Delta_m(x)| < m \qquad (m > N, \ x \in M),$$

and then, by V (see (1.19)), for $m > N$ and for $x \in M$ one of (1.9), (1.11), (1.12), (1.14), (1.15), (1.17) or (1.18) must hold. If (1.9) holds, i.e. if both $f_m(x)$ and $g_m(x)$ are finite for the values of m and x considered, then by (1.8) and (1.25)

$$|f_m(x) - g_m(x)| < \varepsilon. \tag{1.26}$$

Further, if for the values of m and x considered at least one of $f_m(x)$ or $g_m(x)$ is equal to $+\infty$, then (1.11), (1.14) or (1.17) must hold,(13) and consequently with the use of (1.10), (1.13) and (1.25) we obtain

$$0 < \frac{1}{f_m(x)} < \varepsilon, \quad g_m(x) = +\infty,$$

or

$$f_m(x) = +\infty, \quad 0 < \frac{1}{g_m(x)} < \varepsilon,$$

or

$$f_m(x) = +\infty, \quad g_m(x) = +\infty,$$

from which we conclude that

$$f_m(x) > \frac{1}{\varepsilon}, \quad g_m(x) > \frac{1}{\varepsilon}. \tag{1.27}$$

Finally, if for the values of m and x considered at least one of $f_m(x)$ or $g_m(x)$ is equal to $-\infty$, then (1.12), (1.15) or (1.18) must hold, and consequently by (1.10), (1.13) and (1.25) we shall have

$$-\varepsilon < \frac{1}{f_m(x)} < 0, \quad g_m(x) = -\infty,$$

---

(13)When $f_m(x) = +\infty$ or $g_m(x) = +\infty$, none of conditions (1.9), (1.12), (1.15) or (1.18) may hold.

or

$$f_m(x) = -\infty, \quad -\varepsilon < \frac{1}{g_m(x)} < 0,$$

or

$$f_m(x) = -\infty, \quad g_m(x) = -\infty,$$

from which we conclude that

$$f_m(x) < -\frac{1}{\varepsilon}, \quad g_m(x) < -\frac{1}{\varepsilon}. \tag{1.28}$$

Thus for any positive number $\epsilon < 1$ we may choose a natural number N such that for all $x \in M$ one of (1.26), (1.27) or (1.28) will hold, depending on whether both $f_m(x)$ and $g_m(x)$ are finite, or one is equal to $+\infty$ or one is equal to $-\infty$. But this means that (1.3) and (1.4) are uniformly equiconvergent on M (see Definition 2).

Since the measurable set M satisfies (1.6), where σ may be any positive number, Lemma E is proved.

LEMMA F. *Consider functions $\gamma(x)$, $f_m(x)$ and $g_m(x)$, $m = 1, 2, \ldots$, that are defined almost everywhere on some segment $[a, b]$ and are measurable on this segment, and suppose that* (1.3) *and* (1.4) *are equiconvergent almost everywhere on $[a, b]$. Define the functions $\gamma_\alpha(x)$ and $\gamma_{-\alpha}(x)$ by the conditions*

$$\gamma_\alpha(x) = \gamma(x) + \alpha, \quad \gamma_{-\alpha}(x) = \gamma(x) - \alpha, \tag{1.29}$$

if $\gamma(x)$ is finite,

$$\gamma_\alpha(x) = \gamma(x), \quad \gamma_{-\alpha}(x) = \frac{1}{\alpha}, \tag{1.30}$$

if $\gamma(x) = +\infty$, and

$$\gamma_\alpha(x) = -\frac{1}{\alpha}, \quad \gamma_{-\alpha}(x) = \gamma(x), \tag{1.31}$$

if $\gamma(x) = -\infty$, where α is an arbitrary positive number.

Then for any positive number ϵ it is possible to choose a natural number N such that

$$\left.\begin{aligned} &E[f_m(x) > \gamma_\varepsilon(x)] - e \subset E[g_m(x) > \gamma(x)], \\ &\{E[g_m(x) > \gamma(x)] \cdot E[\gamma(x) > -\infty]\} \subset E[f_m(x) > \gamma_{-\varepsilon}(x)] + e \end{aligned}\right\} \tag{1.32}$$

for all $m > N$, where e does not depend on m and

$$\text{meas}\, e < \varepsilon. \tag{1.33}$$

PROOF. We denote by M the set of all points on $[a, b]$ at which all the functions $\gamma(x)$, $f_m(x)$ and $g_m(x)$, $m = 1, 2, \ldots$, are defined. Then

$$M \subset [a, b], \quad \text{meas}\, M = b - a. \tag{1.34}$$

We set

$$\left.\begin{array}{c} E_1=\{M\cdot E[-\infty<\gamma(x)<+\infty]\}, \\ E_2=\{M\cdot E[\gamma(x)=+\infty]\}, \quad E_3=\{M\cdot E[\gamma(x)=-\infty]\}, \end{array}\right\} \tag{1.35}$$

from which it follows that

$$M=E_1+E_2+E_3, \tag{1.36}$$

$$(E_1\cdot E_2)=0, \quad (E_1\cdot E_3)=0, \quad (E_2\cdot E_3)=0 \tag{1.37}$$

and moreover, on the basis of (1.29), (1.30), (1.31) and (1.35),

$$\gamma_\varepsilon(x)=\gamma(x)+\varepsilon, \quad \gamma_{-\varepsilon}(x)=\gamma(x)-\varepsilon \qquad (x\in E_1), \tag{1.38}$$

$$\gamma_\varepsilon(x)=+\infty, \qquad \gamma_{-\varepsilon}(x)=\frac{1}{\varepsilon} \qquad (x\in E_2), \tag{1.39}$$

$$\gamma_\varepsilon(x)=-\frac{1}{\varepsilon}, \qquad \gamma_{-\varepsilon}(x)=-\infty \qquad (x\in E_3), \tag{1.40}$$

where ϵ is some fixed positive number.

Since $\gamma(x)$ is finite at each point of E_1, we may define a measurable set Ω and a positive number ϵ' that will satisfy the conditions

$$\Omega\subset E_1, \quad \text{meas}\,\Omega>\text{meas}\,E_1-\frac{\varepsilon}{2}, \tag{1.41}$$

$$\varepsilon'<\varepsilon, \quad \gamma(x)+\varepsilon<\frac{1}{\varepsilon'}, \quad \gamma(x)-\varepsilon>-\frac{1}{\varepsilon'} \qquad (x\in\Omega), \tag{1.42}$$

from which it follows by (1.38) that

$$-\frac{1}{\varepsilon'}<\gamma_{-\varepsilon}(x)<\gamma(x)<\gamma_\varepsilon(x)<\frac{1}{\varepsilon'} \qquad (x\in\Omega). \tag{1.43}$$

Moreover, taking into consideration (1.35), (1.39) and (1.40), we obtain

$$+\infty=\gamma(x)=\gamma_\varepsilon(x)>\gamma_{-\varepsilon}(x)=\frac{1}{\varepsilon} \qquad (x\in E_2), \tag{1.44}$$

$$-\infty=\gamma(x)=\gamma_{-\varepsilon}(x)<\gamma_\varepsilon(x)=-\frac{1}{\varepsilon} \qquad (x\in E_3). \tag{1.45}$$

Setting

$$\Omega'=\Omega+E_2+E_3, \tag{1.46}$$

we have, on the basis of (1.34), (1.36) and (1.41),

$$\Omega'\subset M, \quad \text{meas}\,\Omega'>b-a-\frac{\varepsilon}{2}. \tag{1.47}$$

By the hypotheses of the lemma, (1.3) and (1.4) are equiconvergent almost everywhere on $[a, b]$, and consequently by (1.34) and (1.47) they are equiconvergent almost everywhere on Ω'. Then by Lemma E we may define a measurable set Ω'' that satisfies the conditions

$$\Omega'' \subset \Omega', \quad \text{meas}\, \Omega'' > \text{meas}\, \Omega' - \frac{\varepsilon}{2} \tag{1.48}$$

and on which (1.3) and (1.4) are *uniformly* equiconvergent.

Putting together (1.34), (1.47) and (1.48), we obtain

$$\Omega'' \subset [a, b], \quad \text{meas}\, \Omega'' > b - a - \varepsilon, \tag{1.49}$$

where Ω'' does not depend on m.

Since (1.3) and (1.4) are uniformly equiconvergent on Ω'', we may by Definition 2 define a natural number N such that the following conditions are satisfied:

α). $$|f_m(x) - g_m(x)| < \varepsilon' \qquad (m > N,\ x \in \Omega''), \tag{1.50}$$

if $f_m(x)$ and $g_m(x)$ are finite,

β). $$f_m(x) > \frac{1}{\varepsilon'}, \quad g_m(x) > \frac{1}{\varepsilon'} \quad (m > N,\ x \in \Omega''), \tag{1.51}$$

if $f_m(x)$ or $g_m(x)$ is equal to $+\infty$, and

γ). $$f_m(x) < -\frac{1}{\varepsilon'}, \quad g_m(x) < -\frac{1}{\varepsilon'} \quad (m > N,\ x \in \Omega''), \tag{1.52}$$

if $f_m(x)$ or $g_m(x)$ is equal to $-\infty$.

Under these conditions ϵ' is a positive number satisfying (1.42), which was defined above.

Setting

$$\left.\begin{aligned} H'_m &= \{\Omega \cdot \Omega'' \cdot E\,[f_m(x) > \gamma_\varepsilon(x)]\}, \\ H''_m &= \{E_2 \cdot \Omega'' \cdot E\,[f_m(x) > \gamma_\varepsilon(x)]\}, \\ H'''_m &= \{E_3 \cdot \Omega'' \cdot E\,[f_m(x) > \gamma_\varepsilon(x)]\} \\ &(m = 1, 2, \ldots) \end{aligned}\right\} \tag{1.53}$$

and taking (1.46) and (1.48) into consideration, we have

$$\left.\begin{aligned} \{\Omega'' \cdot E\,[f_m(x) > \gamma_\varepsilon(x)]\} = H'_m + H''_m + H'''_m \\ (m = 1, 2, \ldots). \end{aligned}\right\} \tag{1.54}$$

Moreover, it is clear that

$$\left.\begin{aligned} H'_m \subset [a, b],\ H''_m \subset [a, b],\ H'''_m \subset [a, b] \\ (m = 1, 2, \ldots). \end{aligned}\right\} \tag{1.55}$$

Let us suppose now that for some $m > N$

$$x \in H'_m. \tag{1.56}$$

Then from (1.41) and the first equality in (1.53) we have

$$x \in \Omega, \ x \in E_1, \ x \in \Omega'' \tag{1.57}$$

and consequently, by (1.43),

$$-\frac{1}{\varepsilon'} < \gamma_\varepsilon(x). \tag{1.58}$$

Moreover, from the first equality of (1.35) and the second relation in (1.57) we see that $\gamma(x)$ is finite at the point x being considered.

We prove now that, for the values of x and m we are considering,

$$g_m(x) > -\infty. \tag{1.59}$$

Indeed, if $g_m(x) = -\infty$ were to hold for these values of x and m, then it would follow from condition γ) in (1.52), the inequality $m > N$ and the relation $x \in \Omega''$ (see (1.57)) that $f_m(x) < -1/\epsilon'$, from which, using (1.58), we have $f_m(x) < \gamma_\epsilon(x)$, and consequently $x \notin H'_m$ by the first equality of (1.53), which contradicts (1.56). Thus (1.59) holds for the values of x and m being considered.

Let us suppose now that for these values of x and m the function $g_m(x)$ is finite. Then if $f_m(x)$ is also finite, on the basis of condition α) (see (1.50)), the third relation in (1.57) and the inequality $m > N$ we have

$$g_m(x) > f_m(x) - \varepsilon', \tag{1.60}$$

from which, taking into account (1.56), the first equality of (1.53) and the inequality $\epsilon' < \epsilon$ (see (1.42)), we obtain

$$g_m(x) > \gamma_\varepsilon(x) - \varepsilon,$$

and consequently, by the second relation in (1.57) and the first equality in (1.38),

$$g_m(x) > \gamma(x). \tag{1.61}$$

Thus, (1.61) holds if both $f_m(x)$ and $g_m(x)$ are finite for the x and m being considered.

Let us suppose now that for the values of x and m being considered $g_m(x)$ is finite and $f_m(x) = +\infty$. Then from condition β) (see (1.51)), the third relation in (1.57) and the inequality $m > N$ it follows that

$$g_m(x) > \frac{1}{\varepsilon'}, \tag{1.62}$$

from which, taking into account the first relation in (1.57) and the second inequality in (1.42), we once again obtain (1.61). Since $f_m(x)$ may not be equal to $-\infty$ if (1.56) holds (see the first equality in (1.53)), from (1.56) for some x and some $m > N$ inequality (1.61) follows, if $g_m(x)$ is finite.

Let us now treat the case in which for the values of x and m being considered we have $g_m(x) = +\infty$.

We have already seen that for the value of x under consideration $\gamma(x)$ is finite, and in this case we once again obtain (1.61). Thus, if x satisfies (1.56) for any $m > N$, then in all possible cases (1.61) holds.[14]

---

[14] We recall that for the values of x and m being considered the equality $g_m(x) = -\infty$ cannot hold (see (1.59)).

Since $H'_m \subset [a, b], m = 1, 2, \ldots$ (see (1.55)), it follows from this that

$$H'_m \subset E[g_m(x) > \gamma(x)]. \tag{1.63}$$

Let us suppose now that for some $m > N$

$$x \in H'''_m. \tag{1.64}$$

Then, on the basis of the third equality in (1.53),

$$x \in E_3, \quad x \in \Omega'' \tag{1.65}$$

and consequently, by (1.45) and the inequality $\epsilon' < \epsilon$ (see (1.42)),

$$\gamma_\varepsilon(x) = -\frac{1}{\varepsilon} > -\frac{1}{\varepsilon'}. \tag{1.66}$$

Moreover, we see from the third equality in (1.35) and from the first relation in (1.65) that at the point x under consideration

$$\gamma(x) = -\infty. \tag{1.67}$$

We shall prove that for the values of x and m being considered (1.59) holds. Indeed, if $g_m(x) = -\infty$ were to hold for these values of x and m, then from condition γ) (see (1.52)) and the second relation in (1.65) it would follow that $f_m(x) < -1/\epsilon'$, and hence, on the basis of (1.66) and the third equality in (1.53),

$$f_m(x) < \gamma_\varepsilon(x), \quad x \notin H'''_m,$$

which contradicts (1.64). Thus, (1.59) holds for the values of x and m under consideration.

Putting together (1.59) and (1.67) we see that for the values of x and m under consideration (1.61) holds, i.e. this inequality holds for all $m > N$ and for all x that satisfy (1.64). Since $H'''_m \subset [a, b]$ (see (1.55)), we have

$$H'''_m \subset E[g_m(x) > \gamma(x)], \quad (m > N). \tag{1.68}$$

If $x \in E_2$, then $\gamma_\epsilon(x) = +\infty$ by (1.39), and consequently the sets E_2 and $E[f_m(x) > \gamma_\epsilon(x)]$ do not have common points for any natural number m; and then, by the second equality in (1.53),

$$H''_m = 0 \qquad (m = 1, 2, \ldots). \tag{1.69}$$

Putting together (1.54), (1.63), (1.68) and (1.69), we obtain

$$\left.\begin{array}{c}\{\Omega'' \cdot E[f_m(x) > \gamma_\varepsilon(x)]\} \subset E[g_m(x) > \gamma(x)] \\ (m > N).\end{array}\right\} \tag{1.70}$$

We set

$$\left.\begin{array}{l} G'_m = \{\Omega \cdot \Omega'' \cdot E[g_m(x) > \gamma(x)]\}, \\ G''_m = \{E_2 \cdot \Omega'' \cdot E[g_m(x) > \gamma(x)]\}, \\ G'''_m = \{E_3 \cdot \Omega'' \cdot E[g_m(x) > \gamma(x)]\} \\ \qquad (m = 1, 2, \ldots), \end{array}\right\} \tag{1.71}$$

from which, taking (1.46) and (1.48) into account, we have

$$\left.\begin{array}{l} \{\Omega'' \cdot E[g_m(x) > \gamma(x)]\} = G'_m + G''_m + G'''_m, \\ \qquad (m = 1, 2, \ldots). \end{array}\right\} \tag{1.72}$$

Moreover, it is clear that

$$\left.\begin{array}{l} G'_m \subset [a, b], \; G''_m \subset [a, b], \; G'''_m \subset [a, b], \\ \qquad (m = 1, 2, \ldots). \end{array}\right\} \tag{1.73}$$

Let us suppose now that for some $m > N$

$$x \in G'_m. \tag{1.74}$$

Then, by (1.41) and the first equality of (1.71), relations (1.57) will hold, and consequently, by (1.43), inequality (1.58) will hold. Moreover, from the first equality of (1.35) and the second relation of (1.57) we see that $\gamma(x)$ is finite at the point x under consideration.

We prove now that for the values of x and m being considered we have

$$f_m(x) > -\infty. \tag{1.75}$$

Indeed, if $f_m(x) = -\infty$ were to hold for these values of x and m, then by condition γ) (see (1.52)), the inequality $m > N$ and the relation $x \in \Omega''$ (see (1.57)) it would follow that $g_m(x) < -1/\epsilon'$, and hence, by (1.43) and the first relation in (1.57), $g_m(x) < \gamma(\epsilon)$; consequently, by the first equality in (1.71), $x \notin G'_m$, which contradicts (1.74). Thus (1.75) holds for the values of x and m we are considering.

Let us suppose now that, for the same values of x and m, $f_m(x)$ is finite. Then, if $g_m(x)$ is also finite, we shall have by condition α) (see (1.50)), the third relation in (1.57) and the inequality $m > N$

$$f_m(x) > g_m(x) - \epsilon', \tag{1.76}$$

from which, by (1.74), the first equality of (1.71) and the inequality $\epsilon' < \epsilon$ (see (1.42)), we obtain

$$f_m(x) > \gamma(x) - \varepsilon \tag{1.77}$$

and consequently, by the second relation in (1.57) and the second equality of (1.38),

$$f_m(x) > \gamma_{-\varepsilon}(x). \tag{1.78}$$

Thus, (1.78) holds if for the values of x and m being considered both $g_m(x)$ and $f_m(x)$ are finite.

Let us suppose now that for the values of x and m being considered $f_m(x)$ is finite and $g_m(x) = +\infty$. Then from condition β) (see (1.51)), the third relation in (1.57) and the inequality $m > N$ it follows that $f_m(x) > 1/\epsilon'$, from which we once again obtain (1.78) with the use of the first relation in (1.57) and (1.43). Since $g_m(x)$ cannot be equal to $-\infty$ if (1.74) holds (see the first equality in (1.71)), it follows that if (1.74) holds for some x and some $m > N$, then (1.78) follows if $f_m(x)$ is finite.

We now treat the case in which for the values of x and m under consideration we have $f_m(x) = +\infty$. We have already seen that $\gamma(x)$ is finite for the value of x being considered, and in this case, by the second equality of (1.38) and the second relation in (1.57), we once again obtain (1.78). Thus if x satisfies (1.74) for any $m > N$, then in all possible cases it satisfies (1.78).[15]

Since $G'_m \subset [a, b]$, $m = 1, 2, \ldots$ (see (1.73)), it follows from this that

$$G'_m \subset E[f_m(x) > \gamma_{-\epsilon}(x)] \quad (m > N). \tag{1.79}$$

Using the second and third equalities of (1.71) and the definition of the sets E_2 and E_3 (see (1.35)) we obtain

$$G''_m = 0, \ \{G'''_m \cdot E[\gamma(x) > -\infty]\} = 0 \quad (m = 1, 2, \ldots),$$

and hence, by (1.72) and (1.79),

$$\left.\begin{array}{c}\{\Omega'' \cdot E[g_m(x) > \gamma(x)] \cdot E[\gamma(x) > -\infty]\} \subset E[f_m(x) > \gamma_{-\epsilon}(x)] \\ (m > N).\end{array}\right\} \tag{1.80}$$

Setting $e = [a, b] - \Omega''$ and using (1.49), (1.70) and (1.80), we obtain (1.33) and (1.32) for $m > N$. Here e and Ω'' do not depend on m. Since ϵ is an arbitrary positive number and N was defined only in terms of ϵ, $\gamma(x)$, $f_m(x)$ and $g_m(x)$, $m = 1, 2, \ldots$, Lemma F is proved.

LEMMA G. *Let the functions $F(x)$, $G(x)$, $f_m(x)$ and $g_m(x)$, $m = 1, 2, \ldots$, be defined almost everywhere on some segment $[a, b]$ and be measurable on this segment. Suppose that*

$$G(x) \leqslant F(x) \tag{1.81}$$

almost everywhere on $[a, b]$, and that (1.3) *and* (1.4) *are equiconvergent on this segment.*

Then, if $F(x)$ and $G(x)$ are the upper and lower limits in measure respectively on $[a, b]$ of (1.3), *these functions are the upper and lower limits in measure respectively on $[a, b]$ of* (1.4).

PROOF. Let us choose an arbitrary measurable function $\varphi(x)$ defined almost everywhere on $[a, b]$. For any positive number α we define the function $\varphi_{-\alpha}(x)$ by the conditions

$$\varphi_{-\alpha}(x) = \varphi(x) - \alpha \tag{1.82}$$

if $\varphi(x)$ is finite,

---

[15] We recall that for the values of x and m being considered we cannot have $f_m(x) = -\infty$ (see (1.75)).

$$\varphi_{-\alpha}(x) = \frac{1}{\alpha} \tag{1.83}$$

if $\varphi(x) = +\infty$, and

$$\varphi_{-\alpha}(x) = \varphi(x) \tag{1.84}$$

if $\varphi(x) = -\infty$.

Then $\varphi_{-\alpha}(x)$ will be measurable and defined almost everywhere on $[a, b]$. Let us take a sequence of positive numbers α_n, $n = 1, 2, \ldots$, that satisfy the conditions

$$\lim_{n\to\infty} \alpha_n = 0, \quad \alpha_n > \alpha_{n+1} \quad (n = 1, 2, \ldots). \tag{1.85}$$

From (1.82)–(1.84) it follows that

$$\lim_{n\to\infty} \varphi_{-\alpha_n}(x) = \varphi(x), \quad \varphi_{-\alpha_n}(x) \leqslant \varphi_{-\alpha_{n+1}}(x) \quad (n = 1, 2, \ldots)$$

for all x on $[a, b]$ for which $\varphi(x)$ is defined. In that case

$$E[\varphi(x) > F(x)] = \sum_{n=1}^{\infty} E[\varphi_{-\alpha_n}(x) > F(x)], \tag{1.86}$$

$$\left.\begin{array}{c} E[\varphi_{-\alpha_n}(x) > F(x)] \subset E[\varphi_{-\alpha_{n+1}}(x) > F(x)] \\ (n = 1, 2, \ldots). \end{array}\right\} \tag{1.87}$$

From these relations we obtain

$$\lim_{n\to\infty} \operatorname{meas} E[\varphi_{-\alpha_n}(x) > F(x)] = \operatorname{meas} E[\varphi(x) > F(x)]. \tag{1.88}$$

Let us take an arbitrary number σ. Using (1.86)–(1.88), we may define a positive number ϵ such that

$$E[\varphi_{-\varepsilon}(x) > F(x)] + R = E[\varphi(x) > F(x)], \tag{1.89}$$

where the set R satisfies

$$\operatorname{meas} R < \sigma. \tag{1.90}$$

Moreover, we may assume that

$$\varepsilon < \sigma. \tag{1.91}$$

Comparing (1.29)–(1.31) (see the statement of Lemma F) with (1.82)–(1.84), we see that $\varphi_{-\alpha}(x)$ is obtained from $\varphi(x)$ in exactly the same way that $\gamma_{-\alpha}(x)$ was obtained from $\gamma(x)$. Since, by the hypotheses of Lemma G, (1.3) and (1.4) are equiconvergent everywhere on $[a, b]$, we may use Lemma F with $\varphi(x)$ taken instead of $\gamma(x)$ to choose a natural number N such that

$$\{E[g_m(x) > \varphi(x)] \cdot E[\varphi(x) > -\infty]\} \subset E[f_m(x) > \varphi_{-\varepsilon}(x)] + e \tag{1.92}$$

for all $m > N$, where e does not depend on m and meas $e < \epsilon$. Consequently, by (1.91),

$$\operatorname{meas} e < \sigma. \tag{1.93}$$

From (1.92) we obtain

$$\left.\begin{aligned} &\{E[g_m(x) > \varphi(x)] \cdot E[\varphi(x) > -\infty] \cdot E[\varphi(x) > F(x)]\} \\ &\quad \subset \{E[f_m(x) > \varphi_{-\varepsilon}(x)] \cdot E[\varphi(x) > F(x)]\} + e \\ &\quad (m > N). \end{aligned}\right\} \tag{1.94}$$

Moreover, it is clear that

$$\begin{aligned} &\{E[g_m(x) > \varphi(x)] \cdot E[\varphi(x) > -\infty] \cdot E[\varphi(x) > F(x)]\} \\ &\quad = \{E[g_m(x) > \varphi(x)] \cdot E[\varphi(x) > F(x)]\} \\ &\quad (m = 1, 2, \ldots), \end{aligned}$$

from which it follows by (1.89) and (1.94) that

$$\left.\begin{aligned} &\{E[g_m(x) > \varphi(x)] \cdot E[\varphi(x) > F(x)]\} \\ &\quad \subset \{E[f_m(x) > \varphi_{-\varepsilon}(x)] \cdot E[\varphi_{-\varepsilon}(x) > F(x)]\} + R + e \\ &\quad (m > N). \end{aligned}\right\} \tag{1.95}$$

We have assumed that $F(x)$ is the limit superior in measure on $[a, b]$ of (1.3). Since $\varphi_{-\epsilon}(x)$ is measurable and defined almost everywhere on $[a, b]$, it follows from condition a) of the definition of the limit superior in measure (see (7) of the Introduction) that

$$\lim_{m\to\infty} \operatorname{meas}\{E[f_m(x) > \varphi_{-\varepsilon}(x)] \cdot E[\varphi_{-\varepsilon}(x) > F(x)]\} = 0.$$

It is then possible to define a natural number N' that satisfies

$$N' > N, \tag{1.96}$$

$$\left.\begin{aligned} &\operatorname{meas}\{E[f_m(x) > \varphi_{-\varepsilon}(x)] \cdot E[\varphi_{-\varepsilon}(x) > F(x)]\} < \sigma \\ &\quad (m > N'), \end{aligned}\right\} \tag{1.97}$$

and hence we may use (1.90), (1.93) and (1.95) to obtain

$$\left.\begin{aligned} &\operatorname{meas}\{E[g_m(x) > \varphi(x)] \cdot E[\varphi(x) > F(x)]\} < 3\sigma \\ &\quad (m > N'). \end{aligned}\right\} \tag{1.98}$$

Thus for any positive number σ we may choose a natural number N' such that (1.98) holds, and then

$$\lim_{m\to\infty} \operatorname{meas}\{E[g_m(x) > \varphi(x)] \cdot E[\varphi(x) > F(x)]\} = 0. \tag{1.99}$$

Since $\varphi(x)$ in (1.99) is an arbitrary measurable function that is defined almost everywhere on $[a, b]$, we thus see that $F(x)$ and the $g_m(x)$, $m = 1, 2, \ldots$, satisfy condition a) (equality (7)) if in this condition we replace the $f_k(x)$ by the $g_m(x)$. Stated differently, $F(x)$ satisfies condition a) in the definition of the limit superior in measure with respect to the sequence (1.4).

We prove now that $F(x)$ satisfies condition b) in the definition of the limit superior in measure (see inequality (8) of the Introduction) if in this definition we take the $g_m(x)$ instead of the $f_k(x)$. To do this we take an arbitrary measurable function $\chi(x)$ that is defined almost everywhere on $[a, b]$ and satisfies the condition

$$\text{meas}\, E[F(x) > \chi(x)] > 0. \tag{1.100}$$

For any positive number α we define a function $\chi_\alpha(x)$ by the equalities

$$\chi_\alpha(x) = \chi(x) + \alpha \tag{1.101}$$

if $\chi(x)$ is finite,

$$\chi_\alpha(x) = \chi(x) \tag{1.102}$$

if $\chi(x) = +\infty$, and

$$\chi_\alpha(x) = -\frac{1}{\alpha} \tag{1.103}$$

if $\chi(x) = -\infty$.

Then $\chi_\alpha(x)$ will be measurable and defined almost everywhere on $[a, b]$. Let us choose a sequence of positive numbers α_n, $n = 1, 2, \ldots$, that satisfy (1.85).

From (1.101)–(1.103) it follows that

$$\lim_{n\to\infty} \chi_{\alpha_n}(x) = \chi(x), \quad \chi_{\alpha_n}(x) \geqslant \chi_{\alpha_{n+1}}(x) \quad (n = 1, 2, \ldots)$$

for all x on $[a, b]$ for which $\chi(x)$ is defined. In this case

$$E[F(x) > \chi(x)] = \sum_{n=1}^{\infty} E[F(x) > \chi_{\alpha_n}(x)], \tag{1.104}$$

$$\left.\begin{array}{c} E[F(x) > \chi_{\alpha_n}(x)] \subset E[F(x) > \chi_{\alpha_{n+1}}(x)] \\ (n = 1, 2, \ldots). \end{array}\right\} \tag{1.105}$$

From these relations we obtain

$$\lim_{n\to\infty} \text{meas} E[F(x) > \chi_{\alpha_n}(x)] = \text{meas}\, E[F(x) > \chi(x)]. \tag{1.106}$$

Using (1.100) and (1.106) we may define a positive number δ for which

$$\text{meas}\, E[F(x) > \chi_\delta(x)] > 0. \tag{1.107}$$

Let us suppose that $F(x)$ is the limit superior in measure on $[a, b]$ of (1.3). Since $\chi_\delta(x)$ is measurable, is defined almost everywhere on $[a, b]$ and satisfies (1.107), it follows that if we set

$$\mu = \overline{\lim_{m\to\infty}} \text{meas}\{E[f_m(x) > \chi_\delta(x)] \cdot E[F(x) > \chi_\delta(x)]\} \tag{1.108}$$

and use condition b) of the definition of the limit superior in measure (see (8) of the Introduction),(16) we shall have

$$\mu > 0. \tag{1.109}$$

(16)In condition b) we take $\chi_\delta(x)$ instead of $\chi(x)$.

Let us take an arbitrary positive number ϵ that satisfies

$$\varepsilon < \delta, \quad \varepsilon < \mu. \tag{1.110}$$

Comparing (1.29)–(1.31) (see the statement of Lemma F) with (1.101)–(1.103), we see that $\chi_\alpha(x)$ is obtained from $\chi(x)$ in exactly the same way as $\gamma_\alpha(x)$ was obtained from $\gamma(x)$. Since by the hypotheses of Lemma G the sequences (1.3) and (1.4) are equiconvergent almost everywhere on $[a, b]$, we may use Lemma F (in which we take $\chi(x)$ instead of $\varphi(x)$) to choose a natural number N_0 such that

$$E\,[f_m(x) > \chi_\varepsilon(x)] - e_0 \subset E\,[g_m(x) > \chi(x)] \tag{1.111}$$

for all $m > N_0$, where e_0 does not depend on m, and

$$\text{meas}\, e_0 < \varepsilon. \tag{1.112}$$

From (1.101)–(1.103) and the first inequality of (1.110) we obtain $\chi_\delta(x) \geqslant \chi_\epsilon(x) \geqslant \chi(x)$ for all x on $[a, b]$ at which $\chi(x)$ is defined. In that case

$$E\,[F(x) > \chi_\delta(x)] \subset E\,[F(x) > \chi(x)],$$
$$E\,[f_m(x) > \chi_\delta(x)] \subset E\,[f_m(x) > \chi_\varepsilon(x)]$$
$$(m = 1, 2, \ldots)$$

and consequently, using (1.111),

$$\left.\begin{aligned} &\{E\,[f_m(x) > \chi_\delta(x)] \cdot E\,[F(x) > \chi_\delta(x)]\} - e_0 \subset \\ &\subset \{E\,[g_m(x) > \chi(x)] \cdot E\,(F(x) > \chi(x)]\} \\ &(m = 1, 2, \ldots). \end{aligned}\right\} \tag{1.113}$$

Taking together (1.108), (1.110), (1.112) and (1.113), we have

$$0 < \mu - \varepsilon < \overline{\lim_{m \to \infty}} \{E\,[g_m(x) > \chi(x)] \cdot E\,[F(x) > \chi(x)]\}. \tag{1.114}$$

Since $\chi(x)$ is an arbitrary measurable function that is defined almost everywhere on $[a, b]$ and satisfies (1.100), it follows from (1.114) that $F(x)$ satisfies condition b) of the definition of the limit superior in measure if the $g_m(x)$ are taken instead of $f_k(x)$ in this definition. We have already shown that $F(x)$ satisfies condition a) in the definition of the limit superior in measure with respect to the sequence $g_m(x)$. It follows from this that $F(x)$ is the limit superior in measure on $[a, b]$ of (1.4).

We prove now that $G(x)$ is the limit inferior in measure on $[a, b]$ of (1.4). Indeed, we have assumed that $G(x)$ is the limit inferior in measure on $[a, b]$ of (1.3). Then, using Theorem D of [2], §2, we have that $-G(x)$ is the limit superior in measure on $[a, b]$ of the sequence of functions $-f_m(x)$, $m = 1, 2, \ldots$. Since, by the hypotheses of this lemma, (1.3) and (1.4) are equiconvergent almost everywhere on $[a, b]$, it follows that the sequences $-f_m(x)$ and $-g_m(x)$, $m = 1, 2, \ldots$, are also equiconvergent almost everywhere on $[a, b]$; and then, by the already proved first part of this lemma, $-G(x)$ is the limit superior in measure on $[a, b]$ of the sequence $-g_m(x)$, $m = 1, 2, \ldots$. Using Theorem D of [2] once again, we see that $G(x)$ is the limit inferior in measure on $[a, b]$ of (1.4). We have shown earlier that $F(x)$ is the limit superior in measure on $[a, b]$ of the same sequence. Consequently, Lemma G is proved.

LEMMA H. *Let F_m, $m = 1, 2, \ldots$, be a sequence of measurable sets, and let p_m, $m = 1, 2, \ldots$, be a sequence of natural numbers that satisfy the following conditions:*

I. $$\text{meas } F_m = (b-a)\lambda_m, \qquad F_m \subset (a, b) \qquad (m = 1, 2, \ldots), \tag{1.115}$$

where (a, b) is some interval that does not depend on m, the numbers λ_m satisfy

$$0 < \lambda_m < 1 \qquad (m = 1, 2, \ldots) \tag{1.116}$$

and the series

$$\sum_{m=1}^{\infty} \lambda_m \tag{1.117}$$

diverges.

II. *Each of the sets F_m, $m = 1, 2, \ldots$, consists of a finite number of nonintersecting open intervals whose end-points lie among the numbers $a + (b-a)t/2p_m$, $t = 0, 1, \ldots, 2p_m$.*

III. *The numbers p_{m+1}/p_m, $m = 1, 2, \ldots$, are integers.*

IV. *For each value $m = 1, 2, \ldots$, all the sets*

$$\left\{F_{m+1} \cdot \left(a + \frac{(b-a)(t-1)}{2p_m}, \quad a + \frac{(b-a)t}{2p_m}\right)\right\}$$
$$(t = 1, 2, \ldots, 2p_m)$$

have the same measure M_m, which in general depends on m but not on t.

Under these conditions

$$\text{meas } \overline{\lim_{m\to\infty}} F_m = b - a. \tag{1.118}$$

The proof of this lemma is given in [3] (Lemma 3.1, p. 629).

LEMMA I. *Set*

$$\left.\begin{array}{c} x_{mi} = a + \dfrac{i}{2^m}(b-a) \\ (i = 0, 1, 2, \ldots 2^m, m = 1, 2, \ldots), \end{array}\right\} \tag{1.119}$$

$$\left.\begin{array}{c} E_m = \displaystyle\sum_{j=1}^{2^{m-1}} (x_{m,2j-2}, x_{m,2j-1}), \quad G_m = \sum_{j=1}^{2^{m-1}} (x_{m,2j-1}, x_{m,2j}) \\ (m = 1, 2, \ldots), \end{array}\right\} \tag{1.120}$$

where $a < b$. Then

$$\text{meas } \overline{\lim_{m\to\infty}} E_m = b - a, \quad \text{meas } \overline{\lim_{m\to\infty}} G_m = b - a, \tag{1.121}$$

$$\overline{\lim_{m\to\infty}} E_m \subset (a, b), \qquad \overline{\lim_{m\to\infty}} G_m \subset (a, b). \tag{1.122}$$

PROOF. From (1.119) and (1.120) we have

$$\left.\begin{array}{c} x_{m0}=a<x_{m1}<\dots<x_{mi}<\dots<x_{m,\,2^m}=b \\ (m=1,\ 2,\ \dots), \end{array}\right\} \tag{1.123}$$

$$x_{mi}-x_{m,\,i-1}=\frac{b-a}{2^m} \quad (i=1,\ 2,\ \dots,\ 2^m,\ m=1,\ 2,\ \dots), \tag{1.124}$$

$$E_m\subset(a,\ b),\quad G_m\subset(a,\ b)\quad (m=1,\ 2,\ \dots), \tag{1.125}$$

from which we obtain (1.122).

To prove (1.121) we prove that each of the sequences of sets

$$E_1,\ E_2,\ \dots,\ E_m,\ \dots \tag{1.126}$$

and

$$G_1,\ G_2,\ \dots,\ G_m,\ \dots \tag{1.127}$$

satisfies the hypotheses of Lemma H, if F_m in that lemma is replaced by E_m or G_m.

From (1.120), (1.123) and (1.124) we have

$$\left.\begin{array}{c} \operatorname{meas} E_m=\dfrac{b-a}{2^m}\cdot 2^{m-1}=\dfrac{b-a}{2},\ \operatorname{meas} G_m=\dfrac{b-a}{2} \\ (m=1,\ 2,\ \dots) \end{array}\right\} \tag{1.128}$$

and consequently, by (1.125), both (1.126) and (1.127) satisfy condition I of Lemma H, in which we set $\lambda_m = \frac{1}{2}$, $m = 1, 2, \dots$.

From (1.120) it follows that each of the sets E_m and G_m, $m = 1, 2, \dots$, consists of a finite number of nonintersecting intervals whose end-points lie among the points x_{mi}, $i = 0, 1, \dots, 2^m$, $m = 1, 2, \dots$. It then follows from (1.119) that both of the sequences of sets (1.126) and (1.127) satisfy condition II of Lemma H, in which we set

$$p_m=2^{m-1}\quad (m=1,\ 2,\ \dots) \tag{1.129}$$

and take E_m or G_m in place of F_m.

From (1.129) it follows that the natural numbers p_m, $m = 1, 2, \dots$, satisfy condition III of Lemma H. We prove now that each of the sequences of sets (1.126) and (1.127) together with the natural numbers p_m defined by (1.129) satisfy condition IV of Lemma H.

From (1.119), (1.120) and (1.124) we have

$$\begin{aligned} &\operatorname{meas}\left\{E_{m+1}\cdot\left(a+\frac{(b-a)(t-1)}{2^m},\ a+\frac{(b-a)t}{2^m}\right)\right\} \\ &=\operatorname{meas}\left\{E_{m+1}\cdot\left(a+\frac{(b-a)(2t-2)}{2^{m+1}},\ a+\frac{(b-a)(2t-1)}{2^{m+1}}\right)\right\} \\ &+\operatorname{meas}\left\{E_{m+1}\cdot\left(a+\frac{(b-a)(2t-1)}{2^{m+1}},\ a+\frac{(b-a)\cdot 2t}{2^{m+1}}\right)\right\} \\ &=\operatorname{meas}\{E_{m+1}\cdot(x_{m+1,\,2t-2},\ x_{m+1,\,2t-1})\}=\frac{b-a}{2^{m+1}} \\ &\qquad (t=1,\ 2,\ \dots,\ 2^m,\ m=1,\ 2,\ \dots) \end{aligned}$$

and in the same way

$$\text{meas}\left\{G_{m+1}\cdot\left(a+\frac{(b-a)(t-1)}{2^m},\ a+\frac{(b-a)t}{2^m}\right)\right\}$$
$$=\text{meas}\,\{G_{m+1}\cdot(x_{m+1,\,2t-1},\ x_{m+1,\,2t})\}=\frac{b-a}{2^{m+1}}$$
$$(t=1,\ 2,\ \ldots,\ 2^m,\ m=1,\ 2,\ \ldots),$$

from which it follows by (1.129) that the sequences (1.126) and (1.127) together with the numbers p_m satisfy condition IV of Lemma H, in which we set $M_m = (b-a)/2^{m+1}$.

Thus the sequences of sets (1.126) and (1.127) together with the numbers p_m defined by (1.129) satisfy all the conditions of Lemma H, and by this lemma we obtain (1.121). Since (1.122) has already been proved, Lemma I is completely proved.

LEMMA J. *Assume that the sequence of measurable sets H_m, $m = 1, 2, \ldots,$ satisfies the following conditions*:

α)
$$\left.\begin{array}{c}\text{meas}\,H_m>\lambda>0,\quad H_m\subset[-\pi,\ \pi]\\ (m=1,\ 2,\ \ldots)\end{array}\right\}\tag{1.130}$$

where λ does not depend on m:

β)
$$\lim_{m\to\infty}\frac{\text{meas}\,(H_m\cdot[c,\ d])}{\text{meas}\,H_m}=\frac{d-c}{2\pi}\tag{1.131}$$

for any interval $[c, d] \subset [-\pi, \pi]$.

In that case
$$\lim_{m\to\infty}\frac{\text{meas}\,(H_m\cdot\Omega)}{\text{meas}\,H_m}=\frac{\text{meas}\,\Omega}{2\pi}\tag{1.132}$$

for any measurable set Ω that lies on $[-\pi, \pi]$.

The proof of this lemma was given in [2] (Lemma 5.1). For a set of measure zero, (1.132) is obvious.

We now prove another lemma, one which is a generalization of a well-known theorem of Egorov on sequences of measurable functions. We first give the following definition.

DEFINITION 3. We shall say that the sequence (1.3) of functions $f_m(x)$, $m = 1, 2, \ldots$, defined(17) on some set E *converges uniformly* to a function $f(x)$ on E if for any positive number ϵ it is possible to find a natural number N such that for any $x \in E$ and any natural $m > N$ one of the following conditions holds:

α)
$$-\infty<f_m(x)<+\infty,\quad |f_m(x)-f(x)|<\varepsilon,$$

if $f(x)$ is finite;

β)
$$f_m(x)>\frac{1}{\epsilon}$$

(17)The functions $f_m(x)$ may be equal to $+\infty$ or $-\infty$ on sets of positive measure.

if $f(x) = +\infty$; or

γ) $$f_m(x) < -\frac{1}{\varepsilon}$$

if $f(x) = -\infty$.

LEMMA K. *If the sequence* (1.3) *of measurable functions* $f_m(x)$, $m = 1, 2, \ldots$, *that are defined almost everywhere on a measurable set E converges to* $f(x)$ *almost everywhere on this set, then for any positive number* σ *it is possible to define a measurable set M that satisfies the conditions*

$$M \subset E, \text{ meas } M > \text{meas } E - \sigma \tag{1.133}$$

and is such that (1.3) *converges uniformly to* $f(x)$ *on M.*

PROOF. We define the functions $g_m(x)$, $m = 1, 2, \ldots$, of the sequence (1.4) by

$$g_m(x) = f(x) \quad (m = 1, 2, \ldots). \tag{1.134}$$

Then the functions of this sequence will be defined almost everywhere on E. Also, since by the hypotheses of the lemma (1.3) converges to $f(x)$ almost everywhere on E, it follows immediately from (1.134) that (1.3) and (1.4) are equiconvergent almost everywhere on E.

Let us set

$$\left.\begin{array}{c} E_1 = \{E \cdot E[-\infty < f(x) < +\infty]\}, \\ E_2 = \{E \cdot E[f(x) = +\infty]\}, \quad E_3 = \{E \cdot E[f(x) = -\infty]\}, \end{array}\right\} \tag{1.135}$$

from which it follows that the sets E_i, $i = 1, 2, 3$, are measurable, with

$$E \supset \sum_{i=1}^{3} E_i, \text{ meas } E = \sum_{i=1}^{3} \text{meas } E_i, \ (E_i \cdot E_j) = 0 \quad (i \neq j, \ i, \ j = 1, 2, 3). \tag{1.136}$$

Let us choose an arbitrary number σ. Since $f(x)$ is measurable and finite almost everywhere on E_1, we can find a positive number C and a measurable set E_4 such that

$$E_4 \subset E_1, \text{ meas } E_4 > \text{meas } E_1 - \frac{\sigma}{2}, \tag{1.137}$$

$$|f(x)| < C \qquad (x \in E_4). \tag{1.138}$$

Let us set

$$E' = \sum_{i=2}^{4} E_i. \tag{1.139}$$

Then, using (1.136) and (1.137),

$$E' \subset E, \text{ meas } E' > \text{meas } E - \frac{\sigma}{2}, \tag{1.140}$$

$$(E_i \cdot E_j) = 0 \quad (i \neq j, \ i, \ j = 2, 3, 4), \tag{1.141}$$

with the sequences (1.3) and (1.4) equiconvergent almost everywhere on E'. We apply

Lemma E to these sequences, using $\sigma/2$ and E' instead of σ and E, to see that there exists a measurable set M that satisfies the conditions

$$M \subset E' \subset E, \text{ meas } M > \text{meas } E' - \frac{\sigma}{2} > \text{meas } E - \sigma \tag{1.142}$$

and is such that (1.3) and (1.4) are uniformly equiconvergent on M.

We prove that (1.3) converges to $f(x)$ uniformly on M. Let us choose an arbitrary positive number ϵ and suppose that the positive number ϵ' satisfies the conditions

$$\varepsilon' < \varepsilon, \quad \frac{1}{\varepsilon'} > C. \tag{1.143}$$

Since (1.3) and (1.4) are uniformly equiconvergent on M, by Definition 2 and equalities (1.134) we may find a natural number N such that for all $x \in M$ and for all natural $m > N$ one of the following three inequalities holds:

$$|f_m(x) - f(x)| < \varepsilon', \tag{1.144}$$

if $f_m(x)$ and $f(x)$ are finite,

$$f_m(x) > \frac{1}{\varepsilon'}, \quad f(x) > \frac{1}{\varepsilon'} \tag{1.145}$$

if at least one of $f_m(x)$ or $f(x)$ is equal to $+\infty$, and

$$f_m(x) < -\frac{1}{\varepsilon'}, \quad f(x) < -\frac{1}{\varepsilon'} \tag{1.146}$$

if at least one of $f_m(x)$ or $f(x)$ is equal to $-\infty$.

From (1.139) and (1.142) we obtain that $M \subset \Sigma_{i=2}^{4} E_i$, and consequently, taking (1.135) and (1.137) together with (1.138) and (1.143), we see that $|f(x)| < 1/\epsilon'$ if $x \in M$ and $f(x)$ is finite, i.e., the second inequalities in (1.145) and (1.146) cannot hold if $x \in M$ and $f(x)$ is finite. In that case, if

$$x \in M, \quad m > N \tag{1.147}$$

and $f(x)$ is finite, then $f_m(x)$ cannot equal $+\infty$ or $-\infty$, since in the contrary case both inequalities in (1.145) or both inequalities in (1.146) would hold, which is impossible.

Thus, if x and m satisfy (1.147) and $f(x)$ is finite, then $f_m(x)$ is also finite and satisfies (1.144). Furthermore, if x and m satisfy (1.147) and $f(x) = +\infty$ or $f(x) = -\infty$, then either the first inequality of (1.145) or the first inequality of (1.146) respectively will hold.

Since $\epsilon' < \epsilon$ (see the first inequality of (1.143)), we come to the following conclusion:

If $x \in M$ and $m > N$, then inequalities α, β or γ of Definition 3 will hold, depending on whether $f(x)$ is finite, equal to $+\infty$ or equal to $-\infty$. Since ϵ may be an arbitrary positive number, it follows from this by Definition 3 that (1.3) converges uniformly to $f(x)$ on M. Thus Lemma K is proved.

§2. Proof of Theorem III stated in the Introduction

Let us assume that $F_1(x)$ and $F_2(x)$ are measurable, defined almost everywhere on $[-\pi, \pi]$ and satisfy inequality (4) (see the Introduction) almost everywhere on this segment. We take an arbitrary trigonometric series of type $A(F_1, F_2)$, i.e., one that is universal with respect to the functions $F_1(x)$ and $F_2(x)$ (see the Introduction, Definition A).

Let us consider any measurable functions $\varphi_1(x)$, $\varphi_2(x)$, $\psi_1(x)$ and $\psi_2(x)$ that are defined almost everywhere on $[-\pi, \pi]$ and satisfy (11) (see Definition 2 in the Introduction) almost everywhere on this segment. We may define the six functions listed above at all points of $[-\pi, \pi]$ at which they were not originally defined in such a way that (4) and (11) will hold everywhere on $[-\pi, \pi]$, with these functions still measurable on the given segment.

Let us assume that m is an arbitrary natural number and partition $[-\pi, \pi]$ into 2^m equal parts. We denote the partition points, numbered in increasing order, by x_{mi}, $i = 0, 1, \ldots, 2^m$. Then

$$\left.\begin{array}{c} x_{mi} = -\pi + \dfrac{2\pi i}{2^m} = -\pi + \dfrac{\pi i}{2^{m-1}} \\ x_{mi} - x_{m,\, i-1} = \dfrac{\pi}{2^{m-1}} \\ (i = 1, 2, \ldots, 2^m,\ m = 1, 2, \ldots), \end{array}\right\} \tag{2.1}$$

$$x_{m0} = -\pi < x_{m1} < \ldots < x_{mi} < \ldots < x_{m,\, 2^m} = \pi. \tag{2.2}$$

We set

$$E_m = \sum_{j=1}^{2^{m-1}} (x_{m,\, 2j-2},\ x_{m,\, 2j-1}),\quad G_m = \sum_{j=1}^{2^{m-1}} (x_{m,\, 2j-1},\ x_{m,\, 2j}), \tag{2.3}$$

from which it follows by (2.2) that

$$\left.\begin{array}{c} E_m \subset (-\pi, \pi),\ G_m \subset (-\pi, \pi),\ (E_m \cdot G_m) = 0 \\ (m = 1, 2, \ldots). \end{array}\right\} \tag{2.4}$$

Considering (2.1) and (2.3), we see that the sequences of sets E_m and G_m, $m = 1, 2, \ldots$, satisfy the hypotheses of Lemma I (§1), in which $(a, b) = (-\pi, \pi)$. Consequently, setting

$$E = \overline{\lim_{m\to\infty}} E_m,\quad G = \overline{\lim_{m\to\infty}} G_m, \tag{2.5}$$

we have on the basis of this lemma

$$\left.\begin{array}{c} \operatorname{meas} E = 2\pi,\ \operatorname{meas} G = 2\pi, \\ E \subset (-\pi, \pi),\ G \subset (-\pi, \pi). \end{array}\right\} \tag{2.6}$$

We define the functions $g_m(x)$, $m = 1, 2, \ldots$, from the conditions

$$\left.\begin{array}{c} g_m(x) = \psi_1(x),\quad (x \in E_m),\quad g_m(x) = \psi_2(x),\quad (x \in G_m), \\ g_m(x_{mi}) = \psi_1(x_{mi}) \\ (i = 0, 1, 2, \ldots, 2^m,\ m = 1, 2, \ldots). \end{array}\right\} \tag{2.7}$$

By (2.2) and (2.3) the functions $g_m(x)$, $m = 1, 2, \ldots$, will be defined almost everywhere on $[-\pi, \pi]$ and will be measurable on this segment. Here, since (11) holds almost everywhere on $[-\pi, \pi]$, we have by (2.7)

$$\left.\begin{array}{c}\psi_1(x) \leqslant g_m(x) \leqslant \psi_2(x)\\ (x \in [-\pi, \pi],\ m = 1, 2, \ldots).\end{array}\right\} \tag{2.8}$$

We assume that ν is a natural number and define the numbers m_ν and j_ν from the conditions

$$2^{m_\nu - 1} \leqslant \nu < 2^{m_\nu},\quad j_\nu = \nu - 2^{m_\nu - 1} + 1. \tag{2.9}$$

It is clear that m_ν and j_ν are uniquely defined for any natural ν, with

$$m_\nu > 0,\ 1 \leqslant j_\nu \leqslant 2^{m_\nu - 1} \quad (\nu = 1, 2, \ldots), \tag{2.10}$$

$$\lim_{\nu \to \infty} m_\nu = \infty. \tag{2.11}$$

Let us set

$$\Phi_\nu(x) = \varphi_1(x) \quad [x \in (x_{m_\nu, 2j_\nu - 2},\ x_{m_\nu, 2j_\nu - 1})], \tag{2.12}$$

$$\Phi_\nu(x) = \varphi_2(x) \quad [x \in (x_{m_\nu, 2j_\nu - 1},\ x_{m_\nu, 2j_\nu})], \tag{2.13}$$

$$\left.\begin{array}{c}\Phi_\nu(x) = g_{m_\nu}(x)\\ (x \in [-\pi, \pi] - (x_{m_\nu, 2j_\nu - 2},\ x_{m_\nu, 2j_\nu - 1}) - (x_{m_\nu, 2j_\nu - 1},\ x_{m_\nu, 2j_\nu})\end{array}\right\} \tag{2.14}$$

Then for any natural number ν the functions $\Phi_\nu(x)$ will be defined almost everywhere on $[-\pi, \pi]$ and will be measurable on this interval. Moreover, by (11), (2.8) and (2.12)–(2.14),

$$\varphi_1(x) \leqslant \Phi_\nu(x) \leqslant \varphi_2(x) \quad (x \in [-\pi, \pi],\ \nu = 1, 2, \ldots). \tag{2.15}$$

We now define the functions $f_k(x)$, $k = 1, 2, \ldots$, from the conditions

$$\left.\begin{array}{c}f_{2m}(x) = g_m(x),\ f_{2m-1}(x) = \Phi_m(x)\\ (x \in [-\pi, \pi],\ m = 1, 2, \ldots),\end{array}\right\} \tag{2.16}$$

from which it follows from (11), (2.8) and (2.15) that

$$\left.\begin{array}{c}\varphi_1(x) \leqslant f_k(x) \leqslant \varphi_2(x)\\ (x \in [-\pi, \pi],\quad k = 1, 2, \ldots),\end{array}\right\} \tag{2.17}$$

with the functions $f_k(x)$, $k = 1, 2, \ldots$, measurable and defined almost everywhere on $[-\pi, \pi]$.

We shall prove that $\psi_1(x)$ and $\psi_2(x)$ are the limits superior and inferior respectively in measure on $[-\pi, \pi]$ of the sequence of functions $f_k(x)$, $k = 1, 2, \ldots$. We choose an arbitrary measurable function $\varphi(x)$ defined almost everywhere on $[-\pi, \pi]$ and consider the sets

$$\left.\begin{array}{c}\Omega_k' = \{E[f_k(x) < \varphi(x)] \cdot E[\varphi(x) < \psi_1(x)]\},\\ \Omega_k'' = \{E[f_k(x) > \varphi(x)] \cdot E[\varphi(x) > \psi_2(x)]\}\\ (k = 1, 2, \ldots).\end{array}\right\} \tag{2.18}$$

If k is even, i.e.

$$k = 2m, \tag{2.19}$$

where m is a natural number, then, on the basis of the first equality in (2.16),

$$f_k(x) = f_{2m}(x) = g_m(x) \quad (x \in [-\pi, \pi]), \tag{2.20}$$

and so, by (2.8),

$$\psi_1(x) \leqslant f_k(x) \leqslant \psi_2(x) \quad (x \in [-\pi, \pi]). \tag{2.21}$$

Then from (2.18) we obtain $\Omega'_k = \Omega''_k = 0$, and consequently

$$\operatorname{meas} \Omega'_k = 0, \quad \operatorname{meas} \Omega''_k = 0, \tag{2.22}$$

if k is even.

Let us suppose now that k is odd, i.e.

$$k = 2\nu - 1, \tag{2.23}$$

where ν is a natural number. Then from the second equality in (2.16) we have

$$f_k(x) = f_{2\nu-1}(x) = \Phi_\nu(x) \quad (x \in [-\pi, \pi])$$

and consequently, by (2.18),

$$\left.\begin{array}{c} \Omega'_k = \{E[\Phi_\nu(x) < \varphi(x)] \cdot E[\varphi(x) < \psi_1(x)]\}, \\ \Omega''_k = \{E[\Phi_\nu(x) > \varphi(x)] \cdot E[\varphi(x) > \psi_2(x)]\} \\ (k = 2\nu - 1, \ \nu = 1, 2, \ldots). \end{array}\right\} \tag{2.24}$$

Keeping (2.8) in mind, we find that

$$\begin{array}{c} \{E[g_m(x) < \varphi(x)] \cdot E[\varphi(x) < \psi_1(x)]\} = 0, \\ \{E[g_m(x) > \varphi(x)] \cdot E[\varphi(x) > \psi_2(x)]\} = 0 \\ (m = 1, 2, \ldots), \end{array}$$

and in that case, by (2.14) and (2.24),

$$\left.\begin{array}{c} \Omega'_k \subset (x_{m_\nu, 2j_\nu-2}, x_{m_\nu, 2j_\nu-1}) + (x_{m_\nu, 2j_\nu-1}, x_{m_\nu, 2j_\nu}), \\ \Omega''_k \subset (x_{m_\nu, 2j_\nu-2}, x_{m_\nu, 2j_\nu-1}) + (x_{m_\nu, 2j_\nu-1}, x_{m_\nu, 2j_\nu}) \\ (k = 2\nu - 1, \ \nu = 1, 2, \ldots), \end{array}\right\} \tag{2.25}$$

where the numbers m_ν and j_ν are defined by (2.9). Taking (2.25) together with the second equality in (2.1), we have

$$\left.\begin{array}{c} \operatorname{meas} \Omega'_k \leqslant \dfrac{2\pi}{2^{m_\nu - 1}} = \dfrac{\pi}{2^{m_\nu - 2}}, \quad \operatorname{meas} \Omega''_k \leqslant \dfrac{\pi}{2^{m_\nu - 2}} \\ (k = 2\nu - 1, \ \nu = 1, 2, \ldots). \end{array}\right\} \tag{2.26}$$

If we consider ν as a function of the natural number $k = 2\nu - 1$, then $\lim_{k\to\infty} \nu = \infty$, and consequently, by (2.11),

$$\lim_{k\to\infty} m_\nu = \infty, \tag{2.27}$$

and, then, by (2.26), meas Ω'_k and meas Ω''_k tend to zero as $k \longrightarrow \infty$ while taking odd values. We see from this and the use of (2.22) that

$$\lim_{k\to\infty} \text{meas } \Omega'_k = 0, \qquad \lim_{k\to\infty} \text{meas } \Omega''_k = 0. \tag{2.28}$$

We define the sets Ω'_k and Ω''_k, $k = 1, 2, \ldots$, by (2.18), where $\varphi(x)$ is an arbitrary measurable function defined almost everywhere on $[-\pi, \pi]$. Then $\psi_1(x)$ will satisfy condition α) of the definition of the limit inferior of the sequence of functions $f_k(x)$, $k = 1, 2, \ldots$, if we take $\psi_1(x)$ instead of $G(x)$ in this condition and set $[a, b] = [-\pi, \pi]$ (see equality (9) of the Introduction). In exactly the same way $\psi_2(x)$ satisfies condition a) in the definition of the limit superior of the sequence of functions $f_k(x)$, $k = 1, 2, \ldots$, if we take $\psi_2(x)$ instead of $F(x)$ and once again set $[a, b] = [-\pi, \pi]$ (see equality (7) of the Introduction).

We prove now that the functions $\psi_1(x)$ and $\psi_2(x)$ taken instead of $G(x)$ and $F(x)$ respectively satisfy conditions β) and b) in the definitions of the limits inferior and superior in measure of the sequence of functions $f_k(x)$, $k = 1, 2, \ldots$, on the interval $[a, b] = [-\pi, \pi]$ (see inequalities (10) and (8) of the Introduction).

We choose an arbitrary segment $[c, d] \subset [-\pi, \pi]$. By the second equality in (2.1) we have

$$\left.\begin{array}{c} x_{mi} - x_{m,i-1} = \dfrac{\pi}{2^{m-1}} \\ (i = 1, 2, \ldots, 2^m, \ m = 1, 2, \ldots). \end{array}\right\} \tag{2.29}$$

Then if m satisfies the inequality

$$\frac{\pi}{2^{m-1}} < \frac{d-c}{2}, \tag{2.30}$$

we may define natural numbers r and s for which

$$\left.\begin{array}{c} x_{m,2r-2} \leqslant c < x_{m,2r} < x_{m,2s} < d \leqslant x_{m,2s+2}, \\ 1 \leqslant r < s < 2^{m-1}. \end{array}\right\} \tag{2.31}$$

In general, r and s depend on m.

Using (2.2) and (2.3), for the m that satisfy (2.30) we have

$$\{E_m \cdot (x_{m,2r}, x_{m,2s})\} = \sum_{j=r+1}^{s} (x_{m,2j-2}, x_{m,2j-1})$$

$$\subset \{E_m \cdot (c, d)\} \subset \{E_m \cdot (x_{m,2r-2}, x_{m,2s+2})\} = \sum_{j=r}^{s+1} (x_{m,2j-2}, x_{m,2j-1}),$$

$$\{G_m \cdot (x_{m,2r}, x_{m,2s})\} = \sum_{j=r+1}^{s} (x_{m,2j-1}, x_{m,2j})$$

$$\subset \{G_m \cdot (c, d)\} \subset \{G_m \cdot (x_{m,2r-2}, x_{m,2s+2})\} = \sum_{j=r}^{s+1} (x_{m,2j-1}, x_{m,2j}),$$

and hence, by (2.1) and (2.2),

$$\left.\begin{array}{l}\displaystyle\sum_{j=r+1}^{s}(x_{m,2j-1}-x_{m,2j-2})=\frac{\pi}{2m-1}(s-r)\leqslant \text{meas}\,\{E_m\cdot[c,\ d]\}\\ \displaystyle\qquad\leqslant\sum_{j=r}^{s+1}(x_{m,2j-1}-x_{m,2j-2})=\frac{\pi}{2m-1}(s-r+2),\\ \displaystyle\sum_{j=r+1}^{s}(x_{m,2j}-x_{m,2j-1})=\frac{\pi}{2m-1}(s-r)\leqslant \text{meas}\,\{G_m\cdot[c,\ d]\}\\ \displaystyle\qquad\leqslant\sum_{j=r}^{s+1}(x_{m,2j}-x_{m,2j-1})=\frac{\pi}{2m-1}(s-r+2)\end{array}\right\}\tag{2.32}$$

for all sufficiently large values of m.

On the other hand, by (2.1)–(2.3)

$$\text{meas}\,E_m=\pi,\quad \text{meas}\,G_m=\pi\qquad (m=1,\ 2,\ \ldots),\tag{2.33}$$

and hence, by (2.32),

$$\left.\begin{array}{l}\displaystyle\frac{s-r}{2m-1}\leqslant\frac{\text{meas}\,\{E_m\,[c,\ d]\}}{\text{meas}\,E_m}\leqslant\frac{s-r+2}{2m-1},\\ \displaystyle\frac{s-r}{2m-1}\leqslant\frac{\text{meas}\,\{G_m\cdot[c,\ d]\}}{\text{meas}\,G_m}\leqslant\frac{s-r+2}{2m-1}\end{array}\right\}\tag{2.34}$$

for all sufficiently large m. Also, it follows from (2.1) and (2.31) that

$$\frac{2\pi}{2m-1}(s-r)=x_{m,2s}-x_{m,2r}\leqslant d-c\leqslant x_{m,2s+2}-x_{m,2r-2}=\frac{2\pi}{2m-1}(s-r+2),$$

from which we have

$$\frac{s-r}{2m-1}\leqslant\frac{d-c}{2\pi}\leqslant\frac{s-r+2}{2m-1}$$

for all sufficiently large m. Consequently,

$$\lim_{m\to\infty}\frac{s-r}{2m-1}=\lim_{m\to\infty}\frac{s-r+2}{2m-1}=\frac{d-c}{2\pi}\ (^{18}).\tag{2.35}$$

Taking (2.34) and (2.35) together, we obtain

$$\lim_{m\to\infty}\frac{\text{meas}\,\{E_m\,[c,\ d]\}}{\text{meas}\,E_m}=\frac{d-c}{2\pi},\quad \lim_{m\to\infty}\frac{\text{meas}\,\{G_m\cdot[c,\ d]\}}{\text{meas}\,G_m}=\frac{d-c}{2\pi}\tag{2.36}$$

for any segment $[c, d]\subset[-\pi, \pi]$.

From (2.4), (2.33) and (2.36) we see that both sequences of sets E_m and G_m, $m = 1, 2, \ldots$, satisfy all the conditions of Lemma J if E_m or G_m is taken instead of H_m in that lemma. Then, by that lemma,

$$\lim_{m\to\infty}\frac{\text{meas}\,(E_m\cdot\Omega)}{\text{meas}\,E_m}=\frac{\text{meas}\,\Omega}{2\pi},\quad \lim_{m\to\infty}\frac{\text{meas}\,(G_m\cdot\Omega)}{\text{meas}\,G_m}=\frac{\text{meas}\,\Omega}{2\pi}\tag{2.37}$$

for any measurable set Ω that lies on $[-\pi, \pi]$.

---

(18) We recall that r and s depend on m and are defined for all sufficiently large m.

We now take any two measurable functions $\chi_1(x)$ and $\chi_2(x)$ that are defined almost everywhere on $[-\pi, \pi]$ and satisfy

$$\text{meas}\, E\,[\psi_1(x) < \chi_1(x)] > 0, \quad \text{meas}\, E\,[\psi_2(x) > \chi_2(x)] > 0. \tag{2.38}$$

Then by (2.37),

$$\lim_{m\to\infty} \frac{\text{meas}\,\{E_m \cdot E\,[\psi_1(x) < \chi_1(x)]\}}{\text{meas}\, E_m} = \frac{\text{meas}\, E\,[\psi_1(x) < \chi_1(x)]}{2\pi} > 0,$$
$$\lim_{m\to\infty} \frac{\text{meas}\,\{G_m \cdot E\,[\psi_2(x) > \chi_2(x)]\}}{\text{meas}\, G_m} = \frac{\text{meas}\, E\,[\psi_2(x) > \chi_2(x)]}{2\pi} > 0$$

and consequently, from (2.33),

$$\lim_{m\to\infty} \text{meas}\,\{E_m \cdot E\,[\psi_1(x) < \chi_1(x)]\} = \frac{1}{2}\, \text{meas}\, E\,[\psi_1(x) < \chi_1(x)] > 0,$$

$$\lim_{m\to\infty} \text{meas}\,\{G_m \cdot E\,[\psi_2(x) > \chi_2(x)]\} = \frac{1}{2}\, \text{meas}\, E\,[\psi_2(x) > \chi_2(x)] > 0,$$

whence we have

$$\left.\begin{aligned} &\text{meas}\,\{E_m \cdot E\,[\psi_1(x) < \chi_1(x)]\} > \frac{1}{4}\, \text{meas}\, E\,[\psi_1(x) < \chi_1(x)] > 0, \\ &\text{meas}\,\{G_m \cdot E\,[\psi_2(x) > \chi_2(x)]\} > \frac{1}{4}\, \text{meas}\, E\,[\psi_2(x) > \chi_2(x)] > 0 \end{aligned}\right\} \tag{2.39}$$

for all sufficiently large m.

From (2.4), (2.7) and (2.16) it follows that

$$f_{2m}(x) = \psi_1(x) \qquad (x \in E_m,\ m = 1, 2, \ldots), \tag{2.40}$$
$$f_{2m}(x) = \psi_2(x) \qquad (x \in G_m,\ m = 1, 2, \ldots), \tag{2.41}$$

and then

$$\{E_m \cdot E\,[\psi_1(x) < \chi_1(x)]\} \subset \{E\,[f_{2m}(x) < \chi_1(x)] \cdot E\,[\psi_1(x) < \chi_1(x)]\},$$
$$\{G_m \cdot E\,[\psi_2(x) > \chi_2(x)]\} \subset \{E\,[f_{2m}(x) > \chi_2(x)] \cdot E\,[\psi_2(x) > \chi_2(x)]\}$$

and consequently, from (2.39),

$$\left.\begin{aligned} &\text{meas}\,\{E\,[f_{2m}(x) < \chi_1(x)] \cdot E\,[\psi_1(x) < \chi_1(x)]\} > \frac{1}{4}\text{meas}\, E[\psi_1(x) < \chi_1(x)] > 0, \\ &\text{meas}\,\{E\,[f_{2m}(x) > \chi_2(x)] \cdot E\,[\psi_2(x) > \chi_2(x)]\} > \frac{1}{4}\text{meas}\, E[\psi_2(x) > \chi_2(x)] > 0, \end{aligned}\right\} \tag{2.42}$$

for all sufficiently large m. From (2.42) we obtain

$$\left.\begin{aligned} &\overline{\lim_{k\to\infty}}\ \text{meas}\,\{E\,[f_k(x) < \chi_1(x)] \cdot E\,[\psi_1(x) < \chi_1(x)]\} > 0, \\ &\overline{\lim_{k\to\infty}}\ \text{meas}\,\{E\,[f_k(x) > \chi_2(x)] \cdot E\,[\psi_2(x) > \chi_2(x)]\} > 0. \end{aligned}\right\} \tag{2.43}$$

Since (2.43) holds for any measurable functions $\chi_1(x)$ and $\chi_2(x)$ defined almost everywhere on $[-\pi, \pi]$ that satisfy (2.38), the functions $\psi_1(x)$ and $\psi_2(x)$ satisfy conditions β) and b) respectively in the definitions of the limit inferior and superior in measure of the sequence of functions $f_k(x)$, $k = 1, 2, \ldots$, if $\psi_1(x)$ is taken instead of $G(x)$ and $\psi_2(x)$ is taken instead of $F(x)$ and we set $[a, b] = [-\pi, \pi]$ (see inequalities (10) and (8) of the Introduction).

We have shown previously that $\psi_1(x)$ and $\psi_2(x)$ satisfy conditions α) and a) respectively in these definitions. From this it follows that $\psi_1(x)$ and $\psi_2(x)$ are the limits inferior and superior respectively in measure on $[-\pi, \pi]$ of the sequence of functions $f_k(x)$, $k = 1, 2, \ldots$.

We prove that almost everywhere on $[-\pi, \pi]$

$$\varliminf_{k\to\infty} f_k(x) = \varphi_1(x), \quad \varlimsup_{k\to\infty} f_k(x) = \varphi_2(x). \tag{2.44}$$

As we already know, $\varphi_1(x)$, $\varphi_2(x)$ and the $f_k(x)$, $k = 1, 2, \ldots$, are defined everywhere on $[-\pi, \pi]$, and it follows from (2.17) that

$$\varphi_1(x) \leqslant \varliminf_{k\to\infty} f_k(x) \leqslant \varlimsup_{k\to\infty} f_k(x) \leqslant \varphi_2(x) \tag{2.45}$$

everywhere on this segment. We prove now that

$$\left.\begin{array}{c} \varliminf_{k\to\infty} f_k(x) \leqslant \varphi_1(x), \quad \varlimsup_{k\to\infty} f_k(x) \geqslant \varphi_2(x) \\ {[x \in (E, G)]}, \end{array}\right\} \tag{2.46}$$

where the sets E and G are defined by (2.5).

Let x be an arbitrary point in the set $(E \cdot G)$. Then by (2.5) we may define two increasing sequences of natural numbers n_s and n'_s, $s = 1, 2, \ldots$, such that

$$x \in E_{n_s}, \quad x \in G_{n'_s} \quad (s = 1, 2, \ldots) \tag{2.47}$$

Taking (2.3) and (2.47) together, we see that there exist two sequences of natural numbers n_s and n'_s, $s = 1, 2, \ldots$, that satisfy the conditions

$$\left.\begin{array}{c} x \in (x_{n_s, 2i_s-2}, x_{n_s, 2i_s-1}), \quad x \in (x_{n'_s, 2i'_s-1}, x_{n'_s, 2i'_s}) \\ (s = 1, 2, \ldots), \end{array}\right\} \tag{2.48}$$

$$1 \leqslant i_s \leqslant 2^{n_s-1}, \quad 1 \leqslant i'_s \leqslant 2^{n'_s-1} \quad (s = 1, 2, \ldots), \tag{2.49}$$

and in that case, setting

$$\left.\begin{array}{c} \nu_s = 2^{n_s-1} + i_s - 1, \quad \nu'_s = 2^{n'_s-1} + i'_s - 1 \\ (s = 1, 2, \ldots) \end{array}\right\} \tag{2.50}$$

and using (2.9), (2.12) and (2.13), we have at the given point x

$$\Phi_{\nu_s}(x) = \varphi_1(x), \quad \Phi_{\nu'_s}(x) = \varphi_2(x) \quad (s = 1, 2, \ldots), \tag{2.51}$$

and so, by (2.16),

$$f_{2\nu_s-1}(x) = \varphi_1(x), \quad f_{2\nu'_s-1}(x) = \varphi_2(x) \quad (s = 1, 2, \ldots). \tag{2.52}$$

Since both sequences n_s and n'_s, $s = 1, 2, \ldots$, are increasing, it follows from (2.49) and (2.50) that the sequences of natural numbers ν_s and ν'_s, $s = 1, 2, \ldots$, are also increasing, and then

$$\lim_{s\to\infty} \nu_s = \infty, \quad \lim_{s\to\infty} \nu'_s = \infty. \tag{2.53}$$

Since by hypothesis x is an arbitrary point of the set $(E \cdot G)$, by taking (2.52) and (2.53) together we obtain (2.46). From (2.6) we find that

$$\text{meas}\,(E \cdot G) = 2\pi, \quad (E \cdot G) \subset (-\pi, \pi),$$

and then, by (2.45) and (2.46), inequality (2.44) holds almost everywhere on $[-\pi, \pi]$.

We have assumed that the trigonometric series (1) (see the Introduction) is an arbitrary trigonometric series of type $A(F_1, F_2)$, where $F_1(x)$ and $F_2(x)$ are measurable and satisfy (4) almost everywhere on $[-\pi, \pi]$; and we have chosen four measurable functions $\varphi_1(x)$, $\psi_1(x)$, $\psi_2(x)$ and $\varphi_2(x)$ for which (11) holds everywhere on $[-\pi, \pi]$. Using these functions, we have defined on $[-\pi, \pi]$ a sequence of measurable functions $f_k(x)$, $k = 1, 2, \ldots$, that has all the properties listed above.

Putting together (11) and (2.17), we obtain

$$F_1(x) \leqslant f_k(x) \leqslant F_2(x) \qquad (x \in [-\pi, \pi], \; k = 1, 2, \ldots). \tag{2.54}$$

Since the trigonometric series (1) is of type $A(F_1, F_2)$, we may by Definition A (see the Introduction) define for any natural number k an increasing sequence of natural numbers m_{ki}, $i = 1, 2, \ldots$, such that

$$\lim_{i \to \infty} S_{m_{ki}}(x) = f_k(x) \tag{2.55}$$

almost everywhere on $[-\pi, \pi]$. Then by Lemma K we may define a measurable set M_k that satisfies the conditions

$$M_k \subset [-\pi, \pi], \; \text{meas}\, M_k > 2\pi - \frac{1}{2k} \quad (k = 1, 2, \ldots) \tag{2.56}$$

and is such that the sequence of functions

$$S_{m_{ki}}(x) \qquad (i = 1, 2, \ldots) \tag{2.57}$$

converges uniformly to $f_k(x)$ on M_k. Then by Definition 3 there exists a natural number N_k such that

$$|S_{m_{ki}}(x) - f_k(x)| < \frac{1}{k} \quad (x \in M_k, \; i > N_k) \tag{2.58}$$

if $f_k(x)$ is finite,

$$S_{m_{ki}}(x) > k \qquad (x \in M_k, \; i > N_k) \tag{2.59}$$

if $f_k(x) = +\infty$, and

$$S_{m_{ki}}(x) < -k \qquad (x \in M_k, \; i > N_k) \tag{2.60}$$

if $f_k(x) = -\infty$.

Since for each natural number k the natural numbers m_{ki}, $i = 1, 2, \ldots$, form an increasing sequence, we may define a sequence of natural numbers j_k, $k = 1, 2, \ldots$, satisfying

$$j_k > N_k, \quad m_{k, j_k} < m_{k+1, j_{k+1}} \qquad (k = 1, 2, \ldots). \tag{2.61}$$

Then, setting

$$n_k = m_{k, j_k} \qquad (k = 1, 2, \ldots) \tag{2.62}$$

and using (2.58)–(2.60), we have

$$|S_{n_k}(x) - f_k(x)| < \frac{1}{k} \qquad (x \in M_k) \tag{2.63}$$

if $f_k(x)$ is finite,

$$S_{n_k}(x) > k \qquad (x \in M_k), \tag{2.64}$$

if $f_k(x) = +\infty$, and

$$S_{n_k}(x) < -k \qquad (x \in M_k) \tag{2.65}$$

if $f_k(x) = -\infty$. Here, by the second inequality of (2.61), the natural numbers n_k, $k = 1, 2, \ldots$, form an increasing sequence.

Setting

$$Q = \varliminf_{k \to \infty} M_k \tag{2.66}$$

and using (2.56), we obtain

$$\text{meas } Q = 2\pi, \quad Q \subset [-\pi, \pi]. \tag{2.67}$$

We shall prove that the sequence of functions

$$S_{n_k}(x) \qquad (k = 1, 2, \ldots) \tag{2.68}$$

and

$$f_k(x) \qquad (k = 1, 2, \ldots) \tag{2.69}$$

are equiconvergent at each point of Q.

Let x be an arbitrary point of Q. Then by (2.66) there exists a natural number k_x such that

$$x \in M_k \qquad (k > k_x). \tag{2.70}$$

Let us suppose that k increases without bound while taking values for which $f_k(x)$ is finite. Then we may use (2.63) and (2.70) to see that $S_{n_k}(x) - f_k(x)$ tends to zero. Let us suppose now that k increases without bound while taking values for which $f_k(x) = +\infty$. Then, by (2.64) and (2.70), $S_{n_k}(x)$ tends to $+\infty$. Finally, if k increases without bound while taking values for which $f_k(x) = -\infty$, then, by (2.65) and (2.70), $S_{n_k}(x)$ tends to $-\infty$.

Since $S_{n_k}(x)$ may take only finite values, it follows from what has been said and Definition 1 (§1) that the sequences of functions (2.68) and (2.69) are equiconvergent at each point of Q, and in that case, using (2.67), we see that the sequences of functions (2.68) and (2.69) are equiconvergent almost everywhere on $[-\pi, \pi]$.

We have already seen that $\psi_1(x)$ and $\psi_2(x)$ are the limits inferior and superior respectively in measure on $[-\pi, \pi]$ of (2.69). Also, we have proved that (2.44) holds almost everywhere on $[-\pi, \pi]$. Since (2.68) and (2.69) are equiconvergent almost everywhere on $[-\pi, \pi]$, it follows from (11) (see the Introduction), Lemma G and

Lemma C that $\psi_1(x)$ and $\psi_2(x)$ are the limits inferior and superior respectively in measure on $[-\pi, \pi]$ of (2.68), and moreover, that

$$\underline{\lim}_{k\to\infty} S_{n_k}(x) = \varphi_1(x), \quad \overline{\lim}_{k\to\infty} S_{n_k}(x) = \varphi_2(x) \tag{2.71}$$

almost everywhere on $[-\pi, \pi]$.

Thus, we have assumed that (1) is an arbitrary series of type $A(F_1, F_2)$, where $F_1(x)$ and $F_2(x)$ are any measurable functions that are defined almost everywhere on $[-\pi, \pi]$ and satisfy (4) almost everywhere on this segment, and we have proved that for any measurable functions $\varphi_1(x)$, $\psi_1(x)$, $\psi_2(x)$ and $\varphi_2(x)$ that are defined almost everywhere on $[-\pi, \pi]$ and satisfy (11) almost everywhere on this segment([19]) we may define an increasing sequence of natural numbers n_k, $k = 1, 2, \ldots$, such that (2.71) holds for the sequence of partial sums $S_{n_k}(x)$, $k = 1, 2, \ldots$, of the series (1) almost everywhere on $[-\pi, \pi]$, and, moreover, with $\psi_1(x)$ and $\psi_2(x)$ equal to the limits inferior and superior respectively in measure on $[-\pi, \pi]$ of this sequence. From this it follows that the series (1) is a series of type $B(F_1, F_2)$ (see Definition B in the Introduction), i.e. an arbitrary trigonometric series of type $A(F_1, F_2)$ is a series of type $B(F_1, F_2)$, as was to be shown.

BIBLIOGRAPHY

1. D. E. Men'šov, *Sur les séries trigonométriques universelles*, C. R. (Dokl.) Acad. Sci. URSS **49** (1945), 79–82. MR **7**, 435.

2. ———, *On convergence in measure of trigonometric series*, Trudy Mat. Inst. Steklov. **32** (1950); English transl., Amer. Math. Soc. Transl. (1) 3 (1962), 196–270. MR **12**, 254.

3. ———, *On limits of indeterminacy of Fourier series*, Mat. Sb. **30 (72)** (1950), 601–650. (Russian) MR **14**, 40.

Translated by D. LEVINE

([19])In the proof we have defined the functions $F_1(x)$, $F_2(x)$, $\varphi_1(x)$, $\psi_1(x)$, $\psi_2(x)$ and $\varphi_2(x)$ at the points where they were originally not defined in such a way that (4) and (11) hold everywhere on $[-\pi, \pi]$.

Amer. Math Soc. Transl.
(2) Vol. 111, 1978

ON LIMIT FUNCTIONS OF TRIGONOMETRIC SERIES*(1)

D. E. MEN'ŠOV

§1. Introduction

We choose a series

$$\sum_{n=0}^{\infty} u_n(x), \tag{1.1}$$

whose terms are measurable functions, finite almost everywhere on some segment $[a, b]$. Put

$$Q_n(x) = \sum_{\nu=0}^{n} u_\nu(x). \tag{1.2}$$

We introduce the following definitions.

DEFINITION 1. The function $\varphi(x) \equiv \varphi(x, E)$, defined almost everywhere on some set $E \subset [a, b]$(2) will be called a *limit function of the series* (1.1) if there exists an increasing sequence of natural numbers ρ_k $(k = 1, 2, \ldots)$ such that

$$\lim_{k\to\infty} Q_{\rho_k}(x) = \varphi(x) \tag{1.3}$$

almost everywhere on E, where $Q_n(x)$ is defined from (1.2).

DEFINITION 2. If $\varphi(x) \equiv \varphi(x, E)$ satisfies the conditions of Definition 1, then we will also say that $\varphi(x)$ is a *limit function of the series* (1.1) on the set E (3)

The following assertion follows directly from Definition 2.

If $E \subset [a, b]$, $\operatorname{meas}(E' - E) = 0$, $\operatorname{meas} E' > 0$, and if $\varphi(x, E)$ is a limit function of the series (1.1) on the set E, then the function $\psi(x, E')$, equal to $\varphi(x, E)$ almost everywhere on E', is a limit function of the same series on E'.

In the present paper we will consider the set of all limit functions of an arbitrary trigonometric series

$$\frac{a_0}{2} + \sum_{n=1}^{\infty} (a_n \cos nx + b_n \sin nx). \tag{1.4}$$

AMS (MOS) subject classifications (1970). Primary 42A20; Secondary 28A20.

*Translation of Trudy Moskov. Mat. Obšč. 7 (1958), 291–334. MR 21 #2152.

(1) The basic contents of this paper were reported at the meeting of the Moscow Mathematical Society on 16 April 1957.

(2) When we say that a measurable function is defined almost everywhere on some measurable set, we do not exclude the possibility that this function may be equal to $+\infty$ or $-\infty$ on sets of positive measure.

(3) This last definition was given in [1] (Definition 3). There is an inessential difference, in that in [1] we did not introduce the requirement that meas $E > 0$.

We introduce the set $M = \{\varphi(x, E)\}$ of measurable functions $\varphi(x, E)$, each of which is defined almost everywhere on some set $E \subset [-\pi, \pi]$ of positive measure; this set may be different for different functions $\varphi(x, E) \in M$. (4)

Our problem is the following.

Find a necessary and sufficient condition for the set M to be the set of all limit functions of some trigonometric series.

In order to formulate this condition, we introduce the following definition.

DEFINITION 3. We choose a sequence of functions

$$\varphi_n(x, E_n) \qquad (n = 1, 2, \dots), \tag{1.5}$$

each of which is defined almost everywhere on the corresponding set E_n. Put

$$Q = \varliminf_{n\to\infty} E_n \tag{1.6}$$

and suppose that the set Q and some other set E satisfy the conditions

$$\text{meas } Q > 0, \text{ meas } E > 0, \text{ meas}(E - Q) > 0. \text{(5)} \tag{1.7}$$

We shall say that the function $\varphi(x, E)$, defined almost everywhere on the set E, is a *limit element in the wide sense of the sequence* (1.5) if

$$\lim_{n\to\infty} \varphi_n(x, E_n) = \varphi(x, E) \tag{1.8}$$

almost everywhere on E. (6)

It is easy to see that a limit element in the wide sense of the sequence (1.5) is defined nonuniquely. Indeed, if $\varphi(x, E)$ is a limit element in the wide sense of the sequence (1.5) and $\text{meas}(E' - E) = 0$, meas $E' > 0$, then a function $\psi(x, E')$ equal to $\varphi(x, E)$ almost everywhere on E' is also a limit element in the wide sense of the sequence (1.5).

DEFINITION 4. Choose some set $M = \{\varphi(x, E)\}$ of functions $\varphi(x, E)$, each of which is defined almost everywhere on the corresponding set E, meas $E > 0$. We shall call the function $f(x, E')$, defined almost everywhere on a set E' of positive measure, a *limit element in the wide sense of the set M* if there exists a sequence of functions $\varphi_n(x, E_n)$ $(n = 1, 2, \dots)$, belonging to M, for which $f(x, E')$ is a limit element in the wide sense.

In this definition the functions $\varphi_n(x, E_n)$ are not necessarily distinct for distinct n. In such a case any function $\varphi(x, E)$ of the set $M = \{\varphi(x, E)\}$ is a limit element in the wide sense of that set. Indeed, if we put $E_n = E$ and $\varphi_n(x, E_n) = \varphi(x, E)$ $(n = 1, 2, \dots)$, then obviously $\varphi(x, E)$ is a limit element in the wide sense of the

(4) In what follows we will suppose that each of the functions $\varphi(x, E)$ is considered only at the points of E and is not defined outside that set.

(5) We will always suppose that all sets lie on some segment.

(6) It makes sense to speak of the limit of the functions $\varphi_n(x, E_n)$ almost everywhere on E, since on the basis of (1.6) and (1.7), at any point $x \in E$, other than possibly a set of measure zero, the function $\varphi_n(x, E_n)$ is defined for all sufficiently large values of n.

The functions $\varphi_n(x, E_n)$ may be equal to $+\infty$ and $-\infty$ on sets of positive measure. It is clear how one should understand equation (1.8) if at the point x in question $\varphi(x, E) = +\infty$ or $-\infty$ and $\varphi_n(x, E_n) = +\infty$ or $-\infty$ for an infinite set of values of n.

sequence of functions $\varphi_n(x, E_n)$, from which our assertion follows.

DEFINITION 5. The set $\overline{M}$ will be said to be the *closure of the set* $M = \{\varphi(x, E)\}$ *in the wide sense* if $\overline{M}$ is gotten from M by adjoining to it all the limit elements in the wide sense.

DEFINITION 6. The set $M = \{\varphi(x, E)\}$ will be said to be *closed in the narrow sense* if it coincides with its closure in the wide sense.

From this definition it follows that the empty set is a closed set in the narrow sense.

We note also the following property of sets closed in the narrow sense.

LEMMA A. *Suppose that* $M = \{\varphi(x, E)\}$ *is a set of functions* $\varphi(x, E)$, *closed in the narrow sense, with* meas $E > 0$ *for any function* $\varphi(x, E) \in M$. *If* $f(x, E) \in M$ *and if*

$$\text{meas}\,(E' - E) = 0, \quad \text{meas}\, E' > 0, \quad E' \subset [a, b] \tag{1.9}$$

(7)

and

$$\psi(x, E') = f(x, E) \tag{1.10}$$

almost everywhere on E', *then* $\psi(x, E') \in M$.

PROOF. Since $f(x, E) \in M$, in view of the remark made after Definition 4 this function is a limit element of the set M. Hence it follows on the basis of (1.9), (1.10) and the remark made after Definition 3 that $\psi(x, E')$ is a limit element of M. In such a case, in view of Definition 5, $\psi(x, E') \in \overline{M}$, where $\overline{M}$ is the closure of M. Since by hypothesis M is closed in the narrow sense, it follows from Definition 6 that $\overline{M} = M$, so that $\psi(x, E') \in M$, as required.

It is possible to prove the following theorem.

THEOREM I. *Suppose that* $M = \{\varphi(x, E)\}$ *is a set of measurable functions* $\varphi(x, E)$, *each of which is defined almost everywhere on some set* E *of positive measure lying on* $[-\pi, \pi]$. *For* M *to be the set of all limit functions of the trigonometric series* (1.4), *it is necessary and sufficient that* M *should be closed in the narrow sense.*

Theorem I can be obtained as a consequence of more general theorems. Before formulating these theorems, we recall some definitions from [2], §1.

DEFINITION 7. Suppose given a sequence of measurable functions $f_n(x)$, defined almost everywhere on some segment $[a, b]$. We shall say that the measurable function $F(x)$, defined almost everywhere on $[a, b]$, is *an upper limit in measure on* $[a, b]$ *of the sequence of functions* $f_n(x)$ $(n = 1, 2, \ldots)$ if $F(x)$ has the following properties:

a°. $$\lim_{n\to\infty} \text{meas}\{E[f_n(x) > \varphi(x)] \cdot E[\varphi(x) > F(x)\} = 0 \tag{1.11}$$

[2])

for any measurable function $\varphi(x)$ defined almost everywhere on $[a, b]$;

---

(7) $[a, b]$ is a segment on which the sets E corresponding to all possible functions $\varphi(x, E)$ are situated.

(8) If $\varphi_1(x)$ and $\varphi_2(x)$ are two arbitrary measurable functions defined almost everywhere on $[a, b]$, then we will denote as usual by $E[\varphi_1(x) > \varphi_2(x)]$ the set of all points on $[a, b]$ for which $\varphi_1(x) > \varphi_2(x)$.

b°. $$\overline{\lim_{n\to\infty}} \text{ meas}\{E[f_n(x) > \psi(x)] \cdot E[F(x) > \psi(x)]\} > 0 \tag{1.12}$$

for any measurable function $\psi(x)$ defined almost everywhere on $[a, b]$ and such that meas $E[F(x) > \psi(x)] > 0$.

DEFINITION 8. Suppose that the functions $f_n(x)$ have the same properties as in Definition 7. We will say that a measurable function $G(x)$, defined almost everywhere on $[a, b]$, is a *lower limit in measure on that segment* of the sequence of functions $f_n(x)$ $(n = 1, 2, \ldots)$, if $G(x)$ satisfies the following conditions:

α°. $$\lim_{n\to\infty} \text{meas}\{E[f_n(x) < \tau(x)] \cdot E[\tau(x) < G(x)]\} = 0 \tag{1.13}$$

for any measurable function $\tau(x)$ defined almost everywhere on $[a, b]$;

β°. $$\overline{\lim_{n\to\infty}} \text{ meas}\{E[f_n(x) < \chi(x)] \cdot E[G(x) < \chi(x)]\} > 0 \tag{1.14}$$

for any measurable function $\chi(x)$ defined almost everywhere on $[a, b]$ and such that meas $E[G(x) \leqslant \chi(x)] > 0$.

It is possible to prove that the upper and lower limits in measure are defined uniquely up to a set of measure zero and have many properties analogous to the corresponding properties of the ordinary upper and lower limits.(9)

Now we choose a series (1.1) whose terms are measurable functions, finite almost everywhere on $[a, b]$. We introduce the following definition.

DEFINITION 9. We will say that the measurable functions $F(x)$ and $G(x)$, defined almost everywhere on $[a, b]$, are *upper* and *lower limits in measure on* $[a, b]$ *of the series* (1.1), if $F(x)$ and $G(x)$ are respectively equal to the upper and lower limits in measure on $[a, b]$ of sequence of partial sums of the same series.(10)

The first part of Theorem I, i.e. the necessity of the condition entering into its formulation, may be obtained from the following theorem.

THEOREM II. *Suppose that the terms of the series* (1.1) *are measurable and finite almost everywhere on some segment* $[a, b]$, *and suppose that* $F(x)$ *and* $G(x)$ *are respectively upper and lower limits in measure on* $[a, b]$ *of that series. Then the set* $M = \{\varphi(x, E)\}$ *of all limit functions of the series* (1.1) *has the following properties*:

α. *M is closed in the narrow sense.*

β. *If* $\varphi(x, E) \in M$, *then* $\varphi(x, E)$ *is measurable on E and*

$$G(x) \leqslant \varphi(x, E) \leqslant F(x) \tag{1.15}$$

almost everywhere on that set.

γ. *If* $\varphi(x, E) \in M$ *and if*

---

(9) [2], §2, Theorems A–E. It was supposed in [2] that the functions $f_n(x)$ $(n = 1, 2, \ldots)$ are finite almost everywhere on $[a, b]$. But the definitions of the upper and lower limits in measure make sense without any restriction, while Theorems A–E remain valid without any restriction and their proofs remain unchanged.

(10) This definition was given in [1] (Definition 5).

$$E_0 = E + E\,[F(x) = G(x)], \tag{1.16}$$

then the function $\varphi_0(x, E_0)$ *defined by*

$$\varphi_0(x, E_0) = \left\{ \begin{array}{lll} \varphi(x, E) & \textit{for} & x \in E, \\ F(x) & \textit{for} & x \in E_0 - E, \end{array} \right\} \tag{1.17}$$

also belongs to M.(11)

The second part of Theorem I, i.e. the sufficiency, is obtained from the following theorem.

THEOREM III. *Suppose that* $M = \{\varphi(x, E)\}$ *is a nonempty set of functions* $\varphi(x, E)$, *each of which is defined almost everywhere on the corresponding set* E, *while*

$$\operatorname{meas} E > 0, E \subset [-\pi, \pi]. \tag{1.18}$$

Suppose that the set M *satisfies conditions* α, β, *and* γ *of Theorem* II, *where* $F(x)$ *and* $G(x)$ *are measurable functions defined almost everywhere on* $[-\pi, \pi]$ *and satisfying the inequality*

$$G(x) \leqslant F(x) \tag{1.19}$$

almost everywhere on that segment.(12)

In this case there exists a trigonometric series (1.4), *having the following properties*:

A. *M is the set of all limit functions of the series* (1.4).

B. $F(x)$ *and* $G(x)$ *are respectively upper and lower limits in measure on* $[-\pi, \pi]$ *of the series* (1.4).

C. $$\lim_{n\to\infty} a_n = 0, \qquad \lim_{n\to\infty} b_n = 0. \tag{1.20}$$

The theorems formulated in [1] may be deduced from Theorems II and III formulated just above. This derivation will be published in another paper.

§2

Before entering into the proof of Theorem II, we make a number of preparatory remarks. We recall first of all the definition of uniform convergence on a set for the case when the limit function may take on the values $+\infty$ or $-\infty$.

DEFINITION 10. We will say that the sequence of functions $f_n(x)$ $(n = 1, 2, \ldots)$ defined on some set E *converges uniformly* to the function $f(x)$ on E, if for any positive number ϵ it is possible to determine a natural number N such that for any $x \in E$ and any natural $n > N$ one of the following conditions is satisfied:

a°. $-\infty < f_n(x) < +\infty$, and $|f_n(x) - f(x)| < \epsilon$, if $f(x)$ is finite;

b°. $f_n(x) > 1/\epsilon$, if $f(x) = +\infty$;

(11)In (1.16) we define $\varphi(x, E)$ and $F(x)$ in any way on a set of measure zero where they were originally not defined.

(12)We are supposing that the conditions α, β and γ are formulated for the segment $[a, b] = [-\pi, \pi]$.

c°. $f_n(x) < -1/\epsilon$, if $f(x) = -\infty$.

Egorov's theorem is valid for this definition of uniform convergence; in fact, the following assertion is true (see, for example, [3], Lemma K).

THEOREM α. *If the sequence of measurable functions $f_n(x)$ $(n = 1, 2, \ldots)$ defined almost everywhere on a set E of positive measure converges almost everywhere on that set to a function $f(x)$, then for any positive number σ it is possible to determine a measurable set L satisfying the conditions*

$$L \subset E, \text{ meas } L > \text{meas } E - \sigma \tag{2.1}$$

and such that the sequence $f_n(x)$ converges uniformly to $f(x)$ on L.

In [2] we gave a definition of convergence in measure of a sequence of measurable functions in the case when the limiting function is equal to $+\infty$ or $-\infty$ on a set of positive measure. In fact we made the following definition.

DEFINITION 11. Suppose that the functions

$$f_1(x), f_2(x), \ldots, f_n(x), \ldots \tag{2.2}$$

are measurable and finite almost everywhere on some set E of positive measure. We shall say that the sequence of functions $f_n(x)$ *converges in measure on the set E to the measurable function* $f(x)$, if the equation

$$f_n(x) = g_n(x) + \alpha_n(x) \ (n = 1, 2, \ldots) \tag{2.3}$$

is satisfied almost everywhere on E, where $g_n(x)$ and $\alpha_n(x)$ are measurable and finite almost everywhere on E, while

$$\lim_{n\to\infty} g_n(x) = f(x) \tag{2.4}$$

almost everywhere on E and the sequence of functions $\alpha_n(x)$ converges in measure to zero on the same set.(13)

Suppose that the functions $f_n(x)$ of the sequence (2.2) are measurable and finite almost everywhere on some segment $[a, b]$. In [2] the following theorem was proved (Theorem J of §3).

If the measurable functions $F(x)$ and $G(x)$ are respectively upper and lower limits in measure on $[a, b]$ of the sequence (2.2) and if $F(x) = G(x)$ almost everywhere on that segment, then the sequence (2.2) converges in measure on $[a, b]$ to the common value of the functions $F(x)$ and $G(x)$.

We may generalize this theorem somewhat. In fact the following theorem holds.

THEOREM β. *If the measurable functions $F(x)$ and $G(x)$ are respectively upper and lower limits in measure on $[a, b]$ of the sequence (2.2) and if $F(x) = G(x)$ on a set E of positive measure, $E \subset [a, b]$, then the sequence (2.2) converges in measure*

(13)See [2], §1. The function $f(x)$ may be equal to $+\infty$ or $-\infty$ on a set of positive measure. In [2] we suppose that the set E is a segment, but the definition remains unchanged for any set of positive measure.

on the set E to the common value of F(x) and G(x).(14)

Theorem β is an immediate consequence of the following fact.

If F(x) and G(x) are defined as in Theorem β, *then* (2.3) *is satisfied almost everywhere on* $[a, b]$, *where* $g_n(x)$ *and* $\alpha_n(x)$ $(n = 1, 2, \ldots)$ *are measurable and finite almost everywhere on* $[a, b]$,

$$\overline{\lim_{n\to\infty}} g_n(x) = F(x), \qquad \underline{\lim_{n\to\infty}} g_n(x) = G(x)$$

almost everywhere on the same segment, and the sequence of functions $\alpha_n(x)$ *converges in measure to zero on* $[a, b]$.(15)

Theorem β follows directly from this theorem. Indeed, if $F(x) = G(x)$ on a set $E \subset [a, b]$ of positive measure, then in view of the preceding theorem

$$\overline{\lim_{n\to\infty}} g_n(x) = \underline{\lim_{n\to\infty}} g_n(x) = F(x) = G(x) = \lim_{n\to\infty} g_n(x)$$

almost everywhere on E. Hence, on the basis of (2.3), the properties of the functions $g_n(x)$ and $\alpha_n(x)$, and Definition 11, the sequence (2.2) converges in measure on E to the common value of $F(x)$ and $G(x)$, as we were required to prove.

For a sequence converging in measure to a function which can take on infinite values, there is a theorem which is well known in the case of a limit function which is finite almost everywhere.

THEOREM γ. *If the sequence* (2.2) *converges in measure to the function* $f(x)$ *on a set E of positive measure, then there exists an increasing sequence of natural numbers* n_k $(k = 1, 2, \ldots)$ *such that*

$$\lim_{k\to\infty} f_{n_k}(x) = f(x) \tag{2.5}$$

almost everywhere on E.

PROOF. Since by hypothesis the sequence (2.2) converges in measure on E to $f(x)$, on the basis of Definition 11 we have equation (2.3), where $g_n(x)$ and $\alpha_n(x)$ are measurable and finite almost everywhere on E, while (2.4) holds almost everywhere on E and the sequence of functions $\alpha_n(x)$ $(n = 1, 2, \ldots)$ converges to zero on E. On the other hand, since it is known (see, for example, [4], Chapter IV) that Theorem γ is valid in the case when the limit function is finite almost everywhere on E, there exists an increasing sequence of natural numbers n_k $(k = 1, 2, \ldots)$ such that

$$\lim_{k\to\infty} \alpha_{n_k}(x) = 0 \tag{2.6}$$

almost everywhere on E. Comparing (2.3), (2.4) and (2.6), we see that (2.5) is satisfied almost everywhere on E, as we were required to prove.

PROOF OF THEOREM II. We suppose that the terms of the series (1.1) are measurable and finite almost everywhere on some segment $[a, b]$. Denote by $M = \{\varphi(x, E)\}$ the set of all limit functions $\varphi(x, E)$ of this series. If M is the empty set,

(14)In this section we will always suppose that the functions $f_n(x)$ of the sequence (2.2) are measurable and finite almost everywhere on $[a, b]$.

(15)[2], §3, Theorem H. In this theorem $F(x)$ does not necessarily have to equal $G(x)$ almost everywhere.

the theorem is proved. If M is nonempty, then each of the functions $\varphi(x, E) \in M$ is measurable and is defined almost everywhere on the corresponding set E, while

$$\text{meas } E > 0, E \subset [a, b] \tag{2.7}$$

(see Definition 1 of §1).

We prove first of all that the set M satisfies condition α (see the formulation of Theorem II of §1), i.e. we prove that M is closed in the narrow sense.

Suppose that the function $f(x, E')$, defined almost everywhere on the point set E', is an arbitrary function belonging to $\overline{M}$, the closure in the wide sense of M (see Definition 5 in §1). Then in view of that definition the function $f(x, E')$ is a limit element in the wide sense of the set M, and therefore, on the basis of Definition 4 in §1, there exists a sequence of functions $\varphi_n(x, E_n)$ $(n = 1, 2, \ldots)$, belonging to M, for which $f(x, E')$ is a limit element in the wide sense.

Hence in view of Definition 3 it follows that

$$\text{meas } E' > 0, \qquad E' \subset [a, b], \qquad \text{meas}\,(E' - \varliminf_{n\to\infty} E_n) = 0 \tag{2.8}$$

and

$$\lim_{n\to\infty} \varphi_n(x, E_n) = f(x, E') \tag{2.9}$$

almost everywhere on E'.(16)

We denote by M the set of all limit functions of the type (1.1). In such a case, since

$$\varphi_n(x, E_n) \in M \qquad (n = 1, 2, \ldots), \tag{2.10}$$

it follows that each of the functions $\varphi_n(x, E_n)$ is a limit function of the series (1.1), and, accordingly, on the basis of Definition 1,

$$\text{meas } E_n > 0, \qquad E_n \subset [a, b] \qquad (n = 1, 2, \ldots), \tag{2.11}$$

and for each value $n = 1, 2, \ldots$ it is possible to define an increasing sequence of natural numbers ρ_{nk} $(k = 1, 2, \ldots)$ such that

$$\lim_{k\to\infty} Q_{\rho_{nk}}(x) = \varphi_n(x, E_n) \qquad (n = 1, 2, \ldots) \tag{2.12}$$

almost everywhere on E_n, where $Q_n(x)$ is defined from (1.2).

From (2.12) it follows in view of Theorem α that for each $n = 1, 2, \ldots$ it is possible to define a measurable set L_n satisfying the conditions

$$L_n \subset E_n, \qquad \text{meas } L_n > \text{meas } E_n - \frac{1}{2^n} \qquad (n = 1, 2, \ldots) \tag{2.13}$$

and such that the sequence of functions $Q_{\rho_{nk}}(x)$ $(k = 1, 2, \ldots)$ converges uniformly to $\varphi_n(x, E_n)$ on L_n. In such a case, on the basis of Definition 10, for each $n = 1, 2, \ldots$ it is possible to define a natural number k_n such that

(16) Each of the functions $\varphi_n(x, E_n)$ is defined almost everywhere on the corresponding set E_n.

$$\lambda_n < \lambda_{n+1} \qquad (n = 1, 2, \ldots), \tag{2.14}$$

where

$$\lambda_n = \rho_{n,\, k_n} \qquad (n = 1, 2, \ldots), \tag{2.15}$$

and, moreover, for any $x \in L_n$ $(n = 1, 2, \ldots)$ one of the following conditions is satisfied:

a°. $$-\infty < Q_{\lambda_n}(x) < +\infty, \qquad |Q_{\lambda_n}(x) - \varphi_n(x, E_n)| < \frac{1}{2^n}, \tag{2.16}$$

if $\varphi_n(x, E_n)$ is finite;

b°. $$Q_{\lambda_n}(x) > 2^n, \tag{2.17}$$

if $\varphi_n(x, E_n) = +\infty$; or

c°. $$Q_{\lambda_n}(x) < -2^n, \tag{2.18}$$

if $\varphi_n(x, E_n) = -\infty$.

Denote by e' the set of all points of E' at which (2.9) is not satisfied. Put

$$e_n = E_n - L_n \qquad (n = 1, 2, \ldots), \tag{2.19}$$

$$e'' = E' - \varliminf_{n\to\infty} E_n, \qquad E''' = E' - \varlimsup_{n\to\infty} e_n - e' - e''. \tag{2.20}$$

Then on the basis of (2.8) and (2.13) we have

$$L_n = E_n - e_n, \qquad \text{meas}\, e_n < \frac{1}{2^n} \qquad (n = 1, 2, \ldots), \tag{2.21}$$

$$\text{meas}\, e' = 0, \qquad \text{meas} \varlimsup_{n\to\infty} e_n = 0, \qquad \text{meas}\, e'' = 0, \tag{2.22}$$

$$E'' \subset E', \qquad \text{meas}\, E'' = \text{meas}\, E'. \tag{2.23}$$

Moreover, in view of (2.20) and (2.21)

$$E'' \subset \varliminf_{n\to\infty} E_n - \varlimsup_{n\to\infty} e_n \subset \varliminf_{n\to\infty} L_n. \tag{2.24}$$

We note further that at all points of E'' the equation (2.9) is satisfied.

We now suppose that x satisfies the condition

$$x \in E''. \tag{2.25}$$

Then on the basis of (2.24) we may define a natural number n_x such that

$$x \in L_n \qquad (n > n_x). \tag{2.26}$$

Accordingly, at the given point x, for any $n > n_x$ one of the conditions a°, b°, or c° is satisfied (see (2.16)–(2.18)).

Now we suppose that n increases unboundedly, taking on values for which $\varphi_n(x, E_n)$ is finite. Then in view of (2.16) $Q_{\lambda_n}(x) - \varphi(x, E_n) \longrightarrow 0$, and accordingly, on the basis of (2.9),

$$Q_{\lambda_n}(x) \longrightarrow f(x, E'). \tag{2.27}$$

Thus, at the point x in question condition (2.27) is satisfied as n increases unboundedly, taking on values for which $\varphi_n(x, E_n)$ is finite.

Now we suppose that n increases unboundedly, taking on values for which $\varphi_n(x, E_n) = +\infty$. Then on the basis of (2.9) we have $f(x, E') = +\infty$. Moreover, $Q_{\lambda_n}(x) \longrightarrow +\infty$, in view of (2.17). Hence it follows that (2.27) is satisfied at the point x in question in the case when n increases unboundedly, taking on values for which $\varphi_n(x, E_n) = +\infty$.

Finally we suppose that n increases unboundedly, taking on values for which $\varphi_n(x, E_n) = -\infty$. In this case, on the basis of (2.9) we have $f(x, E') = -\infty$, and, moreover, $Q_{\lambda_n}(x) \longrightarrow -\infty$, on the basis of (2.18). We thus see that (2.27) is satisfied at the point x in question also for the case when n increases unboundedly taking on values for which $\varphi_n(x, E_n) = -\infty$.

Since we have supposed that x is an arbitrary point satisfying (2.25), it follows from the above that

$$\lim_{n\to\infty} Q_{\lambda_n}(x) = f(x, E') \tag{2.28}$$

at any point x of the set E''; in this case, on the basis of (2.23) equation (2.28) is satisfied almost everywhere on E'.

The functions $Q_n(x)$ are partial sums of the series (1.1) (see (1.2)). Further, as we already know, the set E' satisfies the conditions meas $E' > 0$ and $E' \subset [a, b]$ (see (2.8)), and the natural numbers λ_n $(n = 1, 2, \ldots)$ form an increasing sequence (see (2.14)). In such a case, on the basis of Definition 1 of §1, $f(x, E')$ is a limit function of the series (1.1).

On the other hand, by hypothesis, M is the set of all limit functions of the series (1.1). Accordingly, $f(x, E') \in M$. In such a case, since $f(x, E')$ is an arbitrary function of the set $\overline{M}$, any function of $\overline{M}$ also belongs to M. But $\overline{M}$ is the closure in the wide sense of M. Thus M coincides with its closure in the wide sense. Thus in view of Definition 6 of §1 the set M is closed in the narrow sense.

Thus we have proved that the set M of all limit functions of the series (1.1) is closed in the narrow sense. In other words we have proved that M satisfies condition α of Theorem II.

Now we shall prove that M satisfies condition β of the same theorem. First of all we show that each of the functions $\varphi(x, E) \in M$ is measurable on the corresponding set E.

Indeed, if $\varphi(x, E) \in M$, then this function is a limit function of the series (1.1), and so there exists an increasing sequence of natural numbers n_k $(k = 1, 2, \ldots)$ for which

$$\lim_{k\to\infty} Q_{n_k}(x) = \varphi(x, E) \tag{2.29}$$

almost everywhere on E. We already know that the terms of (1.1) are measurable and finite almost everywhere on $[a, b]$ and that the functions $Q_n(x)$ are defined by (1.2). In such a case, since $E \subset [a, b]$, it follows from (2.29) that $\varphi(x, E)$ is measurable on E.

Now we shall prove that for any function $\varphi(x, E) \in M$

$$\text{meas}\, E\,[\varphi(x, E) > F(x)] > 0.\text{(17)} \tag{2.30}$$

Then $\varphi(x, E) > -\infty$ for points x belonging to $E\,[\varphi(x, E) > F(x)]$, so that, on the basis of (2.30), it is possible to determine a measurable set H_0 such that

$$H_0 \subset E\,[\varphi(x, E) > F(x)],\ \text{meas}\, H_0 > 0, \tag{2.31}$$

while on H_0 either

$$-\infty < \varphi(x, E) < +\infty \tag{2.32}$$

everywhere or

$$\varphi(x, E) = +\infty \tag{2.33}$$

everywhere. Clearly

$$H_0 \subset E. \tag{2.34}$$

It follows from (2.31) that

$$F(x) < +\infty\ (x \in H_0). \tag{2.35}$$

In such a case it is possible to choose a positive constant λ and a measurable set H_0' such that

$$H_0' \subset H_0,\ \text{meas}\, H_0' > 0, \tag{2.36}$$

$$F(x) < \lambda\ \ (x \in H_0'). \tag{2.37}$$

Moreover, if (2.32) is satisfied everywhere on H_0', then H_0' may be chosen in such a way that

$$\varphi(x, E) - \epsilon_0 > F(x)\ (x \in H_0'), \tag{2.38}$$

where ϵ_0 is some positive number. Further, it follows from (2.36) that either (2.32) is satisfied everywhere on H_0' or else (2.33) is satisfied everywhere there.

Since in view of (2.34) and (2.36) $H_0' \subset E$, and since (2.29) is satisfied almost everywhere on E, it follows that (2.29) is satisfied also almost everywhere on H_0'. In such a case, on the basis of Theorem α of §2 it is possible to determine a measurable set H_0'' satisfying the conditions

$$H_0'' \subset H_0,\ \text{meas}\, H_0'' > 0 \tag{2.39}$$

and such that the sequence of functions $Q_{n_k}(x)$ $(k = 1, 2, \ldots)$ converges uniformly

(17) $E[\varphi(x, E) > F(x)] \subset E$, since $\varphi(x, E)$ is not defined outside E.

to $\varphi(x, E)$ on H_0''.

If (2.32) is satisfied everywhere on H_0'', then in view of Definition 10 and (2.39) it is possible to determine a natural number k_0 such that

$$Q_{n_k}(x) > \varphi(x, E) - \epsilon_0 \; (x \in H_0'', k > k_0). \tag{2.40}$$

If now (2.33) is satisfied everywhere on H_0', it follows from Definition 10 and (2.39) that a number k_0 may be determined in such a way that

$$Q_{n_k}(x) > \lambda \; (x \in H_0'', k > k_0). \tag{2.41}$$

Put

$$\varphi(x) = \varphi(x, E) - \epsilon_0 \; (x \in H_0'')$$

if (2.32) is satisfied everywhere on H_0',

$$\varphi(x) = \lambda \; (x \in H_0'')$$

if (2.33) is satisfied everywhere on H_0', and

$$\varphi(x) = 0 \; (x \in [a, b] - H_0'')$$

in both cases. Then $\varphi(x)$ will be measurable and defined everywhere on $[a, b]$, while on the basis of (2.37)–(2.39)

$$\varphi(x) > F(x) \; (x \in H_0''), \tag{2.42}$$

and on the basis of (2.40) and (2.41)

$$Q_{n_k}(x) > \varphi(x) \; (x \in H_0'', k > k_0). \tag{2.43}$$

It follows from (2.42) and (2.43) that

$$H_0'' \subset \{E[Q_{n_k}(x) > \varphi(x)] \cdot E[\varphi(x) > F(x)]\}.$$

Hence in view of (2.39)

$$\overline{\lim_{n\to\infty}} \text{ meas} \{E[Q_{n_k}(x) > \varphi(x)] \cdot E[\varphi(x) > F(x)]\} \geqslant \text{meas } H_0'' > 0.$$

Since the natural numbers n_k $(k = 1, 2, \ldots)$ form an increasing sequence, we have that

$$\overline{\lim_{n\to\infty}} \text{ meas} \{E[Q_n(x) > \varphi(x)] \cdot E[\varphi(x) > F(x)]\} > 0. \tag{2.44}$$

Hence it follows that the function $F(x)$ cannot be the upper limit in measure on $[a, b]$ of the sequence of functions $Q_n(x)$ $(n = 1, 2, \ldots)$ (see Definition 7 of §1, condition a°); and since the $Q_n(x)$ are partial sums of the series (1.1), $F(x)$ cannot be the upper limit in measure on $[a, b]$ of that series, which contradicts the hypotheses of Theorem II. We have arrived at this contradiction on the assumption that for some function $\varphi(x, E) \in M$ inequality (2.30) is satisfied. Therefore

$$\varphi(x, E) \leqslant F(x) \tag{2.45}$$

almost everywhere on E, if $\varphi(x, E) \in M$.

Now we shall prove that for any function $\varphi(x, E) \in M$ the inequality

$$G(x) \leqslant \varphi(x, E) \tag{2.46}$$

is satisfied almost everywhere on E. Indeed, put

$$v_n(x) = - u_n(x) \ (n = 0, 1, 2, \ldots). \tag{2.47}$$

Then, on the basis of (1.2),

$$\sum_{\nu=0}^{n} v_\nu(x) = - Q_n(x) \ (n = 0, 1, 2, \ldots), \tag{2.48}$$

i.e. the quantities $- Q_n(x)$ $(n = 0, 1, 2, \ldots)$ are the partial sums of the series

$$\sum_{n=0}^{\infty} v_n(x). \tag{2.49}$$

By the hypothesis of Theorem II, $G(x)$ is the lower limit in measure on $[a, b]$ of the series (1.1), i.e. it is the lower limit in measure on $[a, b]$ of the sequence of partial sums $Q_n(x)$ of this series (see Definition 9 of §1). Then, in view of Theorem D in §2 of [2], $- G(x)$ is the upper limit in measure on $[a, b]$ of the sequence of functions $- Q_n(x)$ $(n = 0, 1, 2, \ldots)$; and since these functions are the partial sums of (2.49), $- G(x)$ is the upper limit in measure on $[a, b]$ of that series.

Further, if $\varphi(x, E) \in M$. i.e. if $\varphi(x, E)$ is a limit function of the series (1.1), then, on the basis of Definition 1 and (2.47), $-\varphi(x, E)$ is a limit function of the series (2.49). In such a case we may apply (2.45) to the series (2.49), putting $- \varphi(x, E)$ and $- G(x)$ into that inequality in place of $\varphi(x, E)$ and $F(x)$. Thus we get

$$- \varphi(x, E) \leqslant - G(x) \tag{2.50}$$

almost everywhere on E, from which we conclude that (2.46) is satisfied almost everywhere on E.

Comparing (2.45) and (2.46), we see that for any function $\varphi(x, E) \in M$ inequality (1.15) is satisfied almost everywhere on the corresponding set E. Since we already know that any function $\varphi(x, E) \in M$ is measurable on the corresponding set E, the set M of all limit functions of the series (1.1) satisfies hypothesis β of Theorem II.

We have already proved that the set M satisfies hypothesis α. Thus in order to complete the proof of Theorem II, it will suffice for us to prove that M satisfies hypothesis γ.

We suppose that $\varphi(x, E)$ is any function in the set M. For the corresponding point set E we choose the set E_0 defined by (1.16) (see the formulation of Theorem II). Then

$$E_0 - E \subset E[F(x) = G(x)]. \tag{2.51}$$

Since by the hypotheses of Theorem II the functions $F(x)$ and $G(x)$ are upper and lower limits in measure respectively on $[a, b]$ of the series (1.1), i.e. are upper and lower limits in measure on $[a, b]$ of the sequence of its partial sums $Q_n(x)$ $(n = 0, 1, 2, \ldots)$, on the basis of Theorem β and (2.51) it follows that the sequence of functions $Q_n(x)$ $(n = 0, 1, 2, \ldots)$ converges in measure on the set $E_0 - E$ to $F(x)$, while on that set

$$F(x) = G(x).\text{(18)} \tag{2.52}$$

Further, since $\varphi(x, E) \in M$ and is therefore a limit function of the series (1.1), in view of Definition 1 there exists an increasing sequence of natural numbers ρ_k $(k = 1, 2, \ldots)$ such that

$$\lim_{n\to\infty} Q_{\rho_k}(x) = \varphi(x, E) \tag{2.53}$$

almost everywhere on E. Since the sequence of functions $Q_n(x)$ $(n = 0, 1, 2, \ldots)$ converges in measure on $E_0 - E$ to the function $F(x)$, the sequence of functions $Q_{\rho_k}(x)$ $(k = 1, 2, \ldots)$ has the same property. Then it is possible to select from the sequence ρ_k an increasing sequence of numbers ρ'_k $(k = 1, 2, \ldots)$ such that

$$\lim_{k\to\infty} Q_{\rho'_k}(x) = \varphi(x, E) \tag{2.54}$$

almost everywhere on E and

$$\lim_{k\to\infty} Q_{\rho'_k}(x) = F(x) \tag{2.55}$$

almost everywhere on $E_0 - E$.

It follows from (2.54) and (2.55) that

$$\lim_{k\to\infty} Q_{\rho'_k}(x) = \varphi_0(x, E_0) \tag{2.56}$$

almost everywhere on E_0, where $\varphi_0(x, E_0)$ is defined from (1.17). Since the $Q_n(x)$ $(n = 0, 1, 2, \ldots)$ are partial sums of the series (1.1) and since the natural numbers ρ'_k $(k = 1, 2, \ldots)$ form an increasing sequence, it follows from (2.56) that $\varphi_0(x, E)$ is a limit function of the series (1.1), (19) i.e. $\varphi_0(x, E_0) \in M$.

So, if $\varphi(x, E) \in M$ and if E_0 and $\varphi_0(x, E_0)$ are defined respectively by (1.16) and (1.17), then the set M satisfies hypothesis γ of Theorem II, and that theorem is proved.

---

(18) We may apply Theorem β in the case when $\operatorname{meas}(E_0 - E) > 0$. If that measure is zero, we may say that any sequence of functions converges on that set, both in measure and almost everywhere.

(19) Clearly meas $E_0 > 0$, $E_0 \subset [a, b]$, since the set E satisfies the same conditions. Thus $\varphi_0(x, E_0)$ satisfies all the conditions in Definition 1 of § 1.

§3

Before proceeding to the proof of Theorem III, as formulated in §1, we introduce certain definitions and prove a number of lemmas, and also recall certain definitions and lemmas contained in other papers.

First of all we recall the following definition.

DEFINITION 12. We will say that $f(x)$, defined almost everywhere on some set E of positive measure, is a *limit function in the sense of convergence almost everywhere on E for the set $M = \{\varphi(x)\}$ of functions defined almost everywhere on E*, if there exists a sequence of functions $\varphi_m(x) \in M$ ($m = 1, 2, \ldots$) which converges to $f(x)$ almost everywhere on E.(20)

Now we introduce three more definitions.

DEFINITION 13. Suppose that $M = \{\varphi(x)\}$ is a collection of functions $\varphi(x)$ each of which is defined on some set of positive measure.(21) A set M' lying in M will be said to be *everywhere dense on the set M in the sense of convergence almost everywhere on E*, if any function $\varphi(x) \in M$ is a limit function in the sense of convergence almost everywhere on E for the set M'.

DEFINITION 14. Suppose that $M_0 = \{[\varphi(x), \psi(x)]\}$ is a set of pairs of functions $\varphi(x)$ and $\psi(x)$, each of which is defined almost everywhere on some set E of positive measure. The pair of functions $f(x)$ and $g(x)$, defined almost everywhere on E, will be said to be a *limit pair in the sense of convergence almost everywhere on E for the set M_0*, if there exists a sequence of pairs of functions $[\varphi_m(x), \psi_m(x)] \in M_0$ ($m = 1, 2, \ldots$), such that

$$\lim_{m\to\infty} \varphi_m(x) = f(x) \quad \text{and} \quad \lim_{m\to\infty} \psi_m(x) = g(x)$$

almost everywhere on E.(22)

DEFINITION 15. Suppose that $M_0 = \{[\varphi(x), \psi(x)]\}$ has the same meaning as in Definition 14. A set M_0' lying in M_0 will be said to be *everywhere dense on M_0 in the sense of convergence almost everywhere on E*, if any pair $[\varphi(x), \psi(x)] \in M_0$ is a limit pair in the sense of convergence almost everywhere on E for the set M_0'.

We shall prove the following lemmas.

LEMMA 1. *Suppose that $M = \{\varphi(x)\}$ is some set of measurable functions defined almost everywhere on a set E of positive measure. Then it is possible to determine a finite or countable set M', lying in M and everywhere dense on that set in the sense of convergence almost everywhere on E.*

PROOF. Suppose that k is any natural number. For any function $\varphi(x) \in M$, for the points where this function is defined, put

---

(20)See [1], p. 777. The functions $\varphi_m(x)$ are not necessarily different for different m.

(21)We suppose here that the set E does *not* depend on the functions $\varphi(x) \in M$.

(22)The pairs $[\varphi_m(x), \psi_m(x)]$ are not necessarily distinct for different m.

$$[\varphi(x)]_k = \begin{cases} \varphi(x) & (\text{if } |\varphi(x)| \leq k), \\ k & (\text{if } \varphi(x) > k), \\ -k & (\text{if } \varphi(x) < -k). \end{cases} \tag{3.1}$$

Then each of the functions $[\varphi(x)]_k$ is a bounded measurable function defined almost everywhere on E. Accordingly, each of the functions $[\varphi(x)]_k$ has a summable square on E.

Denote by M_k the set of all functions $[\varphi(x)]_k$ corresponding to all possible functions $\varphi(x) \in M$ and a given number k. Then, if we denote by L^2 the set of all distinct measurable functions with summable square on E, (23) we have

$$M_k \subset L^2 \quad (k = 1, 2, \ldots). \tag{3.2}$$

As usual, the norm $\|f\|$ of a function $f(x) \in L^2$ will be defined by

$$\|f\| = +\sqrt{\int_E f^2(x)\,dx}\,. \tag{3.3}$$

It is easy to show that there exists a finite or countable set M'_k lying in M_k and everywhere dense in this set in the sense of the metric defined by the norm (3.3). Suppose that the $\psi_{k\nu}(x)$, $\nu = 1, 2, \ldots$, are all possible functions of the set M'_k. Generally speaking these functions may coincide for distinct ν.(24)

From the definition of M_k it follows that for each function $f(x) \in M_k$ it is possible to define a function $\varphi(x) \in M$ such that

$$[\varphi(x)]_k = f(x) \tag{3.4}$$

almost everywhere on E. Since $M'_k \subset M_k$, for each function $\psi_{k\nu}(x)$ we may define a function $\varphi_{k\nu}(x)$ such that

$$\varphi_{k\nu}(x) \in M, \qquad [\varphi_{k\nu}(x)]_k = \psi_{k\nu}(x) \tag{3.5}$$

almost everywhere on E.

For each natural number k we denote by M''_k the set of all distinct functions $\varphi_{k\nu}(x)$, $\nu = 1, 2, \ldots$. Then

---

(23) Two functions defined almost everywhere on E will be regarded as *different* if they are defined on different sets of total measure on E, or else have different values at at least one point $x \in E$.

(24) The existence of a set M'_k having the property noted above may be proved as follows. Suppose that the functions $f_\nu(x)$, $\nu = 1, 2, \ldots$, form an everywhere dense set on L^2 in the sense of the metric defined by the norm (3.3). Put

$$\rho_\nu = \inf_{f \in M_k} \|f - f_\nu\| \quad (\nu = 1, 2, \ldots)$$

and define for each ν a function $\psi_{k\nu}(x)$ satisfying the conditions

$$\psi_{k\nu}(x) \in M_k, \qquad \|\psi_{k\nu} - f_\nu\| < \rho_\nu + \frac{1}{\nu} \quad (\nu = 1, 2, \ldots).$$

Denote by M'_k the set of all distinct functions $\psi_{k\nu}(x)$, $\nu = 1, 2, \ldots$. Then M'_k is a finite or countable set contained in M_k.

Now choose any function $f(x) \in M_k$. Define an increasing sequence of integers n_s, $s = 1, 2, \ldots$, such that $\lim_{s\to\infty}\|f - f_{n_s}\| = 0$. Then $\lim_{s\to\infty}\rho_{n_s} = 0$, so that $\lim_{s\to\infty}\|\psi_{k,n_s} - f_{n_s}\| = 0$, and hence $\lim_{s\to\infty}\|\psi_{k,n_s} - f\| = 0$. Since $f(x)$ is an arbitrary function from M_k, it therefore follows that M'_k is everywhere dense on M_k in the sense of the metric (3.3).

$$\varphi_{k\nu}(x) \in M_k'', \qquad M_k'' \subset M \qquad (\nu = 1, 2, \ldots, k = 1, 2, \ldots), \tag{3.6}$$

each of the M_k'' being a finite or countable set.

Put

$$M' = \sum_{k=1}^{\infty} M_k''. \tag{3.7}$$

Then M' is a finite or countable set. From (3.6) and (3.7) it follows that

$$M' \subset M. \tag{3.8}$$

We shall show that M' is everywhere dense on M in the sense of convergence almost everywhere on E.

Choose any function $\varphi(x) \in M$. We define an increasing sequence of natural numbers m_k, $k = 1, 2, \ldots$, and a sequence of measurable sets e_k as follows.

Put

$$m_1 = 1, \quad e_1 = E. \tag{3.9}$$

Now we suppose that m_{k-1} and e_{k-1} have already been defined, k being any natural number satisfying the condition

$$k > 1. \tag{3.10}$$

Since the set M_k' is everywhere dense on M_k in the sense of the metric (3.3), since for the given k the functions $\psi_{k\nu}(x)$, $\nu = 1, 2, \ldots$, belong to M_k', and since

$$[\varphi(x)]_k \in M_k, \tag{3.11}$$

where $\varphi(x)$ is the function on M which we have chosen, it follows that there exists an increasing sequence of natural numbers ν_j, $j = 1, 2, \ldots$, such that

$$\lim_{j \to \infty} \| \psi_{k, \nu_j}(x) - [\varphi(x)]_k \| = 0. \text{[25]}$$

Hence, in view of a well-known theorem, it is possible to select from the sequence ν_j a subsequence of integers ρ_j such that

$$\lim_{j \to \infty} \psi_{k, \rho_j}(x) = [\varphi(x)]_k \tag{3.12}$$

almost everywhere on E.

In view of Egorov's theorem, it follows from (3.12) that there exists a measurable set e_k satisfying the conditions

$$e_k \subset E, \text{ meas } e_k < \frac{1}{2^k} \tag{3.13}$$

and such that the sequence $\psi_{k,\rho_j}(x)$ converges uniformly to the function $[\varphi(x)]_k$ on the set $E - e_k$.

(25) The sequence of numbers ν_j, $j = 1, 2, \ldots$, may be considered to be increasing in the case when any function M_k' is repeated a countable number of times in the sequence $\psi_{k\nu}(x)$, $\nu = 1, 2, \ldots$. We will suppose that this condition is satisfied.

Hence we may find a natural number m_k such that

$$m_k > m_{k-1} \tag{3.14}$$

and

$$|\varphi_{k,m_k}(x) - [\varphi(x)]_k| < \frac{1}{2k} \qquad (x \in E - e_k). \tag{3.15}$$

Thus, we have defined a number m_1 and a set e_1 from (3.9). Then we have supposed that the natural number m_{k-1} and the measurable set e_{k-1} have been defined, with $k > 1$, and thereupon we have defined a natural number m_k and a measurable set e_k. Then we may define step by step all the natural numbers m_k and all the measurable sets e_k, conditions (3.13), (3.14), and (3.15) being satisfied for all $k > 1$.(26) It follows from (3.14) that the natural numbers m_k form an increasing sequence.

Comparing (3.5) and (3.15), we have

$$|[\varphi_{k,m_k}(x)]_k - [\varphi(x)]_k| < \frac{1}{2k} \qquad (x \in E - e_k, \ k > 1). \tag{3.16}$$

Moreover, on the basis of (3.6) and (3.7)

$$\varphi_{k,m_k}(x) \in M' \ (k > 1).\text{(27)} \tag{3.17}$$

Put

$$e = \overline{\lim_{k\to\infty}} \, e_k. \tag{3.18}$$

In view of (3.13) it follows that

$$e \subset E, \text{ meas } e = 0. \tag{3.19}$$

We will prove that

$$\lim_{k\to\infty} \varphi_{k,m_k}(x) = \varphi(x) \tag{3.20}$$

for all x satisfying the condition

$$x \in E - e. \tag{3.21}$$

If x satisfies this condition, then in view of (3.18) there exists a natural number k_x such that

$$x \in E - e_k \qquad (k > k_x). \tag{3.22}$$

Then on the basis of (3.16) the functions $[\varphi(x)]_k$ are defined at the point x in question for $k > k_x$, so that the function $\varphi(x)$ is also defined at that point. Thus $\varphi(x)$ is defined for all x satisfying (3.21). In addition, in view of (3.1)

(26) We recall that the numbers m_k and the sets e_k have been defined in terms of the function $\varphi(x)$ which we have chosen.

(27) The functions $\varphi_{k,m_k}(x)$ may coincide for different values of k.

$$\lim_{k \to \infty} [\varphi(x)]_k = \varphi(x) \qquad (x \in E - e). \tag{3.23}$$

Now we suppose that x is any point satisfying (3.21). First we suppose that at this point

$$-\infty < \varphi(x) < +\infty. \tag{3.24}$$

Then the number k_x satisfying (3.22) may be chosen so large that

$$-k + 1 < \varphi(x) < k - 1 \qquad (k > k_x). \tag{3.25}$$

Thus on the basis of (3.1),

$$[\varphi(x)]_k = \varphi(x) \qquad (k > k_x). \tag{3.26}$$

Comparing (3.16), (3.22) and (3.26), we have

$$\varphi(x) - \frac{1}{2^k} < [\varphi_{k,\, m_k}(x)]_k < \varphi(x) + \frac{1}{2^k} \qquad (k > k_x),$$

so that, by (3.25),

$$-k < [\varphi_{k,\, m_k}(x)]_k < k \qquad (k > k_x),$$

and so, in view of the definition of the functions $[\varphi(x)]_k$ (see (3.1)),

$$[\varphi_{k,\, m_k}(x)]_k = \varphi_{k,\, m_k}(x) \qquad (k > k_x). \tag{3.27}$$

Taking into account (3.16), (3.22), (3.26) and (3.27), we see that at the point x in question the inequality

$$|\varphi_{k,\, m_k}(x) - \varphi(x)| < \frac{1}{2^k} \qquad (k > k_x)$$

is satisfied, so that at x equation (3.20) is satisfied if condition (3.24) is fulfilled.

Now we suppose further that at x condition (3.21) is satisfied as before, and that

$$\varphi(x) = +\infty. \tag{3.28}$$

Then, in view of (3.1), $[\varphi(x)]_k = k$ $(k = 1, 2, \ldots)$, so that, on the basis of (3.16) and (3.22),

$$[\varphi_{k,\, m_k}(x)]_k > k - \frac{1}{2^k} \qquad (k > k_x). \tag{3.29}$$

Comparing (3.1) and (3.29), we see that at x

$$\varphi_{k,\, m_k}(x) \geqslant [\varphi_{k,\, m_k}(x)]_k > k - \frac{1}{2^k} \qquad (k > k_x).$$

(3.20) is a consequence of this if (3.21) and (3.28) are satisfied.

Finally we suppose that at the point in question condition (3.21) is satisfied and that

$$\varphi(x) = -\infty. \tag{3.30}$$

Comparing (3.1) and (3.30), we have $[\varphi(x)]_k = -k$ $(k = 1, 2, \ldots)$, so that on the basis of (3.16) and (3.22)

$$[\varphi_{k,\, m_k}(x)]_k < -k + \frac{1}{2^k} < 0 \qquad (k > k_x),$$

and hence, in view of (3.1) and (3.30),

$$\varphi_{k,\, m_k}(x) \leqslant [\varphi_{k,\, m_k}(x)]_k < -k + \frac{1}{2^k} \qquad (k > k_x).$$

From this inequality we obtain (3.20), if (3.21) and (3.30) are satisfied at the point in question.

Thus (3.20) is valid if $x \in E - e$ and if (3.24) or one of the equations (3.28) or (3.30) is satisfied, i.e. (3.20) is satisfied for all $x \in E - e$.

Hence in view of (3.19) it follows that (3.20) is satisfied almost everywhere on E, so that, on the basis of (3.17) and Definition 12 in §3, the function $\varphi(x)$ of interest to us is a limit function in the sense of convergence almost everywhere on E for the set M'.

By hypothesis $\varphi(x)$ is an arbitrary function belonging to M. Accordingly, in view of Definition 13 in §3 the set M' is everywhere dense on M in the sense of convergence almost everywhere on E. Since we have already proved that M' is a finite or countable set, Lemma 1 is completely proved.

LEMMA 2. *Suppose that $M_0 = \{[\varphi(x), \psi(x)]\}$ is some set of pairs of measurable functions $\varphi(x)$ and $\psi(x)$, defined almost everywhere on a segment $[a, b]$. Then it is possible to define a finite or countable set of pairs of functions lying in M_0 and everywhere dense on that set in the sense of convergence almost everywhere on $[a, b]$.*

PROOF. Without loss of generality we may suppose that $\psi(x)$ is not defined for $x = a$. For each pair $[\varphi(x), \psi(x)] \in M_0$ of functions $\varphi(x)$ and $\psi(x)$ we define a function

$$f(x) \equiv f(x;\ \varphi,\ \psi) \tag{3.31}$$

from the conditions that

$$f(x) = \varphi(x) \tag{3.32}$$

at any point $x \in [a, b]$ at which $\varphi(x)$ is defined, and

$$f(x) = \psi(x - b + a) \tag{3.33}$$

at any point x satisfying the condition $b < x \leqslant 2b - a$ and such that $\psi(x - b + a)$ is defined. Then $f(x)$ is measurable and defined almost everywhere on the segment $[a, 2b - a]$.

Denote by $M_1 = \{f(x)\}$ the set of functions corresponding to all possible pairs $[\varphi(x), \psi(x)] \in M_0$. Then (3.32) and (3.33) establish a one-to-one correspondence between the sets M_0 and M_1.

Since M_1 is some collection of measurable functions defined almost everywhere on $[a, 2b - a]$, on the basis of Lemma 1 there exists a finite or countable set M_1' lying in M_1 and everywhere dense on that set in the sense of convergence almost everywhere on $[a, 2b - a]$.

Suppose that the set M_1' consists of the functions $f_n(x)$, $n = 1, 2, \ldots$. (28) Denote by $\varphi_n(x)$ and $\psi_n(x)$, $n = 1, 2, \ldots$, the functions, defined almost everywhere on $[a, b]$, to which the function $f_n(x)$ corresponds under the conditions (3.32) and (3.33). Then

$$\varphi_n(x) = f_n(x), \quad \psi_n(x) = f_n(b - a + x) \qquad (n = 1, 2, \ldots) \tag{3.34}$$

almost everywhere on $[a, b]$.

Denote by M_0' the set of all possible distinct pairs $[\varphi_n(x), \psi_n(x)]$ corresponding to the functions $f_n(x) \in M_1'$, $n = 1, 2, \ldots$. Then M_0' is a finite or countable set, while

$$M_0' \subset M_0. \tag{3.35}$$

We shall show that M_0' is everywhere dense on M_0 in the sense of convergence almost everywhere on $[a, b]$. Indeed, suppose that $[\varphi(x), \psi(x)]$ is any pair lying in M_0. Choose the corresponding function $f(x)$ defined by (3.32) and (3.33). Then $f(x) \in M_1$, and since M_1' is everywhere dense on M_1' in the sense of convergence almost everywhere on $[a, 2b - a]$, on the basis of Definition 13 we see that $f(x)$ is a limit function of the set M_1' in the sense of convergence almost everywhere on $[a, 2b - a]$.

Hence, in view of Definition 12 of §13, and also from the definition of M_1', it follows that there exists an increasing sequence of natural numbers n_k, $k = 1, 2, \ldots$, for which

$$\lim_{k\to\infty} f_{n_k}(x) = f(x) \tag{3.36}$$

almost everywhere on $[a, 2b - a]$.

Comparing (3.32)–(3.34) and (3.36), we see that

$$\lim_{k\to\infty} \varphi_{n_k}(x) = \varphi(x), \quad \lim_{k\to\infty} \psi_{n_k}(x) = \psi(x) \tag{3.37}$$

almost everywhere on $[a, b]$. Since the definition of M_0' implies

$$[\varphi_{n_k}(x), \psi_{n_k}(x)] \in M_0' \qquad (k = 1, 2, \ldots),$$

it follows on the basis of (3.37) and Definition 14 that the pair $[\varphi(x), \psi(x)]$ which we have chosen is a limit pair in the sense of convergence almost everywhere on $[a, b]$ for the set M_0'. But $[\varphi(x), \psi(x)]$ is any pair lying in M_0. Therefore in view of Definition 15 the set M_0' is everywhere dense on M_0. Since M_0' is a finite or countable set lying in M_0, Lemma 2 is proved.

Now we recall some definitions and lemmas.(29)

DEFINITION 16. Choose two sequences of real numbers

(28) We will suppose that each function of M_1' is repeated a countable number of times in the sequence $\{f_n(x)\}_{n=1,2,\ldots}$.

(29) In the present paper we number in succession both the newly proved lemmas as well as those proved in [2] and [3] and which will be needed in the proof of Theorem III. For the convenience of the reader we recall the formulations of all of these lemmas. In the numbering of the definitions we proceed in the same way.

$$U_1, U_2, \ldots, U_m, \ldots \tag{3.38}$$

and

$$V_1, V_2, \ldots, V_m, \ldots, \tag{3.39}$$

supposing that each of the numbers U_m and V_m is either finite or equal to $+\infty$ or $-\infty$.

We will say that the sequences (3.38) and (3.39) are *equiconvergent* if the following conditions are satisfied:

a°. $U_m - V_m$ tends to zero when m increases unboundedly taking on values for which both the numbers U_m and V_m are finite.

b°. U_m and V_m tend to $+\infty$ or $-\infty$ when m increases unboundedly taking on values for which at least one of the numbers U_m or V_m is equal to $+\infty$ or $-\infty$ (see [3], §1, Definition 1).

The following assertions are easily proved for equiconvergent sequences.

LEMMA 3. *If the sequences* (3.38) *and* (3.39) *are equiconvergent and if the natural numbers* n_k, $k = 1, 2, \ldots$, *form an increasing sequence, then the sequences of numbers* U_{n_k} *and* V_{n_k}, $k = 1, 2, \ldots$, *are also equiconvergent.*

LEMMA 4. *Select a sequence of real numbers*

$$W_1, W_2, \ldots, W_n, \ldots \tag{3.40}$$

and suppose that on the one hand the two sequences (3.38) *and* (3.40) *are equiconvergent and, on the other hand, the sequences* (3.39) *and* (3.40) *are equiconvergent. In this case the two sequences* (3.38) *and* (3.39) *are also equiconvergent.*

LEMMA 5. *If the sequences* (3.38) *and* (3.39) *are equiconvergent and if* $\lim_{m\to\infty} U_m = U$, *where* U *is a finite number or an infinity of definite sign, then* $\lim_{m\to\infty} V_m = U$.

The following lemmas also hold.

LEMMA 6 ([3], §1, LEMMA G). *Suppose that the measurable functions* $F(x)$, $G(x)$, $f_m(x)$ *and* $g_m(x)$ $(m = 1, 2, \ldots)$ *are defined almost everywhere on some segment* $[a, b]$. *Suppose that the inequality*

$$G(x) \leqslant F(x) \tag{3.41}$$

is satisfied almost everywhere on $[a, b]$ *and that the sequences*

$$f_1(x), f_2(x), \ldots, f_m(x), \ldots \tag{3.42}$$

and

$$g_1(x), g_2(x), \ldots, g_m(x), \ldots \tag{3.43}$$

of functions converge uniformly almost everywhere on the same segment.

In this case, if $F(x)$ *and* $G(x)$ *are respectively the upper and lower limits in measure on* $[a, b]$ *of the sequence* (3.42), *then the same functions are respectively the upper and lower limits in measure on* $[a, b]$ *of the sequence* (3.43).

LEMMA 7. ([30]) *Suppose given measurable functions* $F(x)$, $G(x)$, *and* $f_m(x)$ $(m = 1, 2, \ldots)$, *defined almost everywhere on* $[-\pi, \pi]$, *such that*

$$G(x) \leqslant f_m(x) \leqslant F(x) \qquad (m = 1, 2, \ldots) \tag{3.44}$$

almost everywhere on that segment. Then there exist a trigonometric series (1.4), *functions* $h_m(x)$ $(m = 1, 2, \ldots)$ *which are continuous on* $[-\pi, \pi]$, *and natural numbers* ν_m $(m = 1, 2, \ldots)$, *which have the following properties*:

1) *The sequences* $\{f_m\}$ *and* $\{h_m\}$ $(m = 1, 2, \ldots)$ *converge uniformly almost everywhere on* $[-\pi, \pi]$ (*see Definition* 16).

2)
$$\nu_m < \nu_{m+1} \qquad (m = 1, 2, \ldots) \tag{3.45}$$

and

$$\lim_{m\to\infty} [h_m(x) - S_{\nu_m}(x)] = 0 \tag{3.46}$$

almost everywhere on $[-\pi, \pi]$, *where*

$$S_n(x) = \frac{a_0}{2} + \sum_{j=1}^{n} (a_j \cos jx + b_j \sin jx) \qquad (n = 1, 2, \ldots). \tag{3.47}$$

3) *The upper and lower limits in measure on* $[-\pi, \pi]$ *of the sequence* $S_n(x)$ $(n = 1, 2, \ldots)$ *are equal respectively to* $F(x)$ *and* $G(x)$.

4) *For any increasing sequence* n_k $(k = 1, 2, \ldots)$ *of natural numbers, either the upper and lower limits in measure on* $[-\pi, \pi]$ *of the sequence* $S_{n_k}(x)$ $(k = 1, 2, \ldots)$ *are equal respectively to* $F(x)$ *and* $G(x)$, *or one may determine a nondecreasing sequence of natural numbers* p_k $(k = 1, 2, \ldots)$ *such that*

$$\lim_{k\to\infty} p_k = 0 \tag{3.48}$$

and the sequence $\{h_{p_k}(x) - S_{n_k}(x)\}$ $(k = 1, 2, \ldots)$ *converges in measure to zero on* $[-\pi, \pi]$.

5)
$$\lim_{n\to\infty} a_n = 0, \quad \lim_{n\to\infty} b_n = 0.\text{[30]}$$

LEMMA 8 ([5], LEMMA 3.6). *Consider two measurable functions* $\varphi(x)$ *and* $\psi(x)$, *defined almost everywhere on some segment* $[a, b]$ *and satisfying the inequality*

$$\varphi(x) \leqslant \psi(x) \tag{3.49}$$

almost everywhere on that segment. Suppose that $\{c_k\}$ $(k = 1, 2, \ldots)$ *is a sequence of points, not necessarily lying on* $[a, b]$, *and let* $\{\delta_k\}$ *be a sequence of numbers satisfying*

---

[30] [2], Lemma 5.2. In the formulation of Lemma 5.2 in [2] there is a misprint: $h_{p_k}(x) - S_{p_k}(x)$ was printed in place of $h_{p_k}(x) - S_{n_k}(x)$.

$$\lim_{k\to\infty} \delta_k = 0. \tag{3.50}$$

Put

$$x_{ki} = c_k + i\delta_k, \quad \Delta_{ki} = (x_{ki}, x_{k,i+1}) \tag{3.51}$$
$$(k = 1, 2, \ldots;\ i = 0, 1, 2, \ldots, -1, -2, \ldots).$$

Moreover, for those points for which $\varphi(x)$ *and* $\psi(x)$ *are defined, choose a sequence of functions*

$$\omega_1(x),\ \omega_2(x),\ \ldots,\ \omega_k(x),\ \ldots, \tag{3.52}$$

defined from the equations

$$\left.\begin{aligned} \omega_k(x) &= \begin{cases} \psi(x) & (\textit{if } x \in \Delta_{k,2j}), \\ \varphi(x) & (\textit{if } x \in \Delta_{k,2j+1}) \end{cases} \\ (k &= 1, 2, \ldots;\ j = 0, 1, 2, \ldots, -1, -2, \ldots). \end{aligned}\right\} \tag{3.53}$$

Then the functions $\omega_k(x)$ $(k = 1, 2, \ldots)$ *will be measurable and defined almost everywhere on* $[a, b]$, *and moreover* $\varphi(x)$ *and* $\psi(x)$ *will be respectively the upper and lower limits in measure on* $[a, b]$ *of the sequence* (3.52).

We shall prove the following lemma.

LEMMA 9. *Suppose that the functions* $\varphi(x)$, $\psi(x)$, $\alpha_k(x)$ *and* $\beta_k(x)$, $k = 1, 2, \ldots$, *are measurable and defined almost everywhere on some segment* $[a, b]$. *Suppose that*

$$\lim_{k\to\infty} \alpha_k(x) = \varphi(x), \quad \lim_{k\to\infty} \beta_k(x) = \psi(x) \tag{3.54}$$

almost everywhere on that segment. [31]. *Suppose that at the points at which* $\varphi(x)$, $\psi(x)$, *and the* $\alpha_k(x)$ *and* $\beta_k(x)$, $k = 1, 2, \ldots$, *are defined, the functions* $\omega_k(x)$ *are defined from* (3.53), *and the functions* $g_k(x)$ *are defined from the equations*

$$\left.\begin{aligned} g_k(x) &= \begin{cases} \beta_k(x) & (\textit{if } x \in \Delta_{k,2j}) \\ \alpha_k(k) & (\textit{if } x \in \Delta_{k,2j+1}) \end{cases} \\ (k &= 1, 2, \ldots;\ j = 0, 1, 2, \ldots, -1, -2, \ldots), \end{aligned}\right\} \tag{3.55}$$

where the Δ_{ki}, x_{ki}, c_k *and* δ_k *satisfy the conditions of Lemma* 8.

In this case the functions $\omega_k(x)$ *and* $g_k(x)$ *are measurable and defined almost everywhere on* $[a, b]$, *and, moreover, these function sequences are equiconvergent almost everywhere on the same segment.*

PROOF. It follows directly from the hypothesized properties of the functions $\varphi(x)$, $\psi(x)$, $\alpha_k(x)$ and $\beta_k(x)$, $k = 1, 2, \ldots$, and the definitions of the intervals Δ_{ki} and of the functions $\omega_k(x)$ and $g_k(x)$ at (3.51), (3.53) and (3.55), that the functions $\omega_k(x)$ and $g_k(x)$ are measurable and defined almost everywhere on $[a, b]$. Therefore, in order to complete the proof of Lemma 9, it remains for us to show that the sequences $\{\omega_k(x)\}$ and $\{g_k(x)\}$ are equiconvergent almost everywhere on $[a, b]$.

---

(31) In Lemma 9 we do not require that (3.49) be satisfied almost everywhere on $[a, b]$.

Denote by H the set of all points on $[a, b]$ at each of which all the functions $\varphi(x)$, $\psi(x)$, $\alpha_k(x)$, $\beta_k(x)$, $\omega_k(x)$ and $g_k(x)$ are defined, and equations (3.54) are satisfied. In this case, in view of our assumptions,

$$\text{meas } H = b - a, \quad H \subset [a, b]. \tag{3.56}$$

Suppose that x is an arbitrary point satisfying the condition

$$x \in H. \tag{3.57}$$

Then at this point all the functions in question are defined, the functions $\omega_k(x)$ and $g_k(x)$ are defined by (3.53) and (3.55), and for any k the point x in question belongs to one of the intervals $\Delta_{k,2j}$ or $\Delta_{k,2j+1}$, where j is some integer, in general depending on x.

If for some k the point x in question belongs to one of the intervals $\Delta_{k,2j}$, then in view of (3.53) and (3.55) we have

$$\omega_k(x) = \psi(x), \quad g_k(x) = \beta_k(x). \tag{3.58}$$

If x belongs to one of the intervals $\Delta_{k,2j+1}$, then, on the basis of the same equations,

$$\omega_k(x) = \varphi(x), \quad g_k(x) = \alpha_k(x). \tag{3.59}$$

Thus, for any x satisfying (3.57), and for any natural number k, either both equations (3.58) are satisfied, or else both equations (3.59). Since both of the function systems $\varphi(x)$, $\alpha_k(x)$, $k = 1, 2, \ldots$, and $\psi(x)$, $\beta_k(x)$, $k = 1, 2, \ldots$, enter equivalently into the conditions of the lemma, it suffices for us to consider the values of k for which only one of the equation pairs (3.58) or (3.59) is satisfied.

For definiteness we consider those values of k for which (3.58) is satisfied. First we suppose that at the point x in question the function $\psi(x)$ is finite. From the definition of the set H it follows that (3.54) is satisfied at this point x, so that it is possible to determine an integer k_0 such that the functions $\beta_k(x)$ will be finite for all k satisfying the condition

$$k > k_0. \tag{3.60}$$

Now if k increases without bound, taking on values for which (3.58) and (3.60) are satisfied at x, then the corresponding values of $\omega_k(x)$ and $g_k(x)$ will be finite, and in view of (3.54) the difference $\omega_k(x) - g_k(x)$ will tend to zero.

Now suppose further that at the point x in question the equation

$$\psi(x) = +\infty \tag{3.61}$$

is satisfied. In this case, if k increases unboundedly, taking on values for which (3.58) holds, then for these values of k the functions $\omega_k(x) = +\infty$, and, on the basis of (3.54), $g_k(x) \longrightarrow +\infty$. Thus, if k increases unboundedly, taking on the above-mentioned values of x, then at x at least one of the functions $\omega_k(x)$ or $g_k(x)$ is equal to $+\infty$, and both of them tend to $+\infty$.

Finally, suppose that at the point x

$$\psi(x) = -\infty. \tag{3.62}$$

Then, if k increases unboundedly, taking on the same values as before, then for these k we have $\omega_k(x) = -\infty$ and, by (3.54), $g_k(x) \longrightarrow -\infty$. Accordingly, at the point x and for the given values of k, at least one of the functions $\omega_k(x)$ or $g_k(x)$ is equal to $-\infty$, and both of these functions tend to $-\infty$, when k increases unboundedly taking on the values under consideration.

Analogous considerations hold when k increases unboundedly, taking on values for which (3.59) is satisfied.

Taking into account Definition 16 and the properties of the functions $\omega_k(x)$ and $g_k(x)$ which we have found, we see that at any point x satisfying (3.57) the function sequences $\{\omega_k(x)\}$ and $\{g_k(x)\}$, $k = 1, 2, \ldots$, are equiconvergent. Hence in view of (3.56) these sequences are equiconvergent almost everywhere on $[a, b]$, as we were required to prove.

§4

We turn to the proof of Theorem III, formulated in §1. Suppose that the functions $F(x)$, $G(x)$, and the set $M = \{\varphi(x, E)\}$ satisfy the hypotheses of Theorem III, i.e. conditions α, β, and γ of Theorem II (see the formulation of Theorem III). To each function $\varphi(x, E) \in M$(32) we assign a pair of functions $\varphi(x)$, $\psi(x)$, defined from the equations

$$\varphi(x) = \psi(x) = \varphi(x, E) \tag{4.1}$$

almost everywhere on E, and

$$\varphi(x) = G(x), \quad \psi(x) = F(x) \tag{4.2}$$

almost everywhere on $[-\pi, \pi] - E$. Then, in view of the definitions of $\varphi(x, E)$, $G(x)$ and $F(x)$ (see the formulation of Theorem III), $\varphi(x)$ and $\psi(x)$ will be measurable and defined almost everywhere on $[-\pi, \pi]$.

We are supposing that the set M satisfies conditions α, β, and γ of Theorem II. In view of condition γ, if $\varphi(x)$ and $\psi(x)$ correspond to the given function $\varphi(x, E) \in M$, and if the set E_0 and the function $\varphi_0(x, E_0)$ are defined by (1.16) and (1.17), then $\varphi_0(x, E_0) \in M$, and, moreover, from (4.2),

$$\varphi(x) = \psi(x) = F(x) = G(x) = \varphi_0(x, E_0)$$

almost everywhere on $E_0 - E$. Accordingly, in view of (1.16), (1.17), (4.1) and (4.2),

$$\varphi(x) = \psi(x) = \varphi_0(x, E_0) \tag{4.3}$$

almost everywhere on E_0, and

(32) In view of the hypotheses of Theorem III, M is a nonempty set. We will suppose that each of the functions $\varphi(x, E)$ is not defined outside of E.

$$\varphi(x) = G(x), \quad \psi(x) = F(x) \tag{4.4}$$

almost everywhere on $[-\pi, \pi] - E_0$.

Thus, if $\varphi(x)$ and $\psi(x)$ are the functions corresponding to $\varphi(x, E) \in M$, then these same functions are those corresponding to the function $\varphi_0(x, E_0)$ defined by (1.17) for the given function $\varphi(x, E)$.(33) Moreover, if $\varphi(x, E) \in M$, then it follows from (1.15) and (1.19) (see the formulation of Theorem III and condition β of Theorem II) that for the functions $\varphi(x)$ and $\psi(x)$ corresponding to $\varphi(x, E)$ under equations (4.1) and (4.2) we have

$$G(x) \leqslant \varphi(x) \leqslant \psi(x) \leqslant F(x) \tag{4.5}$$

almost everywhere on $[-\pi, \pi]$.

Denote by M_0 the set of pairs $[\varphi(x), \psi(x)]$ of functions $\varphi(x)$ and $\psi(x)$ corresponding to all possible functions $\varphi(x, E) \in M$. Then the inequalities (4.5) are satisfied almost everywhere on $[-\pi, \pi]$ for arbitrary functions $\varphi(x)$ and $\psi(x)$ for which $[\varphi(x), \psi(x)] \in M_0$.

Since the functions $\varphi(x)$ and $\psi(x)$ forming the pairs $[\varphi(x), \psi(x)] \in M_0$ are measurable and are defined almost everywhere on $[-\pi, \pi]$, on the basis of Lemma 2 we may find a countable set $M_0' = \{[\varphi_m(x), \psi_m(x)]\}$ of pairs of functions $\varphi_m(x)$, $\psi_m(x)$, $m = 1, 2, \ldots$, lying in M_0 and everywhere dense there in the sense of convergence almost everywhere on $[-\pi, \pi]$.(34) Then the functions $\varphi_m(x)$ and $\psi_m(x)$, $m = 1, 2, \ldots$, will be measurable and defined almost everywhere on $[-\pi, \pi]$, and, moreover, in view of (4.5),

$$G(x) \leqslant \varphi_m(x) \leqslant \psi_m(x) \leqslant F(x) \qquad (m = 1, 2, \ldots) \tag{4.6}$$

almost everywhere on $[-\pi, \pi]$.

If $\varphi(x, E) \in M$ and if the functions $\varphi(x)$ and $\psi(x)$ correspond to it, then, as we already know, the same functions correspond to the function $\varphi_0(x, E_0)$ defined by (1.17), while $\varphi_0(x, E_0) \in M$, and, in view of (1.16),

$$E[F(x) = G(x)] \subset E_0. \tag{4.7}$$

In other words, if $[\varphi(x), \psi(x)] \in M$, then the functions $\varphi(x)$ and $\psi(x)$ correspond to a function $\varphi_0(x, E_0) \in M$ for which (4.7) is satisfied. Since $M_0' \subset M$, it therefore follows that for any $m = 1, 2, \ldots$ the functions $\varphi_m(x)$ and $\psi_m(x)$ correspond to a certain function

$$g_m(x, E_m) \in M, \tag{4.8}$$

for which

$$E[F(x) = G(x)] \subset E_m, \tag{4.9}$$

---

(33) In what follows, when we say that $\varphi(x)$ and $\psi(x)$ correspond to some function $\varphi(x, E) \in M$, we will always mean that $\varphi(x)$ and $\psi(x)$ are defined from (4.1) and (4.2) for the given $\varphi(x, E)$.

(34) The pairs $[\varphi_m(x), \psi_m(x)]$ are not necessarily distinct for distinct m; see footnote (22). In this case M_0' might consist of a finite number of distinct pairs.

E_m being that set on the total measure of which the function $g_m(x, E_m)$ is defined. Here

$$\text{meas } E_m > 0, \quad E_m \subset [-\pi, \pi], \tag{4.10}$$

and moreover, in view of (4.1) and (4.2)

$$\varphi_m(x) = \psi_m(x) = g_m(x, E_m) \tag{4.11}$$

almost everywhere on E_m and

$$\varphi_m(x) = G(x), \quad \psi_m(x) = F(x) \tag{4.12}$$

almost everywhere on $[-\pi, \pi] - E_m$.

Now from a well-known theorem on measurable sets we have

$$E_m = E'_m + e'_m - e''_m, \tag{4.13}$$

where E'_m is the sum of a finite number of segments

$$d_{mk} = [\alpha_{mk}, \beta_{mk}], \; k = 1, \ldots, q_m, \tag{4.14}$$

which are disjoint,

$$E'_m \subset [-\pi, \pi]; \quad e'_m \subset [-\pi, \pi], \quad e''_m \subset [-\pi, \pi] \tag{4.15}$$

and

$$\text{meas } e'_m < \frac{1}{2m}, \quad \text{meas } e''_m < \frac{1}{2m} \tag{4.16}$$

If necessary by making some of the segments d_{mk} somewhat smaller, we may always assume that

$$d_{mk} \subset (-\pi, \pi), \; k = 1, \ldots, q_m, \tag{4.17}$$

From the definition of the sets E'_m it follows moreover that

$$E'_m = \sum_{k=1}^{q_m} d_{mk} \subset (-\pi, \pi). \tag{4.18}$$

For each $m = 1, 2, \ldots$, we now enumerate the segments d_{mk}, $k = 1, \ldots, q_m$, from left to right and put

$$\beta_{m0} = -\pi, \quad \alpha_{m, q_m+1} = \pi. \tag{4.19}$$

Then from (4.18)

$$(-\pi, \pi) - E'_m = \sum_{k=0}^{q_m} d'_{mk}, \tag{4.20}$$

where

$$d'_{mk} = (\beta_{mk}, \alpha_{m, k+1}), \; k = 0, \ldots, q_m; \tag{4.21}$$

$$\omega_m = \min_{0 \leq k \leq q_m} \operatorname{meas} d'_{mk}. \tag{4.22}$$

For each $m = 1, 2, \ldots$, we choose a natural number ρ_m sufficiently large that the inequality

$$\frac{\pi}{2^{\rho_m}} < \frac{\omega_m}{3} \tag{4.23}$$

is satisfied. We may suppose that

$$\rho_1 < \rho_2 < \cdots < \rho_m < \rho_{m+1} < \cdots, \tag{4.24}$$

from which it follows that

$$\lim_{m \to \infty} \rho_m = \infty. \tag{4.25}$$

For those points $x \in [-\pi, \pi]$ for which $\varphi_m(x)$ and $\psi_m(x)$ are defined for all m, we define functions $f_m(x)$, $m = 1, 2, \ldots$, from the conditions

$$f_m(x) = \begin{cases} \psi_m(x) & (\text{if } x \in D_{m, 2j}), \\ \varphi_m(x) & (\text{if } x \in D_{m, 2j+1}) \end{cases} \quad (m = 1, 2, \ldots; \ j = 0, 1, 2, \ldots, -1, -2, \ldots), \tag{4.26}$$

where

$$D_{mi} = (\xi_{mi}, \xi_{m, i+1}), \quad \xi_{mi} = i \cdot \sigma_m, \quad \sigma_m = \frac{\pi}{2^{\rho_m}} \quad (m = 1, 2, \ldots; \ i = 0, 1, 2, \ldots, -1, -2, \ldots). \tag{4.27}$$

Since the functions $\varphi_m(x)$ and $\psi_m(x)$ are measurable and defined almost everywhere on $[-\pi, \pi]$, the functions $f_m(x)$ have the same property. Moreover, from (4.6) and (4.20) we have

$$G(x) \leqslant \varphi_m(x) \leqslant f_m(x) \leqslant \psi_m(x) \leqslant F(x), \tag{4.28}$$

$m = 1, 2, \ldots$, almost everywhere on $[-\pi, \pi]$.

By hypothesis the functions $F(x)$ and $G(x)$ are measurable and defined almost everywhere on $[-\pi, \pi]$ (see the formulation of Theorem III in §1). It follows in view of (4.28) that the functions $F(x)$, $G(x)$ and $f_m(x)$, $m = 1, 2, \ldots$, satisfy all the hypotheses of Lemma 7. Therefore, by that lemma, there exist a trigonometric series (1.4), functions $h_m(x)$ continuous on $[-\pi, \pi]$, and natural numbers ν_m which have properties 1)–5) formulated in Lemma 7.

In view of Definition 9 in §1, properties 3) and 5) of the series (1.4) coincide with properties B and C of that series, appearing in the formulation of Theorem III. Therefore, in order to complete the proof of that theorem, it remains for us to verify that the series (1.4) has property A appearing in that formulation.

Choose any function $\varphi(x, E)$ satisfying the condition

$$\varphi(x, E) \in M, \tag{4.29}$$

where M is the set of functions satisfying the hypotheses of Theorem III, i.e. conditions α, β and γ of Theorem II. Suppose that $\varphi(x)$ and $\psi(x)$ are the functions corresponding to the given function $\varphi(x, E)$, i.e. defined from (4.1) and (4.2). Then, in view of the definition of the set M_0, for pairs $[\varphi(x), \psi(x)]$ of these functions the condition

$$[\varphi(x), \psi(x)] \in M_0 \tag{4.30}$$

will be satisfied.

Since the set $M_0' = \{[\varphi_m(x), \psi_m(x)]\}$ of pairs of functions $\varphi_m(x), \psi_m(x)$, $m = 1, 2, \ldots$, lies in M_0 and is everywhere dense in that set in the sense of convergence almost everywhere on $[-\pi, \pi]$, in view of Definition 15 and (4.30) the pair $[\varphi(x), \psi(x)]$ of functions which we are considering is a limit pair for the set M_0' in the sense of convergence almost everywhere on $[-\pi, \pi]$. Hence, on the basis of Definition 14 it is possible to determine an increasing sequence of natural numbers m_k, $k = 1, 2, \ldots$, such that

$$\lim_{k\to\infty} \varphi_{m_k}(x) = \varphi(x), \quad \lim_{k\to\infty} \psi_{m_k}(x) = \psi(x) \tag{4.31}$$

almost everywhere on $[-\pi, \pi]$.(35)

For those points x at which the functions $\varphi(x)$ and $\psi(x)$ are defined, put

$$\left.\begin{aligned} \chi_m(x) = \begin{cases} \psi(x) & (\text{if } x \in D_{m,2j}), \\ \varphi(x) & (\text{if } x \in D_{m,2j+1}) \end{cases} \\ (m = 1, 2, \ldots;\ i = 0, 1, 2, \ldots, -1, -2, \ldots), \end{aligned}\right\} \tag{4.32}$$

where D_{mi}, ξ_{mi} and σ_m are defined from (4.27). Then the functions $\chi_m(x)$, $m = 1, 2, \ldots$, will be measurable and defined almost everywhere on $[-\pi, \pi]$. Moreover, in view of (4.25) and (4.27) $\lim_{m\to\infty} \sigma_m = 0$, so that

$$\lim_{k\to\infty} \sigma_{m_k} = 0. \tag{4.33}$$

Taking into account (4.5), (4.27), (4.32) and (4.33), we see that the functions $\varphi(x)$, $\psi(x)$ and $\chi_{m_k}(x)$, the numbers $\xi_{m_k,i}$ and σ_{m_k}, and the intervals $D_{m_k,i}$ ($k = 1, 2, \ldots;\ i = 0, 1, 2, \ldots, -1, -2, \ldots$) satisfy all the hypotheses of Lemma 8, if in this lemma one puts $\delta_k = \sigma_{m_k}$, $\Delta_{k_i} = D_{m_k,i}$, $c_k = 0$ and $[a, b] = [-\pi, \pi]$ (see (3.49)–(3.51) and (3.53)). In this case, on the basis of Lemma 8, the functions $\psi(x)$ and $\varphi(x)$ will be respectively the upper and lower limit functions, relative to convergence in measure on $[-\pi, \pi]$, of the sequence $\{\chi_{m_k}(x)\}$, $k = 1, 2, \ldots$.

Further, we see from (4.26), (4.27) and (4.31)–(4.33) that the functions $\varphi(x)$, $\psi(x)$, $\varphi_{m_k}(x)$, $\psi_{m_k}(x)$, $f_{m_k}(x)$, $\chi_{m_k}(x)$, the numbers $\xi_{m_k i}$, σ_{m_k}, and the intervals $D_{m_k,i}$ ($k = 1, 2, \ldots;\ i = 0, 1, 2, \ldots, -1, -2, \ldots$) satisfy all the hypotheses of Lemma 9 if in this lemma one puts

(35) The sequence of numbers m_k may be regarded as increasing in the case when any pair belonging to M_0' is repeated in the succession of pairs $[\varphi_m(x), \psi_m(x)]$ a countable number of times. We will suppose that this condition is satisfied.

$$\alpha_k(x)=\varphi_{m_k}(x),\ \beta_k(x)=\psi_{m_k}(x),\ g_k(x)=f_{m_k}(x),\ \omega_k(x)=\chi_{m_k}(x),\ x_{ki}=\xi_{m_k,i},$$
$$\delta_k=\sigma_{m_k},\ \Delta_{ki}=D_{m_k,i},\ c_k=0 \text{ and } [a,\ b]=[-\pi,\ \pi]$$

(see (3.50), (3.51) and (3.53)–(3.55). In this case, on the basis of Lemma 9, the sequences $\{\chi_{m_k}(x)\}$ and $\{f_{m_k}(x)\}$, $k = 1, 2, \ldots$, are equiconvergent almost everywhere on $[-\pi, \pi]$.

Since the trigonometric series (1.4) and the functions $h_m(x)$, $m = 1, 2, \ldots$, satisfy all the conditions of Lemma 7, it follows that the functions $h_m(x)$ are continuous on $[-\pi, \pi]$, and moreover, in view of property 1) of these functions and of the functions $f_m(x)$, $m = 1, 2, \ldots$ (see the formulation of Lemma 7), the sequences $\{h_m(x)\}$ and $\{f_m(x)\}$ are equiconvergent almost everywhere on $[-\pi, \pi]$. By their definition the natural numbers m_k, $k = 1, 2, \ldots$, form an increasing sequence, and in such a case it follows from Lemma 3 and the properties of the functions $f_m(x)$ and $h_m(x)$ which we have found that the sequences $\{f_{m_k}(x)\}$ and $\{h_{m_k}(x)\}$ are equiconvergent almost everywhere on $[-\pi, \pi]$.

On the other hand, we have already proved that the sequences $\{\chi_{m_k}(x)\}$ and $\{f_{m_k}(x)\}$ are equiconvergent almost everywhere on $[-\pi, \pi]$, from which it follows on the basis of Lemma 4 of §3 that the sequences $\{\chi_{m_k}(x)\}$ and $\{h_{m_x}(x)\}$ are equiconvergent as $k \longrightarrow \infty$ almost everywhere on that segment.

Since the functions $h_m(x)$, $m = 1, 2, \ldots$, and the partial sums $S_n(x)$, $n = 1, 2, \ldots$ of the trigonometric series (1.4) are continuous, and therefore finite everywhere on $[-\pi, \pi]$, it follows, on taking into account the definition of equiconvergence of series (Definition 16 of §3) and property 2) of the functions $h_m(x)$ and of the trigonometric series (1.4) (see the formulation of Lemma 7, equation (3.46)), that the sequences $\{h_m(x)\}$ and $\{S_{\nu_m}(x)\}$, $m = 1, 2, \ldots$, are equiconvergent almost everywhere on $[-\pi, \pi]$. Hence, on the basis of Lemma 3, the sequences $\{h_{m_k}(x)\}$ and $\{S_{\nu'_k}(x)\}$, $k = 1, 2, \ldots$, are also equiconvergent almost everywhere on that segment. Here we have written

$$\nu'_k=\nu_{m_k}\quad (k=1,\ 2,\ \ldots). \tag{4.34}$$

Applying Lemma 4 once again, we are able to assert that the sequences $\{\chi_{m_k}(x)\}$ and $\{S_{\nu'_k}(x)\}$ are equiconvergent as $k \longrightarrow \infty$ almost everywhere on $[-\pi, \pi]$. Since we have already proved that the $\psi(x)$ and $\varphi(x)$ corresponding to our function $\varphi(x, E) \in M$ are upper and lower limit functions relative to convergence in measure on $[-\pi, \pi]$ for the sequence $\{\chi_{m_k}(x)\}$, $k = 1, 2, \ldots$, on the basis of (4.5) and Lemma 6 we find that those functions are also upper and lower limit functions relative to convergence in measure on $[-\pi, \pi]$ for the sequence $\{S_{\nu'_k}(x)\}$, $k = 1, 2, \ldots$. Hence, in view of (4.1) and Theorem β of §2 it follows that the sequence $\{S_{\nu'_k}(x)\}$ converges in measure on the set E to the function $\varphi(x, E)$ in question.[36] In such a case, on the basis of Theorem γ of §2, we can find an increasing sequence of natural numbers k_j, $j = 1, 2, \ldots$, such that

---

[36] By the hypothesis of Theorem III, meas $E > 0$, $E \subset [-\pi, \pi]$, if $\varphi(x, E) \in M$, so that all the hypotheses of Theorem β are satisfied.

$$\lim_{j\to\infty} S_{\nu_j''}(x) = \varphi(x, E) \tag{4.35}$$

almost everywhere on E, where

$$\nu_j'' = \nu_{k_j}' \quad (j = 1, 2, \ldots). \tag{4.36}$$

Recalling property 2) of the natural numbers ν_m, $m = 1, 2, \ldots$ (see the formulation of Lemma 7, inequality (3.45)), we see that these numbers form an increasing sequence. On the other hand, by definition, the sequences $\{m_k\}$ and $\{k_j\}$ $(k = 1, 2, \ldots; j = 1, 2, \ldots)$ are increasing, so that in view of (4.34) and (4.36) it follows that the sequence $\{\nu_j''\}$ is also increasing.

Thus, we have chosen an arbitrary function $\varphi(x, E)$ satisfying (4.29) and have constructed for this function an increasing sequence of natural numbers ν_j'' which satisfies (4.35) almost everywhere on E. In addition, in view of the properties of the set M we have meas $E > 0$, $E \subset [-\pi, \pi]$ (see the formulation of Theorem III, condition (1.18)). Since the $S_n(x)$ are partial sums of the trigonometric series (1.4), it therefore follows that any function $\varphi(x, E)$ lying in M is a limit function of that series (see Definition 1). In such a case, in order to prove that the trigonometric series (1.4) has the property A figuring in the formulation of Theorem III, it suffices for us to show that any limit function of the series (1.4) lies in M.

Suppose that $\varphi(x, E)$ is any limit function of (1.4). Then, on the basis of Definition 1, that function is defined almost everywhere on E, while

$$\text{meas } E > 0, \quad E \subset [-\pi, \pi]. \tag{4.37}$$

Moreover, there exists an increasing sequence of natural numbers n_k such that

$$\lim_{k\to\infty} S_{n_k}(x) = \varphi(x, E) \tag{4.38}$$

almost everywhere on E, where the $S_n(x)$ are the partial sums of (1.4).

Since the natural numbers n_k, $k = 1, 2, \ldots$, form an increasing sequence, on the basis of property 4) of the functions $h_m(x)$, $m = 1, 2, \ldots$, and of trigonometric series (1.4) (see the formulation of Lemma 7) one of the following two assertions must be satisfied:

A°. The upper and lower limits in measure on $[-\pi, \pi]$ of the sequence $\{S_{n_k}(x)\}$ are equal respectively to the functions $F(x)$ and $G(x)$ figuring in the formulation of Theorem III.

B°. It is possible to find a nondecreasing sequence of natural numbers p_k such that

$$\lim_{k\to\infty} p_k = \infty \tag{4.39}$$

and the sequence of functions $h_{p_k}(x) - S_{n_k}(x)$ converges in measure to zero on $[-\pi, \pi]$.

We suppose first that condition A° is satisfied. Then, as is known,

$$\underline{\lim}_{k\to\infty} S_{n_k}(x) \leqslant G(x) \leqslant F(x) \leqslant \overline{\lim}_{k\to\infty} S_{n_k}(x)$$

almost everywhere on $[-\pi, \pi]$ (see [2], §3, Theorems F and G). Accordingly, on the basis of (4.37) and (4.38),

$$G(x) = F(x) = \varphi(x, E) = \lim_{k\to\infty} S_{n_k}(x) \tag{4.40}$$

almost everywhere on the set E, so that

$$\text{meas}\,(E - E[F(x) = G(x)]) = 0. \tag{4.41}$$

We suppose that M is a nonempty set satisfying all the hypotheses of Theorem III, in particular conditions α, β, and γ of Theorem II (see the formulation of Theorem III in §1). Hence there exists a function $f(x, E')$, defined almost everywhere on a set E', and such that $f(x, E') \in M$, where meas $E' > 0$ and $E' \subset [-\pi, \pi]$. We introduce the notation

$$E'_0 = E' + E[F(x) = G(x)]. \tag{4.42}$$

At the points of E'_0 where $F(x)$, $G(x)$ and $f(x, E')$ are defined, we put

$$f_0(x, E'_0) = \begin{cases} f(x, E') & (\text{if } x \in E'), \\ F(x) & (\text{if } x \in E'_0 - E'). \end{cases} \tag{4.43}$$

Then $f_0(x, E')$ is defined almost everywhere on E'_0, and, moreover, in view of condition γ,

$$f_0(x, E'_0) \in M. \tag{4.44}$$

From (4.44) and β (see the formulation of Theorem II in §1) it follows that $G(x) \leqslant f_0(x, E'_0) \leqslant F(x)$ almost everywhere on E'_0. Hence, on the basis of (4.42),

$$f_0(x, E'_0) = F(x) = G(x) \tag{4.45}$$

almost everywhere on the set $E[F(x) = G(x)]$. Now comparing this last equation with (4.40) and (4.41), we see that

$$\rho(x, E) = f_0(x, E'_0) \tag{4.46}$$

almost everywhere on E.

Since by the hypothesis of Theorem III the set M is closed in the narrow sense (see condition α of Theorem II), and since in view of (4.37), (4.41) and (4.42)

$$\text{meas}\,(E - E'_0) = 0, \quad \text{meas}\, E > 0, \quad E \subset [-\pi, \pi],$$

on the basis of (4.44) and Lemma A (in which we choose $\varphi(x, E)$ and $f_0(x, E'_0)$ in place of $\psi(x, E')$ and $f(x, E)$) we will have $\varphi(x, E) \in M$.

Thus, if condition A° is satisfied, then the function $\varphi(x, E)$ which we are considering belongs to M.

Now suppose that condition B° is satisfied.

We shall show that in this case $\varphi(x, E) \in M$. Recalling (4.39) and the known properties of sequences converging in measure (see for example §2, Theorem γ), we may find an increasing sequence of natural numbers k_s, $s = 1, 2, \dots$, such that

$$p'_s < p'_{s+1} \quad (s = 1, 2, \dots) \tag{4.47}$$

and

$$\lim_{s\to\infty} [h_{p'_s}(x) - S_{n'_s}(x)] = 0 \tag{4.48}$$

almost everywhere on $[-\pi, \pi]$, where

$$p'_s = p_{k_s}, \quad n'_s = n_{k_s} \quad (s = 1, 2, \dots). \tag{4.49}$$

In addition, since the sequences $\{n_k\}$, $k = 1, 2, \dots$, and $\{k_s\}$, $s = 1, 2, \dots$, are increasing, we have

$$n'_s < n'_{s+1} \quad (s = 1, 2, \dots). \tag{4.50}$$

We already know that the functions $h_m(x)$ and $S_n(x)$ are continuous and therefore finite on $[-\pi, \pi]$. Hence, recalling (4.48) and the definition of equiconvergence of series (Definition 16 of §3), we see that the sequences $\{h_{p'_s}(x)\}$ and $\{S_{n'_s}(x)\}$, $s = 1, 2, \dots$, are equiconvergent almost everywhere on $[-\pi, \pi]$.

By definition, the functions $h_m(x)$, $m = 1, 2, \dots$, have along with the functions $f_m(x)$ and the series (1.4) all the properties 1)–5) figuring in the formulation of Lemma 7. Hence, on the basis of property 1) the sequences $\{f_m(x)\}$ and $\{h_m(x)\}$ are equiconvergent almost everywhere on the segment $[-\pi, \pi]$. Since in view of (4.47) the succession of the p'_s is increasing, it follows from Lemma 3 (§3) that the sequences $\{f_{p'_s}(x)\}$ and $\{h_{p'_s}(x)\}$ are also equiconvergent on that segment, almost everywhere. Further, we have already proved that the sequences $\{h_{p'_s}(x)\}$ and $\{S_{n'_s}(x)\}$ are equiconvergent almost everywhere on the same segment. Hence, in view of Lemma 4 the sequences $\{S_{n'_s}(x)\}$ and $\{f_{p'_s}(x)\}$ are equiconvergent almost everywhere on $[-\pi, \pi]$.

Recalling (4.49) and (4.50), we see that the natural numbers n'_s, $s = 1, 2, \dots$, form an increasing sequence extracted from the sequence of numbers n_k, $k = 1, 2, \dots$, so that it follows on the basis of (4.38) that

$$\lim_{s\to\infty} S_{n'_s}(x) = \varphi(x, E) \tag{4.51}$$

almost everywhere on E. Further, recalling Lemma 5, we see from (4.51) that

$$\lim_{s\to\infty} f_{p'_s}(x) = \varphi(x, E) \tag{4.52}$$

almost everywhere on E.

Put

$$\underline{\lim}_{s\to\infty} E_{p'_s} = Q, \quad \overline{\lim}_{s\to\infty} e'_{p'_s} = e', \quad \overline{\lim}_{s\to\infty} e''_{p'_s} = e'', \tag{4.53}$$

where E_m, e'_m and e''_m are the sets defined above (see the definition of the functions $g_m(x, E_m)$ and (4.13)). On the basis of (4.10), (4.15) and (4.16) we will have

$$\left.\begin{array}{c} E_{p'_s} \subset [-\pi, \pi], \quad e'_{p'_s} \subset [-\pi, \pi], \quad e''_{p'_s} \subset [-\pi, \pi], \\ \text{meas } e'_{p'_s} < \frac{1}{2^{p'_s}}, \quad \text{meas } e''_{p'_s} < \frac{1}{2^{p'_s}} \\ (s = 1, 2, \ldots), \end{array}\right\} \tag{4.54}$$

so that in view of (4.53)

$$Q \subset [-\pi, \pi]. \tag{4.55}$$

Further, since the natural numbers p'_s, $s = 1, 2, \ldots$, form an increasing sequence (see (4.47)), we have that

$$\sum_{s=1}^{\infty} \frac{1}{2^{p'_s}} < +\infty,$$

and so from (4.53) and (4.54) we get

$$\text{meas } e' = 0, \quad \text{meas } e'' = 0, \quad e' \subset [-\pi, \pi], \quad e'' \subset [-\pi, \pi]. \tag{4.56}$$

From the definition of the functions $f_m(x)$ given at (4.26), it follows that for any $x \in [-\pi, \pi]$ (with the possible exception of a set of measure zero) the function $f_{p'_s}(x)$ is equal either to $\varphi_{p'_s}(x)$, or to $\psi_{p'_s}(\psi)$, $s = 1, 2, \ldots$. On the other hand, in view of (4.11) and (4.12), almost everywhere on E_m each of the functions $\varphi_m(x)$ and $\psi_m(x)$ can take on only the value $g_m(x, E_m)$. Furthermore, almost everywhere on $[-\pi, \pi] - E_m$ each of those two functions can take on only one of the values $F(x)$ or $G(x)$. In this situation, for any $s = 1, 2, \ldots$ we have

$$f_{p'_s}(x) = g_{p'_s}(x, E_{p'_s}) \tag{4.57}$$

almost everywhere on $E_{p'_s}$, and

$$f_{p'_s}(x) = F(x) \tag{4.58}$$

or

$$f_{p'_s}(x) = G(x) \tag{4.59}$$

almost everywhere on $[-\pi, \pi] - E_{p'_s}$.

Denoting by e_0 the set of all points lying in E for which (4.52) is satisfied, we will have

$$e_0 \subset E \subset [-\pi, \pi], \quad \text{meas } e_0 = 0. \tag{4.60}$$

Further, we denote by e'_s, $s = 1, 2, \ldots$, the set of all points on $E_{p'_s}$ at which (4.57) is not satisfied, and by e''_s the set of all points lying in $[-\pi, \pi] - E_{p'_s}$ at which neither (4.58) nor (4.59) is satisfied. Put

$$e'_0 = e_1 + \sum_{s=1}^{\infty} (e'_s + e''_s), \quad Q' = E - Q - e' - e'' - e_0 - e'_0, \tag{4.61}$$

where e_1 is the set of points on $[-\pi, \pi]$ at which (4.6) is not satisfied for any value $m = 1, 2, \ldots$. Then

$$\text{meas } e_0' = 0 \tag{4.62}$$

and moreover

$$f_{p_s'}(x) = g_{p_s'}(x, E_{p_s'}) \qquad (x \in E_{p_s'} - e_0';\ s = 1, 2, \ldots), \tag{4.63}$$

$$\left.\begin{array}{c} f_{p_s'}(x) = F(x) \quad \text{or} \quad f_{p_s'}(x) = G(x) \\ (x \in [-\pi, \pi] - E_{p_s'} - e_0';\ s = 1, 2, \ldots), \end{array}\right\} \tag{4.64}$$

$$G(x) \leqslant F(x) \qquad (x \notin e_0'). \tag{4.65}$$

Here all the functions $f_{p_s'}(x)$ $(s = 1, 2, \ldots)$, $F(x)$ and $G(x)$ are defined on $[-\pi, \pi] - e_0'$. We note further that on the basis of (4.37) and (4.61)

$$Q' \subset E \subset [-\pi, \pi], \qquad Q' \subset [-\pi, \pi] - e_0', \tag{4.66}$$

from which it follows that the functions $f_{p_s'}(x)$, $s = 1, 2, \ldots$, $F(x)$ and $G(x)$ are defined on the set Q'.

Now suppose that x is any point satisfying the condition

$$x \in Q'. \tag{4.67}$$

Then it follows from (4.61) that

$$x \notin Q, \quad x \notin e_0'. \tag{4.68}$$

Thus, in view of the first equation (4.53), there exists an increasing sequence of natural numbers s_j, $h = 1, 2, \ldots$, satisfying the condition

$$x \notin E_{p_j''} + e_0' \qquad (j = 1, 2, \ldots), \tag{4.69}$$

where

$$p_j'' = p_{s_j}' \qquad (j = 1, 2, \ldots). \tag{4.70}$$

It follows from (4.9) that

$$E[F(x) = G(x)] \subset E_{p_j''} \qquad (j = 1, 2, \ldots), \tag{4.71}$$

and since (4.69) is satisfied for any point $x \in Q'$, on the basis of (4.65) and (4.71) we have

$$G(x) < F(x) \qquad (x \in Q'). \tag{4.72}$$

Recalling (4.53) and (4.61), we see that for any $x \in Q'$ one may determine a natural number s_x such that

$$x \notin e_{p_s'}', \quad x \notin e_{p_s'}'' \qquad (s > s_x,\ x \in Q'). \tag{4.73}$$

Now we shall prove that this number s_x may be determined for each $x \in Q'$ in such a way that not only (4.73), but also one of the following conditions a° or b° must be satisfied:

$$\text{a}^\circ. \quad f_{p_s'}(x) = F(x) \quad \text{or} \quad f_{p_s'}(x) = g_{p_s'}(x, E_{p_s'}) \quad (s > s_x), \tag{4.74}$$

$$\text{b}^\circ. \quad f_{p_s'}(x) = G(x) \quad \text{or} \quad f_{p_s'}(x) = g_{p_s'}(x, E_{p_s'}) \quad (s > s_x). \tag{4.75}$$

Indeed, it follows from (4.61), (4.63) and (4.64) that for any $x \in Q'$ and for any $s = 1, 2, \ldots$, only one of the following three equations may be satisfied:

$$f_{p_s'}(x) = g_{p_s'}(x, E_{p_s'}), \qquad f_{p_s'}(x) = F(x), \qquad f_{p_s'}(x) = G(x). \tag{4.76}$$

Further, if for some $x \in Q'$ the second and third equations (4.76) were satisfied simultaneously for an infinite set of different values of s, then, from (4.72), at that point x the function values $f_{p_s'}(x)$ would not tend to any limit as $s \to \infty$. Since, on the other hand, from the definition of the set e_0 and the second equation (4.61) it follows that equation (4.52) is satisfied at the point x in question, the preceding hypothesis is not valid. Hence at any point $x \in Q'$ either the second equation or the third equation in (4.76) must be satisfied only at a finite number of values of s. Then for each $x \in Q'$ the number s_x may be chosen in such a way that (4.73) is satisfied, and also one of the conditions (4.74) or (4.75), i.e. either a° or b°.

The basic portion of the following arguments will consist in the proof of the equation

$$\text{meas}\, Q' = 0. \tag{4.77}$$

To this end we suppose that

$$\text{meas}\, Q' > 0, \tag{4.78}$$

and prove that one obtains a contradiction.

Denote by H' the set of all points belonging to Q' and at which (4.74) is satisfied, and by H'' the set of all points lying in Q' at which (4.75) is satisfied. Since at each point $x \in Q'$ either (4.74) or (4.75) is satisfied, we have

$$Q' = H' + H'', \tag{4.79}$$

In the case at hand, in view of (4.78), at least one of the sets H' or H'' has positive measure.

Denote by H that one of the sets H' or H'' which has positive measure.(37) Then on the basis of (4.66)

$$H \subset Q' \subset [-\pi, \pi], \qquad \text{meas}\, H > 0. \tag{4.80}$$

Moreover, either (4.74) is satisfied at all the points of H, or else (4.75) is satisfied at all of its points.

Put

(37) If both H' and H'' have positive measure, we put $H = H'$.

$$\Phi_1(x) = F(x), \qquad \Phi_2(x) = G(x) \tag{4.81}$$

if $H = H'$, and

$$\Phi_1(x) = G(x), \qquad \Phi_2(x) = F(x), \tag{4.82}$$

if $H = H''$. Then on the basis of (4.73), (4.75), and (4.80), and the definitions of the sets H', H'', and H,

$$x \notin e'_{p'_s}, \quad x \notin e''_{p'_s} \qquad (s > s_x, \ x \in H), \tag{4.83}$$

$$f_{p'_s}(x) = \Phi_1(x) \quad \text{or} \quad f_{p'_s}(x) = g_{p'_s}(x, E_{p'_s}) \qquad (s > s_x, \ x \in H). \tag{4.84}$$

Moreover, in view of (4.72) and (4.80)

$$\Phi_1(x) \neq \Phi_2(x) \qquad (x \in H).\text{(}^{38}\text{)} \tag{4.85}$$

For any natural number k, we denote by Ω_k the set of all points $x \in H$ for which $s_x \leqslant k$. Then $\Omega_k \subset \Omega_{k+1}$, $k = 1, 2, \ldots$. Moreover, in view of (4.80),

$$H = \sum_{k=1}^{\infty} \Omega_k \subset [-\pi, \pi]. \tag{4.86}$$

It therefore follows that

$$\lim_{k \to \infty} \operatorname{meas} \Omega_k = \operatorname{meas} H. \tag{4.87}$$

Recalling (4.80) and (4.87), we see that there exists a natural number k_0 satisfying

$$\operatorname{meas} \Omega_{k_0} > 0. \tag{4.88}$$

Denote by P a perfect set satisfying the conditions

$$P \subset \Omega_{k_0}, \quad \operatorname{meas} P > 0. \tag{4.89}$$

Then on the basis of the definition of the sets Ω_k and conditions (4.83)–(4.86) and (4.89) we have

$$P \subset [-\pi, \pi], \quad x \notin e'_{p'_s}, \quad x \notin e''_{p'_s} \qquad (s > k_0, \ x \in P), \tag{4.90}$$

$$f_{p'_s}(x) = \Phi_1(x) \quad \text{or} \quad f_{p'_s}(x) = g_{p'_s}(x, E_{p'_s}) \qquad (s > k_0, \ x \in P), \tag{4.91}$$

$$\Phi_1(x) \neq \Phi_2(x) \qquad (x \in P).\text{(}^{39}\text{)} \tag{4.92}$$

In addition, since the functions $g_{p'_s}(x, E_{p'_s})$, $s = 1, 2, \ldots$, are defined only at the points of $E_{p'_s}$ (see the definition of the functions $g_m(x, E_m)$ from conditions (4.8) and (4.9) and the remark at the beginning of §4), in view of (4.91) we have

$$f_{p'_s}(x) = \Phi_1(x) \qquad (s > k_0, \ x \in P - E_{p'_s}). \tag{4.93}$$

(38) Since the functions $f_{p'_s}(x)$, $s = 1, 2, \ldots$, $F(x)$ and $G(x)$ are defined on the set Q', on the basis of (4.80)–(4.82) the functions $f_{p'_s}(x)$, $\Phi_1(x)$ and $\Phi_2(x)$ are defined on H.

(39) Since $P \subset H$ ((4.86) and (4.89)), by the remark to (4.85) the functions $f_{p'_s}(x)$, $s = 1, 2, \ldots$, $\Phi_1(x)$ and $\Phi_2(x)$ are defined on P.

Put

$$D'_{mi} = D_{m,\,2i}, \qquad D''_{mi} = D_{m,\,2i+1}$$
$$(i = 0,\ 1,\ 2,\ \ldots,\ -1,\ -2,\ \ldots;\ m = 1,\ 2,\ \ldots), \tag{4.94}$$

if $H = H'$, and

$$D'_{mi} = D_{m,\,2i+1}, \qquad D''_{mi} = D_{m,\,2i}$$
$$(i = 0,\ 1,\ 2,\ \ldots,\ -1,\ -2,\ \ldots;\ m = 1,\ 2,\ \ldots), \tag{4.95}$$

if $H = H''$, where D_{mi} are the intervals defined by (4.27). Then on the basis of these equations

$$\text{meas}\ D'_{mi} = \frac{\pi}{2^{2m}}, \quad \text{meas}\ D''_{mi} = \frac{\pi}{2^{2m}}$$
$$(i = 0,\ 1,\ 2,\ \ldots,\ -1,\ -2,\ \ldots;\ m = 1,\ 2,\ \ldots). \tag{4.96}$$

Moreover, at the points where the functions $f_m(x)$, $m = 1, 2, \ldots, F(x)$ and $G(x)$ are defined, we find on comparing (4.12) and (4.26) that

$$f_m(x) = \begin{cases} F(x) & (\text{if } x \in D_{m,\,2j} - E_m), \\ G(x) & (\text{if } x \in D_{m,\,2j+1} - E_m) \end{cases}$$
$$(m = 1,\ 2,\ \ldots;\ j = 0,\ 1,\ 2,\ \ldots,\ -1,\ -2,\ \ldots).$$

Hence in view of (4.81), (4.82), (4.94) and (4.95) we have, for the same points x,

$$\left.\begin{array}{l} f_m(x) = \begin{cases} \Phi_1(x) & (\text{if } x \in D'_{mi} - E_m), \\ \Phi_2(x) & (\text{if } x \in D''_{mi} - E_m) \end{cases} \\ (m = 1,\ 2,\ \ldots;\ i = 0,\ 1,\ 2,\ \ldots,\ -1,\ -2,\ \ldots). \end{array}\right. \tag{4.97}$$

Since the perfect set P has positive measure (see (4.89)), there exists a point x_0 of density of that set, which does not coincide with any of the endpoints of the intervals D'_{mi} or D''_{mi} ($m = 1, 2, \ldots; i = 0, 1, 2, \ldots, -1, -2, \ldots$), and satisfies the condition

$$x_0 \in P. \tag{4.98}$$

Then from the definition of D'_{mi} and D''_{mi} (see (4.27), (4.94) and (4.95)) it follows that for any $m = 1, 2, \ldots$ it is possible to determine an integer $i_0 = i_0(x_0, m)$ such that

$$x_0 \in D'_{m,\,i_0} \quad \text{or} \quad x_0 \in D''_{m,\,i_0}. \tag{4.99}$$

Moreover, it follows from (4.80), (4.86) and (4.89) that $P \subset Q'$, so that on the basis of (4.80) and (4.98) we have

$$x_0 \in Q' \subset [-\pi,\ \pi]. \tag{4.100}$$

We have already seen that for any point $x \in Q'$ it is possible to determine an increasing sequence of natural numbers s_j for which (4.69) is satisfied, where the p''_j are defined from (4.70). Then for the given point x_0 it is possible to determine a natural number r satisfying the conditions

$$x_0 \notin E_{p'_r}, \qquad r > k_0, \tag{4.101}$$

it being possible to choose r arbitrarily large.

Hence it follows in view of (4.90) and (4.98) that

$$x_0 \notin e'_{p'_r}, \quad x_0 \notin e''_{p'_r}. \tag{4.102}$$

From (4.13) we have $E'_m - e''_m \subset E_m$, $m = 1, 2, \ldots$. Therefore on the basis of (4.101) and (4.102) the condition

$$x_0 \notin E'_{p'_r}. \tag{4.103}$$

must be satisfied. Further, since x_0 is a point of density of the set P, there exists a positive number δ_0 such that

$$\text{meas}\,(P \cdot \Delta) > \frac{1}{2}\text{meas}\,\Delta \tag{4.104}$$

for any segment Δ satisfying the conditions

$$x_0 \in \Delta, \quad \text{meas}\,\Delta \leqslant \delta_0. \tag{4.105}$$

Put

$$\rho'_s = \rho_{p'_s} \qquad (s = 1, 2, \ldots), \tag{4.106}$$

where the ρ_m satisfy (4.23). Then on the basis of (4.24) and (4.46) we have

$$\rho'_s < \rho'_{s+1} \qquad (s = 1, 2, \ldots), \tag{4.107}$$

so that

$$\lim_{s\to\infty} \rho'_s = \infty. \tag{4.108}$$

Since the natural number r satisfying (4.101) and (4.102) may be chosen to be arbitrarily large, we may assume, in view of (4.108), that

$$\frac{\pi}{2^{\rho'_r}} < \frac{1}{2}\delta_0. \tag{4.109}$$

From the definition of D_{mi}, D'_{mi} and D''_{mi} (see (4.27), (4.94) and (4.95)) it follows that for any natural number m and for any integer i either the intervals D'_{mi}, D''_{mi}, $D'_{m,i+1}$, $D''_{m,i+1}$ are laid out from left to right, or else the intervals D''_{mi}, D'_{mi}, $D''_{m,i+1}$, $D'_{m,i+1}$ are laid out from left to right, while these intervals do not overlap and successive intervals have common endpoints. Hence it follows that for any value $m = 1, 2, \ldots$, any point x of the real axis either lies in one of the intervals D'_{mi} or D''_{mi}, or else it coincides with the endpoint of one of these intervals. Moreover, to each interval D'_{mi} there is adjacent, at each of the left and right endpoints, one interval D''_{mj}, where $j = i$, $i - 1$, or $i + 1$,

Comparing (4.93), (4.98), and (4.101), we see that

$$f_{p'_r}(x_0) = \Phi_1(x_0). \tag{4.110}$$

Further, choosing in (4.99) the integer i_0 for $m = p'_r$, we may assert that one of the following conditions is satisfied:

$$x_0 \in D'_{p'_r,\, i_0} \tag{4.111}$$

or

$$x_0 \in D''_{p'_r,\, i_0}, \tag{4.112}$$

where $i_0 = i_0(x_0, p'_r)$ is a positive integer depending only on x_0 and p'_r.

We shall prove that only condition (4.111) can be satisfied. Indeed, suppose that (4.112) is satisfied. Then on the basis of (4.101) we will have $x_0 \in D''_{p'_r, i_0} - E_{p'_r}$ and therefore, in view of (4.97),

$$f_{p'_r}(x_0) = \Phi_2(x_0). \tag{4.113}$$

On the other hand, comparing (4.92) and (4.98), we get $\Phi_1(x_0) \neq \Phi_2(x_0)$, which contradicts (4.110) and (4.113).

Thus the supposition that (4.112) is satisfied leads to a contradiction, and so (4.111) must be satisfied, where i_0 is some integer.

Since the ρ_m, $m = 1, 2, \ldots$, are natural numbers, it follows from (4.27), (4.94), and (4.95) that for any natural number m the points $x = -\pi$ and $x = \pi$ will be endpoints of the intervals D'_{mi} for some integer value of i. Since for distinct integers i the intervals D'_{mi} do not overlap, on the basis of (4.111) we have

$$x_0 \neq -\pi, \quad x_0 \neq \pi. \tag{4.114}$$

On the other hand, in view of (4.100), $x_0 \in [-\pi, \pi]$. Hence, recalling (4.20), (4.103), and (4.114), we see that

$$x_0 \in d'_{p'_r,\, \lambda_r} \subset (-\pi, \pi), \tag{4.115}$$

where λ_r is some integer satisfying

$$0 \leqslant \lambda_r \leqslant q_{p'_r}. \tag{4.116}$$

We shall show that there exists an interval $D''_{p'_r, j_0}$ which directly adjoins $D'_{p'_r, i_0}$ (see (4.111)) and satisfies

$$D''_{p'_r,\, j_0} \subset d'_{p'_r,\, \lambda_r}. \tag{4.117}$$

Indeed, in the contrary case, in view of (4.111) and (4.115) the inequality

$$\operatorname{meas} d'_{p'_r,\, \lambda_r} \leqslant \operatorname{meas} D'_{p'_r,\, i_0} + \operatorname{meas} D''_{p'_r,\, j'_0} + \operatorname{meas} D''_{p'_r,\, j''_0} \tag{4.118}$$

would be satisfied, where $D''_{p'_r, j'_0}$ and $D''_{p'_r, j''_0}$ directly adjoin the interval $D'_{p'_r, i_0}$ from the two sides. Since on the basis of (4.22), (4.23), (4.106), and (4.116)

$$\frac{\pi}{2^{\rho'_r}} < \frac{\omega_{p'_r}}{3} \leqslant \frac{1}{3} \operatorname{meas} d'_{p'_r,\, \lambda_r},$$

we would find from (4.96), (4.106), and (4.118), that

$$\frac{3\pi}{2^{\rho'_r}} < \operatorname{meas} d''_{p'_r, \lambda_r}, \qquad \operatorname{meas} d'_{p'_r, \lambda_r} \leqslant \frac{3\pi}{2^{\rho'_r}}, \tag{4.119}$$

i.e. we would have a contradiction.

Thus (4.117) must be satisfied, where $D''_{p'_r, j_0}$ directly adjoins $D'_{p'_r, i_0}$ and λ_r satisfies (4.116). Then in view of (4.18), (4.20) and (4.16) we have

$$(E'_{p'_r} \cdot D''_{p'_r, j_0}) = 0. \tag{4.120}$$

Denote by Δ_n a segment one endpoint of which coincides with x_0 and whose other endpoint coincides with the endpoint of $D''_{p'_r, j_0}$ which is farthest from x_0. Since $D''_{p'_r, j_0}$ directly adjoins the interval $D'_{p'_r, i_0}$, on the basis of (4.111) we have

$$D''_{p'_r, j_0} \subset \Delta_0 \subset \overline{D}'_{p'_r, i_0} + \overline{D}''_{p'_r, j_0}, \tag{4.121}$$

where $\overline{D}'_{mi}$ and $\overline{D}''_{mi}$ are the closures of D'_{mi} and D''_{mi}. In this case, since $\overline{D}'_{p'_r, i_0}$ and $\overline{D}''_{p'_r, j_0}$ do not have common interior points, it follows from (4.121) that

$$\operatorname{meas} D''_{p'_r, j_0} \leqslant \operatorname{meas} \Delta_0 \leqslant \operatorname{meas} D'_{p'_r, i_0} + \operatorname{meas} D''_{p'_r, j_0},$$

so that in view of (4.96) and (4.106) we have

$$\frac{\pi}{2^{\rho'_r}} \leqslant \operatorname{meas} \Delta_0 \leqslant \frac{2\pi}{2^{\rho'_r}} = 2 \operatorname{meas} D''_{p'_r, j_0}.$$

Therefore, on the basis of (4.109)

$$\frac{1}{2} \operatorname{meas} \Delta_0 \leqslant \operatorname{meas} D''_{p'_r, j_0}, \quad \operatorname{meas} \Delta_0 < \delta_0. \tag{4.122}$$

Since x_0 is an endpoint of the segment Δ_0, we have $x_0 \in \Delta_0$, so that in view of (4.104), (4.105) and the second inequality (4.122)

$$\operatorname{meas} (P \cdot \Delta_0) > \frac{1}{2} \operatorname{meas} \Delta_0.$$

Hence, recalling the first inequality (4.122) and (4.121), we will have

$$\operatorname{meas} (P \cdot D''_{p'_r, j_0}) > 0.$$

Thus there exists a point x_1 satisfying the conditions(40)

$$x_1 \in P, \quad x_1 \in D''_{p'_r, j_0}. \tag{4.123}$$

Comparing (4.120) and (4.123), we get

$$x_1 \notin E'_{p'_r}. \tag{4.124}$$

Further, it follows from (4.13) that $E_m \subset E'_m + e'_m$, $m = 1, 2, \ldots$. From (4.90), (4.101) and (4.123) we have $x_1 \notin e'_{p'_r}$, so that in view of (4.124)

(40) On the basis of the remark to (4.92), the functions $f_{p'_s}(x)$, $s = 1, 2, \ldots$, $\Phi_1(x)$ and $\Phi_2(x)$ are defined at x_1.

$$x_1 \notin E_{p'_r}. \tag{4.125}$$

Comparing (4.123) and (4.125), we finally get

$$x_1 \in P - E_{p'_r}, \quad x_1 \in D''_{p'_r, j_0} - E_{p'_r}. \tag{4.126}$$

Therefore, recalling (4.97), we see that

$$f_{p'_r}(x_1) = \Phi_2(x_1). \tag{4.127}$$

On the other hand, recalling (4.93), (4.101) and (4.126), we get

$$f_{p'_r}(x_1) = \Phi_1(x_1). \tag{4.128}$$

Since, moreover, on the basis of (4.92) and (4.123) we have $\Phi_1(x_1) \neq \Phi_2(x_1)$, the equations (4.127) and (4.128) lead to a contradiction.

Thus we have arrived at a contradiction under the supposition that (4.78) is satisfied. Therefore (4.77) must be satisfied.

Comparing (4.56), (4.60), (4.61), (4.62) and (4.77), we have

$$\text{meas}\,(E - Q) = 0, \quad \text{meas}\,[(E \cdot Q) - e'_0] = \text{meas}\, E. \tag{4.129}$$

Further, if $x \in (E \cdot Q) - e'_0$, then on the basis of the first equation (4.53) there exists a natural number σ_x such that $x \in E_{p'_s} - e'_0$, $s > \sigma_x$. Therefore, in view of (4.63),

$$f_{p'_s}(x) = g_{p'_s}(x, E_{p'_s}) \qquad [x \in (E \cdot Q) - e'_0, \ s > \sigma_x], \tag{4.130}$$

Hence, taking (4.52) and (4.129) into account, we see that

$$\lim_{s \to \infty} g_{p'_s}(x, E_{p'_s}) = \varphi(x, E) \tag{4.131}$$

almost everywhere on E, where $\varphi(x, E)$ is the limit function in question of the series (1.4).

Since conditions (4.37) are satisfied for the set E corresponding to the function $\varphi(x, E)$ in quesiton, on recalling the first equation (4.129) we get

$$\text{meas}\, Q > 0, \qquad \text{meas}\, E > 0, \qquad \text{meas}\,(E - Q) = 0,$$

the set Q being defined by the first equation (4.53). Thus, on the basis of (4.131) and Definition 3 (§1), in which we take $g_{p'_s}(x, E_{p'_s})$ and $E_{p'_s}$, $s = 1, 2, \ldots$ in place of $\varphi_n(x, e_n)$ and E_n, $n = 1, 2, \ldots$, we see that $\varphi(x, E)$ is a limit element in the wide sense of the sequence

$$g_{p'_s}(x, E_{p'_s}) \, (s = 1, 2, \ldots).$$

Thus, recalling (4.8) and Definition 4, we see that $\varphi(x, E)$ is a limit element in the wide sense of the set M.

Denote by $\overline{M}$ the closure in the wide sense of M (see Definition 5). We then conclude that

$$\varphi(x, E) \in \overline{M}. \tag{4.132}$$

By the hypothesis of Theorem III the set M satisfies conditions α, β, and γ. Then on the basis of condition α (see the formulation of Theorem II) the set M is closed in the narrow sense. Therefore, in view of Definition 6, $\bar{M} = M$. Hence, recalling (4.132), we see that $\varphi(x, E)$ belongs to M.

Thus we suppose that $\varphi(x, E)$ is any limit function of the trigonometric series (1.4), i.e. a function for which (4.38) is satisfied almost everywhere on E, where the $S_n(x)$, $n = 1, 2, \ldots$, are the partial sums of the series (1.4) and the natural numbers n_k form an increasing sequence. Supposing further that condition B° is satisfied (the formulations of conditions A° and B° directly follow (4.38)), we have proved that $\varphi(x, E)$ belongs to M.

Earlier we have already proved that $\varphi(x, E)$ belongs to M if it is a limit function of the trigonometric series (1.4) and if condition A° is satisfied.

Thus we have proved that any limit function of the trigonometric series (1.4) belongs to M, i.e. we have proved that the series (1.4) has property A. As we have already seen, this completes the proof of Theorem III (see the remarks preceding (4.37) and (4.38), and preceding (4.29)).

§5. Derivation of Theorem I from Theorems II and III

If $M = \{\varphi(x, E)\}$ is the set of all limit functions of the trigonometric series (1.4), it follows immediately from Thereom II that M is closed in the narrow sense. Thus the necessity of the condition formulated in Theorem I is proved.

Now choose any set $M = \{\varphi(x, E)\}$ of measurable functions $\varphi(x, E)$, each of which is defined almost everywhere on a corresponding set $E \subset [-\pi, \pi]$ of positive measure. We suppose that M is closed in the narrow sense. In order to complete the proof of Theorem I, it suffices to prove that M is the set of all limit functions of some trigonometric series.

As is well known, there exists a trigonometric series for which any subsequence of partial sums with increasing indices does not tend almost everywhere to any limit, finite or infinite.(41) In such a case our assertion is proved if M is the empty set.

Now suppose that M is nonempty and closed in the narrow sense. Put

$$F(x) = +\infty, \qquad G(x) = -\infty \qquad (-\pi \leqslant x \leqslant \pi). \tag{5.1}$$

Then the set M and the functions $F(x)$ and $G(x)$ satisfy the conditions α, β, and γ of Theorem II (§1). Indeed, M is a closed set in the narrow sense, so that it satisfies condition α. Further, by hypothesis each of the functions $\varphi(x, E)$ belonging to M is measurable and defined almost everywhere on the corresponding set E, while in view of (5.1) we have $G(x) \leqslant \varphi(x, E) \leqslant F(x)$ almost everywhere on that set. Hence condition β of Theorem II is satisfied.

(41) This is a property of any trigonometric series $\Sigma_1^\infty (\alpha_k \cos n_k x + \beta_k \sin n_k x)$ for which $\Sigma_1^\infty (\alpha_k^2 + \beta_k^2)$ diverges and the natural numbers n_k satisfy the condition $(n_k + 1)/n_k > \lambda > 1$, where λ does not depend on k.

Finally, on the basis of (5.1), $E[F(x) = G(x)] = 0$, so that the set E_0 defined in (1.16) coincides with E. Then the function $\varphi_0(x, E_0)$ defined by (1.17) may be regarded as coincident with $\varphi(x, E)$. Therefore, if $\varphi(x, E) \in M$, then the corresponding function $\varphi_0(x, E_0)$ also belongs to M, and this means that M satisfies condition γ.

Thus the set M and the functions $F(x)$ and $G(x)$ defined by (5.1) satisfy conditions α, β, and γ of Theorem II, while by hypothesis each of the functions $\varphi(x, E)$ belonging to M is measurable and defined almost everywhere on a corresponding set $E \subset [-\pi, \pi]$ of positive measure. Moreover, on the basis of (5.1) the functions $F(x)$ and $G(x)$ satisfy (1.19) everywhere on $[-\pi, \pi]$. Then the set M and the functions $F(x)$ and $G(x)$ satisfy all the conditions of Theorem III. Thus there exists a trigonometric series (1.4) having the properties A, B, and C figuring in the formulation of Theorem III. Since by property A the set M is the set of all limit functions of the series (1.4), the proof of Theorem I is complete.

BIBLIOGRAPHY

1. D. E. Men'šov, *On limits of sequences of partial sums of trigonometric series*, Dokl. Akad. Nauk SSSR **106** (1956), 777–780. (Russian) MR **17**, 1080.

2. ———, *On convergence in measure of trigonometric series*, Trudy Mat. Inst. Steklov. 32 (1950); English transl., Amer. Math. Soc. Transl. (1) 3 (1962), 196–270. MR **12**, 254.

3. ———, *On limits of indeterminacy of partial sums of universal trigonometric series*, Moskov. Gos. Univ. Učen. Zap. **165**, Mat. 7 (1954), 3–33; English transl., Amer. Math Soc. Transl. (2) 111 (1978), 1–33. MR **16**, 467.

4. A. Zygmund, *Trigonometrical series*, PWN, Warsaw, 1935; reprint, Chelsea, New York, 1952. MR **17**, 844.

5. D. E. Men'šov, *On limits of indeterminacy in measure of partial sums of universal trigonometric series*, Mat. Sb. **34**(76) (1954), 557–574. (Russian) MR **15**, 952.

Translated by J. M. DANSKIN

Amer. Math. Soc. Transl.
(2) 111, 1978

UNIVERSAL SEQUENCES OF FUNCTIONS*

D. E. MEN'ŠOV

§1. Introduction

Consider the series

$$\sum_{n=0}^{\infty} u_n(x), \tag{1.1}$$

the terms of which are measurable functions a.e. finite on an interval $[a, b]$. Let $S_m(x)$ $(m = 0, 1, \ldots)$ denote the its partial sums, i.e. set

$$S_m(x) = \sum_{k=0}^{m} u_k(x), \tag{1.2}$$

and let us introduce the following definitions.

DEFINITION 1. The series (1.1) is called *universal on* $[a, b]$ if for any measurable function $f(x)$ defined a.e. on $[a, b]$ ($f(x)$ can take on the values $\pm\infty$ on sets of positive measure) one can find an increasing sequence of natural numbers m_k such that

$$\lim_{k\to\infty} S_{m_k}(x) = f(x) \tag{1.3}$$

a.e. on $[a, b]$.

Consider a sequence of measurable functions

$$f_0(x), f_1(x), \ldots, f_m(x), \ldots, \tag{1.4}$$

defined a.e. on $[a, b]$, which in general can take on the values $\pm\infty$ on sets of positive measure.

DEFINITION 2. We shall call the sequence (1.4) *universal on* $[a, b]$ if for any measurable function $f(x)$ defined a.e. on $[a, b]$ one can find an increasing sequence of natural numbers m_k such that

$$\lim_{k\to\infty} f_{m_k}(x) = f(x) \tag{1.5}$$

a.e. on $[a, b]$ (in the sequel we shall not always indicate on which segment a series or sequence is universal).

It is clear that the sequence of partial sums $S_m(x)$ of a series (1.1) which is universal is also universal. Further, if each of the $f_m(x)$ is finite a.e. on $[a, b]$, then it is clear that a series (1.1) whose terms are defined from the equalities

$$u_0(x) = f_0(x), \quad u_m(x) = f_m(x) - f_{m-1}(x) \quad (m = 1, 2, \ldots), \tag{1.6}$$

AMS (MOS) subject classifications (1970). Primary 28A20, 42A20.

* Translation of Mat. Sb. **65 (107)** (1964), 272–312. MR **31** #305.

will be universal on $[a, b]$ if the sequence (1.4) is universal there.

Nevertheless, if the functions $f_m(x)$ of a sequence (1.4) can take on the values $\pm\infty$ on sets of positive measure, then the functions $u_m(x)$ may not always be defined a.e. on $[a, b]$, and hence the series (1.1) will not always be meaningful, so that in this case a universal series does not always correspond to a universal sequence.

In [1] there were considred trigonometric series which are universal relative to two measurable functions $F_1(x)$ and $F_2(x)$ satisfying the inequality

$$F_1(x) \leqslant F_2(x) \tag{1.7}$$

a.e. on an interval $[a, b]$ (see [1], the Introduction, Definition A). This definition can be extended to any sequence (1.4) of measurable functions defined a.e. on $[a, b]$. Then in place of Definition A of [1] we shall have

DEFINITION 3. Let measurable functions $F_1(x)$ and $F_2(x)$ satisfy (1.7) a.e. on $[a, b]$, and let the functions of (1.4) be measurable and defined a.e. on $[a, b]$ ($F_1(x)$, $F_2(x)$ and the functions in (1.4) can take on the values $\pm\infty$ on sets of positive measure). We shall call the sequence (1.4) *universal relative to* $F_1(x)$ *and* $F_2(x)$ if for every measurable function $f(x)$ defined a.e. on $[a, b]$ and satisfying

$$F_1(x) \leqslant f(x) \leqslant F_2(x) \tag{1.8}$$

a.e. on $[a, b]$, one can find an increasing sequence of natural numbers m_k such that (1.5) will hold a.e. on $[a, b]$.

A sequence (1.4) universal relative to $F_1(x)$ and $F_2(x)$ will be called a *sequence of type* $A(F_1, F_2)$. If $F_1(x) = -\infty$ and $F_2(x) = +\infty$ for all x on $[a, b]$, it is clear that a sequence of type $A(F_1, F_2)$ will be universal in the original sense (see Definition 2 above).

In [1] the properties of universal trigonometric series were considered. In order to formulate the main theorem of this paper let us recall the definition of the upper and lower limits of a sequence of functions with respect to a measure.

DEFINITION 4. Let the functions $f_m(x)$ $(m = 0, 1, 2, \ldots)$ of a sequence (1.4) be measurable and defined a.e. on $[a, b]$. We say that a measurable function $F(x)$ defined a.e. on $[a; b]$ is the *upper limit in measure on* $[a, b]$ of a sequence (1.4) if $F(x)$ satisfies the following two conditions:

a°.

$$\lim_{m\to\infty} \operatorname{meas}\{E[f_m(x) > \varphi(x)] \cdot E[\varphi(x) > F(x)]\} = 0 \tag{1.9}$$

for any measurable function $\varphi(x)$ defined a.e. on $[a, b]$. (If $\varphi_1(x)$ and $\varphi_2(x)$ are any two measurable functions defined a.e. on $[a, b]$, then we shall write, as usual,

$$E[\varphi_1(x) > \varphi_2(x)]$$

for the set of points in $[a, b]$ such that $\varphi_1(x) > \varphi_2(x)$.)

b°.

$$\overline{\lim_{m\to\infty}} \operatorname{meas}\{E[f_m(x) > \psi(x)] \cdot E[F(x) > \psi(x)]\} > 0 \tag{1.10}$$

for any measurable function $\psi(x)$ defined a.e. on $[a, b]$ and such that

$$\text{meas}\, E\,[F(x) > \psi(x)] > 0 \tag{1.11}$$

(see, for example, [2], §1, Definition 7).

DEFINITION 5. Let the sequence (1.4) satisfy the same conditions as in Definition 4. We say that a measurable function $G(x)$ defined a.e. on $[a, b]$ is the *lower limit in measure on* $[a, b]$ of the sequence (1.4) if $-G(x)$ is the upper limit in measure on $[a, b]$ of the sequence $-f_m(x)$. [In [2] there was given another definition of lower limit in measure equivalent to Definition 5 above (see [2], §1, Definition 8). The proof of the equivalence of these two definitions is given in [3] (Theorem D, §2).]

The upper and lower limits in measure of a sequence (1.4) will be called its *limits of indeterminacy in measure.*

One can show that the limits of indeterminacy in measure of any sequence of measurable functions are uniquely defined up to a set of measure zero and enjoy many properties analogous to properties of ordinary upper and lower limits. (See [3], §2, Theorems A–E. In [3] it is assumed that the functions in (1.4) are finite a.e. on $[a, b]$. However, Theorems A–E are valid without this restriction, and the proofs do not need changes.)

Let us further introduce the notion of a sequence of type $B(F_1, F_2)$.

DEFINITION 6. Let the measurable functions $F_1(x)$ and $F_2(x)$ be defined and satisfy (1.7) a.e. on $[a, b]$, and let the functions in (1.4) satisfy the previous conditions. We shall call the sequence (1.4) a *sequence of type* $B(F_1, F_2)$ if for any four measurable $\varphi_1(x)$, $\psi_1(x)$, $\psi_2(x)$ and $\varphi_2(x)$, which are defined a.e. on $[a, b]$ and satisfy

$$F_1(x) \leqslant \varphi_1(x) \leqslant \psi_1(x) \leqslant \psi_2(x) \leqslant \varphi_2(x) \leqslant F_2(x) \tag{1.12}$$

a.e. on $[a, b]$, one can find an increasing sequence of natural numbers m_k such that $\psi_1(x)$ and $\psi_2(x)$ will be, respectively, the lower and upper limits in measure on $[a, b]$ of the sequence of functions

$$f_{m_k}(x) \quad (k = 0, 1, 2, \ldots) \tag{1.13}$$

and moreover

$$\varliminf_{k\to\infty} f_{m_k}(x) = \varphi_1(x), \quad \varlimsup_{k\to\infty} f_{m_k}(x) = \varphi_2(x) \tag{1.14}$$

a.e. on $[a, b]$.

If $F_1(x) \equiv -\infty$ and $F_2(x) \equiv +\infty$, then we shall call a sequence of type $B(F_1, F_2)$ a *sequence of type B.* Further, if the terms of the series (1.1) are measurable and finite a.e. on $[a, b]$, then we shall call this series a series of type $A(F_1, F_2)$ or $B(F_1, F_2)$ if its sequence of partial sums is correspondingly a sequence of type $A(F_1, F_2)$ or type $B(F_1, F_2)$.

In [1] (Theorem 1), we proved

THEOREM A. *Whatever be the measurable functions* $F_1(x)$ *and* $F_2(x)$ *defined a.e.*

on $[-\pi, \pi]$ *and satisfying* (1.7) *a.e. thereon, the classes of trigonometric series of type* $A(F_1, F_2)$ *and type* $B(F_1, F_2)$ *coincide.*

This theorem is easily extendible to an arbitrary sequence (1.4) of measurable functions defined a.e. on any $[a, b]$. For these sequences one can even prove a more general theorem.

For the formulation of this more general theorem we shall require some definitions, a few of which were introduced in [2].

DEFINITION 7. Suppose that a sequence (1.4) satisfies the same conditions as previously (see the remark before Definition 2). We shall say that a function $f(x, E)$ defined a.e. on any set E, meas $E > 0$, $E \subset [a, b]$, is the *limit function* of (1.4) on this set if there is an increasing sequence of natural numbers m_k such that

$$\lim_{k\to\infty} f_{m_k}(x) = f(x, E) \tag{1.15}$$

a.e. on E.

DEFINITION 8. Let the terms of (1.1) be measurable functions which are finite a.e. on $[a, b]$. We shall say that a function $f(x, E)$, defined a.e. on a set E, meas $E > 0$, $E \subset [a, b]$, is the *limit function* of the series (1.1) on E if $f(x, E)$ is the limit function of the sequence of partial sums of (1.1) on E. (This definition coincides with Definition 1 in [2], §1. In the sequel we shall simply say that $f(x, E)$ is the limit function of the sequence (1.4) or series (1.1), without mentioning the set E.)

DEFINITION 9. Let $G(x)$ and $F(x)$ be two measurable functions defined a.e. on some $[a, b]$ and satisfying

$$G(x) \leqslant F(x) \tag{1.16}$$

a.e. thereon. Further, suppose that $M = \{\varphi(x, E)\}$ is a set of functions $\varphi(x, E)$, each of which is defined a.e. on E, meas $E > 0$, $E \subset [a, b]$. We shall say that M has *property* $h(G, F)$ if one can find a sequence (1.4) of measurable functions $f_m(x)$ defined a.e. on $[a, b]$ for which M is the set of all limit functions and $F(x)$ and $G(x)$ are, respectively, its upper and lower limits in measure on $[a, b]$.

One can give another definition of property $h(G, F)$ having a structural nature and equivalent to Definition 9. We shall not do so here, as we shall not require it.

As we remarked, one can prove a theorem of which Theorem A is a special case. From this new theorem it will follow, in particular, that a sequence of measurable functions universal in the usual sense is not only a sequence of type B but also possesses certain other properties; namely, it is a sequence of type C (see the remark after Definition 10).

In order to formulate this new theorem we make

DEFINITION 10. Let measurable functions $F_1(x)$ and $F_2(x)$ satisfy (1.7) a.e. on $[a, b]$, and let the functions in (1.4) be measurable and defined a.e. thereon. We shall say that (1.4) is a *sequence of type* $C(F_1, F_2)$ if for any measurable functions $\varphi_1(x)$, $\psi_1(x)$, $\psi_2(x)$ and $\varphi_2(x)$ defined a.e. on $[a, b]$ and satisfying (1.12) thereon a.e., and for any set $M = \{\varphi(x, E)\}$ having property $h(\psi_1, \psi_2)$, one can find an increasing sequence of natural numbers m_k so that the following conditions will hold.

a°. $\psi_1(x)$ and $\psi_2(x)$ are, respectively, the lower and upper limits in measure on $[a, b]$ of the sequence of functions

$$f_{m_k}(x) \quad (k = 0, 1, 2, \ldots); \tag{1.17}$$

b°.

$$\underline{\lim}_{k\to\infty} f_{m_k}(x) = \varphi_1(x), \quad \overline{\lim}_{k\to\infty} f_{m_k}(x) = \varphi_2(x) \tag{1.18}$$

a.e. on $[a, b]$,

c°. M is the set of all limit functions of the sequence (1.17). (When we say that $\varphi(x, E)$ is a limit function of the sequence (1.17), we mean a limit function on the set E.)

If $F_1(x) \equiv -\infty$ and $F_2(x) \equiv +\infty$, then the sequence (1.4) of type $C(F_1, F_2)$ will be called a *sequence of type C.*

One can prove the following

THEOREM 1. *Whatever be the measurable functions $F_1(x)$ and $F_2(x)$ defined a.e. on $[a, b]$ and satisfying* (1.7) *a.e. thereon, the classes of sequences* (1.4) *of type $A(F_1, F_2)$ and type $C(F_1, F_2)$ coincide.* (*See Definition* 10 *and the remark after Definition* 3.)

In particular,

THEOREM 2. *The class of universal sequences* (1.4) (*see Definition* 2) *coincides with the class of sequences of type C.*

It is clear that every sequence (1.4) of type $C(F_1, F_2)$ is a sequence of type $B(F_1, F_2)$ (see Definition 6) and of type $A(F_1, F_2)$, and every sequence of type $B(F_1, F_2)$ is a sequence of type $A(F_1, F_2)$.

In such a case we also have the following theorems.

THEOREM 3. *If $F_1(x)$ and $F_2(x)$ satisfy the same conditions as in Theorem* 1, *then the classes $A(F_1, F_2)$ and $B(F_1, F_2)$ coincide.*

THEOREM 4. *The class of universal sequences* (1.4) *coincides with the class of sequences of type B* (*the definition of a sequence of type B is given after Definition* 6).

From Theorem 3 one derives Theorem A on trigonometric series as a corollary.

From the remark after the formulation of Theorem 2 it follows at once that to prove Theorems 1, 2, 3 and 4 it suffices to prove the next theorem.

THEOREM 5. *If $F_1(x)$ and $F_2(x)$ satisfy the conditions of Theorem* 1, *then every sequence of type $A(F_1, F_2)$ is a sequence of type $C(F_1, F_2)$.*

Let us make some further definitions.

DEFINITION 11. Let the functions $f_m(x)$ $(m = 0, 1, 2, \ldots)$ of the sequence (1.4) be measurable and finite a.e. on $[a, b]$. We shall say that the sequence (1.4) is *universal relative to a linear matrix method T defined by the matrix*

$$\| a_{mk} \| \quad (m, k = 0, 1, 2, \ldots), \tag{1.19}$$

if for any measurable $f(x)$ defined a.e. on $[a, b]$ there is an increasing sequence of natural numbers

$$m_k \quad (k = 0, 1, 2, \ldots), \tag{1.20}$$

such that the sequence

$$f_{m_k}(x) \quad (k = 0, 1, 2, \ldots) \tag{1.21}$$

sums by method T to $f(x)$ a.e. on $[a, b]$.

DEFINITION 12. The linear method T is called *pure* if any sequence converging to zero sums by this method also to zero.

It is clear that every regular method of summation is a pure method.

As is known (see [4], §3.5, Theorem 4), the following theorem is valid.

In order that a linear matrix method T defined by the matrix (1.19) *be pure it is necessary and sufficient that the following two conditions hold*:

1°.

$$\lim_{m\to\infty} a_{mk} = 0 \quad (k = 0, 1, 2, \ldots),$$

2°.

$$\sum_{k=0}^{\infty} | a_{mk} | < M < +\infty,$$

where M does not depend on m.

DEFINITION 13. A linear method T defined by the matrix (1.19) is called *normal* or *strictly triangular* if

$$a_{mm} \neq 0 \quad (m = 0, 1, 2, \ldots), \tag{1.22}$$

$$a_{mk} = 0 \quad (k > m, \ m = 0, 1, 2, \ldots). \tag{1.23}$$

A normal method, by definition, is a linear matrix method.

We have

THEOREM 6. *Let the $f_m(x)$ in* (1.4) *be measurable and finite a.e. on $[a, b]$, and let the pure normal method T be defined by a matrix $\|a_{mk}\|$ with real entries.*

Then, if the sequence (1.4) *is universal* (*in the usual sense*), *it is also universal relative to the method T.*

In the case of a completely regular method T, Theorem 6 follows at once from the definition of a universal sequence.

§2. Properties of equiconvergent sequences and limits of indeterminacy in measure

DEFINITION 14. Consider two numerical sequences

$$U_0, U_1, U_2, \dots, U_m, \dots \tag{2.1}$$

and

$$V_0, V_1, V_2, \dots, V_m, \dots, \tag{2.2}$$

and assume that each of the U_m and V_m is finite, or $+\infty$, or $-\infty$. We shall say that the sequences (2.1) and (2.2) are *equiconvergent* if the following conditions hold (see [1], Lemmas A, B and C):

α°. $U_m - V_m \longrightarrow 0$ as $m \longrightarrow \infty$ taking on the values for which both U_m and V_m are finite.

β°. U_m and V_m tend to $+\infty$ as $m \longrightarrow \infty$ taking on the values for which at least one of U_m or V_m is $+\infty$.

γ°. U_m and V_m tend to $-\infty$ as $m \longrightarrow \infty$ taking on the values for which at least one of U_m or V_m is $-\infty$.

It is easy to prove the following assertions (see [1], §1, Definition 1).

LEMMA 1. *If the sequences* (2.1) *and* (2.2) *are equiconvergent and if for some nondecreasing sequence of natural numbers* p_k *satisfying* $\lim_{k\to\infty} p_k = +\infty$ *one of the two sequences* U_{p_k} *or* V_{p_k} $(k = 0, 1, 2, \dots)$ *converges to a definite limit L, then the other converges to the same limit* (*the number L can equal* $\pm\infty$).

LEMMA 2. *If* (2.1) *and* (2.2) *are equiconvergent, then their sets of limit points coincide.*

LEMMA 3. *If* (2.1) *and* (2.2) *are equiconvergent, then*

$$\varliminf_{m\to\infty} U_m = \varliminf_{m\to\infty} V_m, \quad \varlimsup_{m\to\infty} U_m = \varlimsup_{m\to\infty} V_m.$$

In the sequel we shall need several lemmas on upper and lower limits in measure of a sequence of functions. Moreover, we shall recall certain definitions necessary for formulating these lemmas.

LEMMA 4. *Let the functions* $F(x)$, $G(x)$,

$$f_m(x) \quad (m = 0, 1, 2, \dots) \tag{2.3}$$

and

$$g_m(x) \quad (m = 0, 1, 2, \dots) \tag{2.4}$$

be defined a.e. on $[a, b]$ *and measurable thereon. Suppose that*

$$G(x) \leqslant F(x) \tag{2.5}$$

a.e. on $[a, b]$, *and that* (2.3) *and* (2.4) *are equiconvergent a.e. on* $[a, b]$. *Then, if* $F(x)$ *and* $G(x)$ *are respectively the upper and lower limits in measure on* $[a, b]$ *of* (2.3), *they are also those of* (2.4) (*see* [1], *Lemma* G).

LEMMA 5. *Let the functions* $f_m(x)$ *of the sequence* (2.3) *be measurable and defined a.e. on* $[a, b]$, *and let* $F(x)$ *and* $G(x)$ *be the upper and lower limits in measure on* $[a, b]$ *of this sequence. Take any increasing sequence of natural numbers* m_k $(k = 0, 1, \ldots)$, *and let* $F^{(1)}(x)$ *and* $G^{(1)}(x)$, *respectively, denote the upper and lower limits in measure on* $[a, b]$ *of the sequence*

$$f_{m_k}(x) \quad (k = 0, 1, 2, \ldots). \tag{2.6}$$

Then

$$G(x) \leqslant G^{(1)}(x) \leqslant F^{(1)}(x) \leqslant F(x) \tag{2.7}$$

a.e. on $[a, b]$.

PROOF. First, it is known that

$$G(x) \leqslant F(x), \quad G^{(1)}(x) \leqslant F^{(1)}(x) \tag{2.8}$$

a.e. on $[a, b]$ (see [3], §2, Theorem C). Further, without loss of generality we can assume that $F(x)$, $G(x)$, $F^{(1)}(x)$, $G^{(1)}(x)$, and $f_m(x)$ $(m = 0, 1, \ldots)$ are defined everywhere on $[a, b]$ and that the inequalities (2.8) hold on $[a, b]$.

Now let us prove that

$$F^{(1)}(x) \leqslant F(x) \tag{2.9}$$

a.e. on $[a, b]$. Suppose not, i.e. suppose that

$$\text{meas}\, E\,[F^{(1)}(x) > F(x)] > 0. \tag{2.10}$$

Set

$$E' = E\,[F^{(1)}(x) > F(x) > -\infty], \quad E'' = E\,[F^{(1)}(x) > F(x) = -\infty]. \tag{2.11}$$

Then $F(x)$ is finite at each $x \in E'$, and

$$E' + E'' = E\,[F^{(1)}(x) > F(x)], \quad E' \subset [a, b], \quad E'' \subset [a, b]. \tag{2.12}$$

Take any positive ϵ and define $\varphi_\epsilon(x)$ from the conditions

$$\varphi_\varepsilon(x) = F(x) + \varepsilon \quad (x \in E'), \tag{2.13}$$

$$\varphi_\varepsilon(x) = -\frac{1}{\varepsilon} \quad (x \in E''), \tag{2.14}$$

$$\varphi_\varepsilon(x) = F(x) \quad (x \in [a, b] \setminus E' \setminus E''). \tag{2.15}$$

Then from the definitions of E' and E'' (see (2.11)) we have

$$\varphi_\varepsilon(x) > F(x) > -\infty\ (x \in E'), \quad \varphi_\varepsilon(x) > F(x) = -\infty\ (x \in E'') \tag{2.16}$$

and hence by (2.12) and (2.15)

$$E\,[F^{(1)}(x) > F(x)] = E\,[\varphi_\varepsilon(x) > F(x)]. \tag{2.17}$$

Since $F(x)$ is the upper limit in measure on $[a, b]$ of the sequence (2.3) of functions $f_m(x)$, by condition a° in Definition 4 (see (1.9)), in which we take $\varphi_\epsilon(x)$ in place of $\varphi(x)$, we have

$$\lim_{m\to\infty} \operatorname{meas}\{E[f_m(x) > \varphi_\varepsilon(x)] \cdot E[\varphi_\varepsilon(x) > F(x)]\} = 0, \tag{2.18}$$

and since the m_k form an increasing sequence, it follows from (2.18) that

$$\lim_{k\to\infty} \operatorname{meas}\{E[f_{m_k}(x) > \varphi_\varepsilon(x)] \cdot E[\varphi_\varepsilon(x) > F(x)]\} = 0. \tag{2.19}$$

Moreover, by (2.15) and (2.16),

$$\varphi_\varepsilon(x) \geqslant F(x) \quad (x \in [a, b]). \tag{2.20}$$

In (2.19) and (2.20) ϵ can be any positive number.

Now if for some $x \in [a, b]$ and for some $\epsilon > 0$ the inequality $F^{(1)}(x) > \varphi_\epsilon(x)$ holds, then, by (2.20), $F^{(1)}(x) > F(x)$ for the same x, whence $E[F^{(1)}(x) > \varphi_\epsilon(x)] \subset E[F^{(1)}(x) > F(x)]$, and in this case, by (2.17),

$$E[F^{(1)}(x) > \varphi_\varepsilon(x)] \subset E[\varphi_\varepsilon(x) > F(x)].$$

Hence we obtain, using (2.19), that

$$\lim_{k\to\infty} \operatorname{meas}\{E[f_{m_k}(x) > \varphi_\varepsilon(x)] \cdot E[F^{(1)}(x) > \varphi_\varepsilon(x)]\} = 0, \tag{2.21}$$

where ϵ is any positive number.

From (2.11) we have

$$\operatorname{meas}\{E' \cdot E[F^{(1)}(x) > F(x) + \varepsilon]\} \geqslant \frac{1}{2} \operatorname{meas} E',$$

$$\operatorname{meas}\left\{E'' \cdot E\left[F^{(1)}(x) > -\frac{1}{\varepsilon}\right]\right\} \geqslant \frac{1}{2} \operatorname{meas} E''$$

for all sufficiently small positive ϵ, and hence from (2.13) and (2.14) we obtain that

$$\operatorname{meas} E[F^{(1)}(x) > \varphi_\varepsilon(x)] \geqslant \frac{1}{2} \operatorname{meas} E', \quad \operatorname{meas} E[F^{(1)}(x) > \varphi_\varepsilon(x)] \geqslant \frac{1}{2} \operatorname{meas} E'' \tag{2.22}$$

for the same ϵ. On the other hand, from (2.10) and (2.12) we see that either meas E' > 0 or meas $E'' > 0$, and then using (2.22) we have

$$\operatorname{meas} E[F^{(1)}(x) > \varphi_\varepsilon(x)] > 0 \tag{2.23}$$

for all sufficiently small positive ϵ.

Since $F^{(1)}(x)$ is the upper limit in measure on $[a, b]$ of the sequence (2.6), on using (2.23) and condition b° of Definition 4 (see (1.10) and (1.11)), in which we take $f_{m_k}(x)$, $F^{(1)}(x)$ and $\varphi_\epsilon(x)$ in place of $f_m(x)$, $F(x)$ and $\psi(x)$, we see that

$$\overline{\lim_{k\to\infty}} \operatorname{meas}\{E[f_{m_k}(x) > \varphi_\varepsilon(x)] \cdot E[F^{(1)}(x) > \varphi_\varepsilon(x)]\} > 0$$

for all sufficiently small positive ϵ, and this contradicts (2.21).

Thus (2.9) holds a.e. on $[a, b]$.

Now let us show that a.e. on $[a, b]$ we have

$$G(x) \leqslant G^{(1)}(x). \tag{2.24}$$

Indeed, since by the hypothesis of Lemma 5 $G(x)$ and $G^{(1)}(x)$ are, respectively, the lower limits in measure on $[a, b]$ of the sequences (2.3) and (2.6), then, by Definition 5, $-G(x)$ and $-G^{(1)}(x)$ are, respectively, the upper limits in measure on $[a, b]$ of the sequences $-f_m(x)$ and $-f_{m_k}(x)$, so it follows from the part of Lemma 5 already proved that $-G^{(1)}(x) \leqslant -G(x)$ a.e. on $[a, b]$, and in this case (2.24) holds a.e. thereon.

Combining (2.8) with (2.9) and (2.24), we see that (2.7) holds a.e. on $[a, b]$, and the proof of Lemma 5 is complete.

One can also prove the following assertions.

LEMMA 6. *For any sequence* (2.3) *of measurable functions $f_m(x)$, defined a.e. on $[a, b]$ (the $f_m(x)$ can take on the values $\pm\infty$ on sets of positive measure), one can find a sequence of functions $h_m(x)$, continuous on $[a, b]$, which have the following properties*:

1°. *The sequences $f_m(x)$ and $h_m(x)$ are equiconvergent a.e. on $[a, b]$.*

2°. $h_m(a) = h_m(b)$, $m = 0, 1, \ldots$.

See [3], Lemma 3.3.

THEOREM B. *If $F(x)$ and $G(x)$ are, respectively, the upper and lower limits in measure on $[a, b]$ of a sequence* (2.3) *of measurable functions $f_m(x)$ defined a.e. on $[a, b]$, then*

$$\varliminf_{m\to\infty} f_m(x) \leqslant G(x) \leqslant F(x) \leqslant \varlimsup_{m\to\infty} f_m(x) \tag{2.25}$$

a.e. on $[a, b]$.

PROOF. If each of the $f_m(x)$ is finite a.e. on $[a, b]$, then Theorem B follows at once from Theorems F and G in [3].

In the general case, using Lemma 6, we can assert that there is a sequence of functions

$$h_m(x) \quad (m = 0, 1, 2, \ldots), \tag{2.26}$$

continuous on $[a, b]$ and which is equiconvergent with (2.3) a.e. on $[a, b]$. Then, by the hypotheses of Theorem B and Lemma 4, $F(x)$ and $G(x)$ are, respectively, the upper and lower limits in measure on $[a, b]$ of (2.26), and by Lemma 3 we have

$$\varlimsup_{m\to\infty} h_m(x) = \varlimsup_{m\to\infty} f_m(x), \quad \varliminf_{m\to\infty} h_m(x) = \varliminf_{m\to\infty} f_m(x) \tag{2.27}$$

a.e. on $[a, b]$.

Since the $h_m(x)$ are continuous and hence finite on $[a, b]$, by the part of Theorem B already proved we have

$$\varliminf_{m\to\infty} h_m(x) \leqslant G(x) \leqslant F(x) \leqslant \varlimsup_{m\to\infty} h_m(x)$$

a.e. on $[a, b]$, so it follows from (2.27) that (2.25) holds a.e. thereon. The proof is complete.

As a corollary of Theorem B we have

LEMMA 7. *Let the functions $f_m(x)$ of the sequence* (2.3) *satisfy the conditions of Theorem* B. *Then, if*

$$\lim_{m\to\infty} f_m(x) = f(x) \tag{2.28}$$

a.e. on some set E with

$$\text{meas } E > 0, \quad E \subset [a, b], \tag{2.29}$$

it follows that

$$F(x) = G(x) = f(x) \tag{2.30}$$

a.e. on E, where $F(x)$ and $G(x)$ are the upper and lower limits in measure on $[a, b]$ of (2.3).

PROOF. By Theorem B and the hypotheses of Lemma 7 (see (2.28)) we have

$$f(x) = \varliminf_{m\to\infty} f_m(x) \leqslant G(x) \leqslant F(x) \leqslant \varlimsup_{m\to\infty} f_m(x)$$

a.e. on E, whence (2.30) holds a.e. on E. The proof is complete.

The following lemmas are also valid.

LEMMA 8. *Consider a sequence* (2.3) *of measurable functions $f_m(x)$ defined a.e. on some $[a, b]$, and let $F(x)$ and respectively $G(x)$ denote the upper and lower limits in measure on $[a, b]$ of the sequence* (2.3). *Suppose that $\varphi(x, E)$ is the limit function of the sequence* (2.3) *on the set E (see Definition* 7 *in* § 1). *Then*

$$G(x) \leqslant \varphi(x, E) \leqslant F(x) \tag{2.31}$$

a.e. on E.

PROOF. From Definition 7 it follows that there is an increasing sequence of natural numbers m_k $(k = 0, 1, 2, \ldots)$ such that

$$\lim_{k\to\infty} f_{m_k}(x) = \varphi(x, E) \tag{2.32}$$

a.e. on E, and meas $E > 0$, $E \subset [a, b]$. Let $F^{(1)}(x)$ and $G^{(1)}(x)$ denote the upper and lower limits in measure on $[a, b]$ of the sequence (2.6) of functions $f_{m_k}(x)$. Then, by the definition of $F(x)$ and $G(x)$ and Lemma 5, the inequality (2.7) will hold a.e. on $[a, b]$.

Further, from Lemma 7, in which we take $f_{m_k}(x)$, $\varphi(x, E)$, $F^{(1)}(x)$ and $G^{(1)}(x)$ in place of $f_m(x)$, $f(x)$, $F(x)$ and $G(x)$, we see that $F^{(1)}(x) = G^{(1)}(x) = \varphi(x, E)$ a.e. on E, so that, by (2.7) and the inclusion $E \subset [a, b]$, (2.31) holds a.e. on E. Q.E.D.

LEMMA 9. *Let the functions $f_m(x)$ $(m = 0, 1, \ldots)$, $F(x)$ and $G(x)$ satisfy the same conditions as in Lemma* 8, *where $[a, b]$ is some interval. Suppose that the* *measurable functions $F_0(x)$ and $G_0(x)$ satisfy*

$$G_0(x) \leqslant G(x) \leqslant F(x) \leqslant F_0(x) \tag{2.33}$$

a.e. on $[a, b]$, and at the points x of $[a, b]$ where $f_m(x)$ $(m = 0, 1, \ldots)$, $F_0(x)$ and $G_0(x)$ are defined let us set

$$g_m(x) = \begin{cases} f_m(x), & \text{if} \quad G_0(x) \leqslant f_m(x) \leqslant F_0(x), \\ F_0(x), & \text{if} \quad f_m(x) > F_0(x) \geqslant G_0(x), \\ G_0(x), & \text{if} \quad f_m(x) < G_0(x) \leqslant F_0(x) \end{cases} \tag{2.34}$$
$$(m = 0, 1, 2, \ldots).$$

(From the definition of the functions $g_m(x)$ it follows that if $F_0(x)$, $G_0(x)$ and $g_m(x)$ are defined for some x and m, then for the same x and m the function $f_m(x)$ is also defined.)

Then the functions

$$g_m(x) \quad (m = 0, 1, 2, \ldots) \tag{2.35}$$

are measurable and defined a.e. on $[a, b]$; and, moreover, $F(x)$ and $G(x)$ are the upper and lower limits in measure on $[a, b]$ of the sequence (2.35).

PROOF. Since the functions $f_m(x)$ $(m = 0, 1, \ldots)$, $F_0(x)$ and $G_0(x)$ are measurable and defined a.e. on $[a, b]$, from (2.34) and (2.33) it follows that the functions (2.35) have the same property.

Now let us show that $F(x)$ and $G(x)$ are the upper and lower limits in measure on $[a, b]$ of (2.35).

From (2.33) and (2.34) it follows that

$$G_0(x) \leqslant g_m(x) \leqslant F_0(x) \quad (m = 0, 1, 2, \ldots) \tag{2.36}$$

a.e. on $[a, b]$. Moreover, if at any $x \in [a, b]$ and for any m all the functions $F_0(x)$, $G_0(x)$, $f_m(x)$ and $g_m(x)$ are defined and if for these x and m the inequality $f_m(x) < g_m(x)$ holds, then for the same x and m we have $f_m(x) < G_0(x) = g_m(x)$. Hence if $g_m(x) > G_0(x)$ for any x and m for which all the four above-mentioned functions are defined, then for these x and m we have $f_m(x) \geqslant g_m(x)$, and hence from (2.33) for any measurable function $\varphi(x)$ defined a.e. on $[a, b]$ the inclusion

$$\{E[g_m(x) > \varphi(x)] \cdot E[\varphi(x) > F(x)]\} \subset \{E[f_m(x) > \varphi(x)] \cdot E[\varphi(x) > F(x)]\} + e_m$$
$$(m = 0, 1, 2, \ldots), \tag{2.37}$$

holds, where e_m is the set of all x in $[a, b]$ at which at least one of the functions $F(x)$, $G(x)$, $F_0(x)$, $G_0(x)$, $f_m(x)$ or $g_m(x)$ is not defined, or (2.33) does not hold.

Since (2.33) holds a.e. on $[a, b]$ and since all the above functions are defined a.e. on $[a, b]$, it follows that meas $e_m = 0$ $(m = 0, 1, \ldots)$, so that by (2.37)

$$\text{meas}\{E\,[g_m(x)>\varphi(x)]\cdot E\,[\varphi(x)>F(x)]\}\leqslant\text{meas}\{E\,[f_m(x)>\varphi(x)]\cdot E\,[\varphi(x)>F(x)]\}$$
$$(m=0,1,2,\ldots). \tag{2.38}$$

Since by assumption $F(x)$ is the upper limit in measure on $[a, b]$ of the sequence (2.3) of functions $f_m(x)$, and since $\varphi(x)$ is measurable and defined a.e. on $[a, b]$, by condition a° in the definition of the upper limit in measure (see Definition 4, (1.9)) we see that the right side of (2.38) tends to zero as $m \longrightarrow \infty$, and in this case

$$\lim_{m\to\infty}\text{meas}\{E\,[g_m(x)>\varphi(x)]\cdot E\,[\varphi(x)>F(x)]\}=0. \tag{2.39}$$

Since we assumed that $\varphi(x)$ is an arbitrary measurable function defined a.e. on $[a, b]$, it follows from (2.39) that $F(x)$ satisfies condition a° in the definition of upper limit in measure if in this definition we take (2.35) in place of (2.3).

Now suppose that $\psi(x)$ is an arbitrary measurable function defined a.e. on $[a, b]$ and satisfying (1.11). If for some $x \in [a, b]$ and for some natural m the inequalities

$$f_m(x)>\psi(x),\quad F(x)>\psi(x) \tag{2.40}$$

and inequality (2.33) hold, then, by (2.34), for the same x and m one of the following three conditions holds:

$$g_m(x)=f_m(x)>\psi(x), \tag{2.41}$$

if $G_0(x) \leqslant f_m(x) \leqslant F_0(x)$;

$$g_m(x)=F_0(x)\geqslant F(x)>\psi(x), \tag{2.42}$$

if $f_m(x) > F_0(x)$; or

$$g_m(x)=G_0(x)>f_m(x)>\psi(x), \tag{2.43}$$

if $f_m(x) < G_0(x)$.

It follows in all possible cases that (2.33) and (2.40) imply $g_m(x) > \psi(x)$ and $F(x) > \psi(x)$, and in this case

$$\{E\,[f_m(x)>\psi(x)]\cdot E\,[F(x)>\psi(x)]\}\subset\{E\,[g_m(x)>\psi(x)]\cdot E\,[F(x)>\psi(x)]\}+e'_m$$
$$(m=0,\ 1,\ 2,\ldots), \tag{2.44}$$

where e'_m is the set of x in $[a, b]$ where either (2.33) does not hold or at least one of $F(x)$, $G(x)$, $F_0(x)$, $G_0(x)$, $\psi(x)$, $f_m(x)$ or $g_m(x)$ is not defined. Since all these functions are defined a.e. on $[a, b]$, it follows that meas $e'_m = 0$ $(m = 0, 1, \ldots)$, whence from (2.44) we obtain

$$\begin{aligned}&\overline{\lim_{m\to\infty}}\,\text{meas}\{E\,[f_m(x)>\psi(x)]\cdot E\,[F(x)>\psi(x)]\}\\&\leqslant\overline{\lim_{m\to\infty}}\,\text{meas}\{E\,[g_m(x)>\psi(x)]\cdot E\,[F(x)>\psi(x)]\}.\end{aligned} \tag{2.45}$$

Since by the hypothesis of the present lemma $F(x)$ is the upper limit in measure on $[a, b]$ of the sequence (2.3), and since $\psi(x)$ is any measurable function defined a.e. on $[a, b]$ and satisfying (1.11), by condition b° in the definition of upper limit in measure we see that the left side of (2.45) is positive, and hence from this inequality we have

$$\overline{\lim_{m\to\infty}}\operatorname{meas}\{F[g_m(x)>\psi(x)]\cdot E[F(x)>\psi(x)]\}>0. \tag{2.46}$$

Since $\psi(x)$ is an arbitrary measurable function defined a.e. on $[a, b]$ and satisfying (1.11), it follows from (2.46) that $F(x)$ satisfies condition b°, if therein we take (2.35) in place of (2.3).

We proved earlier that for the function $F(x)$ and the sequence (2.35) the condition a° holds. Hence, by Definition 4, $F(x)$ is the upper limit in measure on $[a, b]$ of the sequence (2.35). Hence to end the proof of Lemma 9 we need to show that $G(x)$ is the lower limit in measure on $[a, b]$ of the sequence (2.35).

By the hypothesis of Lemma 9, $G(x)$ is the lower limit in measure on $[a, b]$ of the sequence (2.3). Then, by Definition 5, $-G(x)$ is the upper limit in measure on $[a, b]$ of the sequence of functions

$$-f_m(x) \quad (m=0,\ 1,\ 2,\ \ldots). \tag{2.47}$$

Hence from (2.33) and (2.34) it follows that a.e. on $[a, b]$

$$-F_0(x)\leqslant -F(x)\leqslant -G(x)\leqslant -G_0(x) \tag{2.48}$$

and

$$-g_m(x)=\begin{cases}-f_m(x), & \text{if}\quad -F_0(x)\leqslant -f_m(x)\leqslant -G_0(x),\\ -G_0(x), & \text{if}\quad -f_m(x)>-G_0(x)\geqslant -F_0(x),\\ -F_0(x), & \text{if}\quad -f_m(x)<-F_0(x)\leqslant -G_0(x)\end{cases}$$

$$(m=0,\ 1,\ 2,\ \ldots).$$

It follows that all the conditions of Lemma 9 hold if we replace $F(x)$, $G(x)$, $F_0(x)$, $G_0(x)$, $f_m(x)$ and $g_m(x)$ by $-G(x)$, $-F(x)$, $-G_0(x)$, $-F_0(x)$, $-f_m(x)$ and $-g_m(x)$. Hence by the part of Lemma 9 already proved we can assert that $-G(x)$ is the upper limit in measure on $[a, b]$ of the sequence of functions

$$-g_m(x) \quad (m=0,\ 1,\ 2,\ \ldots),$$

and in this case, taking Definition 5 into account, we see that $G(x)$ is the lower limit in measure on $[a, b]$ of (2.35). This completes the proof of Lemma 9.

LEMMA 10. *Let the functions $f_m(x)$ $(m = 0, 1, \ldots)$, $F(x)$ and $G(x)$ satisfy the same conditions as in Lemma 8, where $[a, b]$ is some interval. Then one can find an increasing sequence of natural numbers m_k $(k = 0, 1, \ldots)$ such that*

$$G(x)\leqslant \underline{\lim_{k\to\infty}} f_{m_k}(x)\leqslant \overline{\lim_{k\to\infty}} f_{m_k}(x)\leqslant F(x) \tag{2.49}$$

a.e. on $[a, b]$.

PROOF. Let us set

$$H_1 = E[-\infty < F(x) < +\infty], \tag{2.50}$$

$$H_2 = E[F(x) = +\infty], \tag{2.51}$$

$$H_3 = E[F(x) = -\infty], \tag{2.52}$$

$$\varphi_\varepsilon(x) = F(x) + \varepsilon \quad (x \in H_1), \tag{2.53}$$

$$\varphi_\varepsilon(x) = F(x) = +\infty \quad (x \in H_2), \tag{2.54}$$

$$\varphi_\varepsilon(x) = -\frac{1}{\varepsilon} \quad (x \in H_3), \tag{2.55}$$

where ϵ is any positive number.

Since $F(x)$ is measurable and defined a.e. on $[a, b]$ (see the conditions in the formulation of Lemma 8), we have

$$H_1 + H_2 + H_3 \subset [a, b], \tag{2.56}$$

$$\text{meas}\,(H_1 + H_2 + H_3) = b - a, \tag{2.57}$$

and in this case for any $\epsilon > 0$ the function $\varphi_\epsilon(x)$ is measurable and defined a.e. on $[a, b]$.

Further, since $F(x)$ is the upper limit in measure on $[a, b]$ of the sequence (2.3) of functions $f_m(x)$ (see the conditions in the formulation of Lemma 8) and since for any $\epsilon > 0$ the function $\varphi_\epsilon(x)$ is measurable and defined a.e. on $[a, b]$, by condition a° (see Definition 4) we have for any natural number r

$$\lim_{m\to\infty} \text{meas}\,\{E[f_m(x) > \varphi_{\frac{1}{r}}(x)] \cdot E[\varphi_{\frac{1}{r}}(x) > F(x)]\} = 0. \tag{2.58}$$

Since by (2.50)–(2.55)

$$E[\varphi_{\frac{1}{r}}(x) > F(x)] = H_1 + H_3 \quad (r = 1, 2, \ldots), \tag{2.59}$$

it follows from (2.58) that

$$\lim_{m\to\infty} \text{meas}\,\{E[f_m(x) > \varphi_{\frac{1}{r}}(x)] \cdot (H_1 + H_3)\} = 0, \tag{2.60}$$

and hence for any $r = 1, 2, \ldots$ we can find a set Ω_r and a natural number ν_r which satisfy the conditions

$$\text{meas}\,\Omega_r > \text{meas}(H_1 + H_3) - \frac{1}{2^r}, \quad \Omega_r \subset H_1 + H_3 \quad (r = 1, 2, \ldots), \tag{2.61}$$

$$f_{\nu_r}(x) \leqslant \varphi_{\frac{1}{r}}(x) \quad (x \in \Omega_r, \ r = 1, 2, \ldots). \tag{2.62}$$

Hence we can always assume that the natural numbers

$$\nu_r \ (r = 1, 2, \ldots) \tag{2.63}$$

form an increasing sequence.

From (2.61) and (2.62), by (2.54) we have

$$f_{\nu_r}(x) \leqslant \varphi_{\frac{1}{r}}(x) \quad (x \in E_r, \ r = 1, 2, \ldots), \tag{2.64}$$

and

$$E_r = \Omega_r + H_2 \setminus e_r \quad (r = 1, 2, \ldots), \tag{2.65}$$

where meas $e_r = 0$ (e_r is the set of $x \in [a, b]$ where $f_{\nu_r}(x)$ is not defined). Moreover, taking (2.56), (2.57), (2.61) and (2.65) into account,

$$E_r \subset H_1 + H_2 + H_3 \subset [a, b], \text{ meas } E_r > b - a - \frac{1}{2^r} \quad (r = 1, 2, \ldots). \tag{2.66}$$

Setting

$$E = \varliminf_{r \to \infty} E_r, \tag{2.67}$$

we get, from (2.66),

$$\text{meas } E = b - a, \quad E \subset [a, b]. \tag{2.68}$$

From (2.50)–(2.55) we have

$$\lim_{r \to \infty} \varphi_{\frac{1}{r}}(x) = F(x) \quad (x \in H_1 + H_2 + H_3). \tag{2.69}$$

Further, from (2.64) and (2.67) it follows that

$$f_{\nu_r}(x) \leqslant \varphi_{\frac{1}{r}}(x) \quad (x \in E, \; r > r(x)), \tag{2.70}$$

where $r(x)$ generally depends on x and, moreover, by (2.66) and (2.67), $E \subset H_1 + H_2 + H_3$; and in this case from (2.68)–(2.70) we have

$$\varlimsup_{r \to \infty} f_{\nu_r}(x) \leqslant F(x) \tag{2.71}$$

a.e. on $[a, b]$.

Since $G(x)$ is the lower limit in measure on $[a, b]$ of the sequence (2.3) of functions $f_m(x)$ (see the hypotheses of Lemma 10 and Lemma 8), on letting $G^{(1)}(x)$ denote the lower limit in measure on $[a, b]$ of the sequence of functions

$$f_{\nu_r}(x) \quad (r = 1, 2, \ldots) \tag{2.72}$$

and noting that (2.63) is an increasing sequence of natural numbers, we can assert by Lemma 5 that

$$G(x) \leqslant G^{(1)}(x) \tag{2.73}$$

a.e. on $[a, b]$.

From Definition 5 it follows that $-G_1(x)$ is the upper limit in measure on $[a, b]$ of the sequence

$$-f_{\nu_r}(x) \quad (r = 1, 2, \ldots). \tag{2.74}$$

We have already proved that from a sequence (2.3) of functions $f_m(x)$ one can extract a subsequence (2.72) with increasing indices ν_r for which (2.71) holds a.e. on $[a, b]$. Hence $F(x)$ is the upper limit in measure on $[a, b]$ of a sequence (2.3). Taking the sequence (2.74) in place of (2.3) and $-G^{(1)}(x)$ in place of $F(x)$ we can assert that there is an increasing sequence of natural numbers

$$m_k \quad (k=0,\ 1,\ 2,\ \ldots) \tag{2.75}$$

in the sequence (2.63) for which

$$\overline{\lim_{k\to\infty}}\,[-f_{m_k}(x)] \leqslant -G^{(1)}(x)$$

a.e. on $[a, b]$, and hence by (2.73)

$$G(x) \leqslant \underline{\lim}_{k\to\infty} f_{m_k}(x) \tag{2.76}$$

a.e. on $[a, b]$.

Further, since the increasing sequence (2.75) is contained in the increasing sequence (2.63), by (2.71) we have

$$\overline{\lim_{k\to\infty}} f_{m_k}(x) \leqslant \overline{\lim_{r\to\infty}} f_{\nu_r}(x) \leqslant F(x) \tag{2.77}$$

a.e. on $[a, b]$.

From (2.76) and (2.77) it follows that (2.49) holds a.e. on $[a, b]$. This completes the proof of Lemma 10.

LEMMA 11. *Let the functions* $f_m(x)$ $(m = 0, 1, \ldots)$, $F(x)$ *and* $G(x)$ *satisfy all the conditions in Lemma* 8, *where* $[a, b]$ *is some interval, and let the functions* $g_m(x)$ *be defined from equalities* (2.34), *where* $F_0(x)$ *and* $G_0(x)$ *are measurable and satisfy* (2.33) *a.e. on* $[a, b]$.

Then the set of all limit functions of the sequence (2.3) *coincides with the set of all limit functions of the sequence* (2.4) (*or, what is the same, the sequence* (2.35)).

PROOF. First, from Lemma 9 it follows that the functions $g_m(x)$ are measurable and defined a.e. on $[a, b]$, and, moreover, $F(x)$ and $G(x)$ are the upper and lower limit in measure on $[a, b]$ of the sequence (2.4).

Take any measurable function $\varphi(x, E)$ defined a.e. on some set E and satisfying the inequality

$$G(x) \leqslant \varphi(x, E) \leqslant F(x) \tag{2.78}$$

a.e. on E, and

$$\operatorname{meas} E > 0, \quad E \subset [a, b]. \tag{2.79}$$

For this $\varphi(x, E)$, set

$$E_1 = E\,[G_0(x) < \varphi(x, E) < F_0(x)], \tag{2.80}$$

$$E_2 = E\,[G_0(x) < \varphi(x, E) = F_0(x)], \tag{2.81}$$

$$E_3 = E\,[G_0(x) = \varphi(x, E) < F_0(x)], \tag{2.82}$$

$$E_4 = E\,[G_0(x) = \varphi(x, E) = F_0(x)]. \tag{2.83}$$

By the hypothesis of this lemma, the functions $F(x)$, $G(x)$, $F_0(x)$ and $G_0(x)$ are measurable and satisfy (2.33) a.e. on $[a, b]$; moreover, $\varphi(x, E)$ is defined a.e. on E and is not defined off E. In this case from (2.78)–(2.83) we have

$$E_j \subset E \quad (j = 1, 2, 3, 4), \tag{2.84}$$

$$E = E_1 + E_2 + E_3 + E_4 + e, \tag{2.85}$$

where

$$\text{meas}\, e = 0. \tag{2.86}$$

Suppose now that $\varphi(x, E)$ is any limit function of (2.3) (see Definition 7 of §1). Then one can find an increasing sequence

$$m_k \quad (k = 0, 1, 2, \ldots), \tag{2.87}$$

such that

$$\lim_{k\to\infty} f_{m_k}(x) = \varphi(x, E) \quad (x \in H), \tag{2.88}$$

where

$$\text{meas}\, H = \text{meas}\, E > 0, \quad H \subset E \subset [a, b]. \tag{2.89}$$

Moreover, from Lemma 8 it follows that for a given $\varphi(x, E)$ (2.78) holds a.e. on $[a, b]$, and in this case for $\varphi(x, E)$ we can find from (2.80)–(2.83) sets E_j $(j = 1, 2, 3, 4)$ satisfying (2.84)–(2.86).

Suppose at the outset that $x \in (E_1 \cdot H)$. Then, by (2.80) and (2.88), $G_0(x) < f_{m_k}(x) < F_0(x)$ for all sufficiently large k, and hence, by (2.34), $g_{m_k}(x) = f_{m_k}(x)$ for the same k, and in this case from (2.88) we find that

$$\lim_{k\to\infty} g_{m_k}(x) = \varphi(x, E) \quad (x \in (E_1 \cdot H)). \tag{2.90}$$

Further, suppose that

$$x \in (E_2 \cdot H). \tag{2.91}$$

Suppose, moreover, that for this x and for an infinite set of k we have

$$f_{m_k}(x) \leqslant F_0(x). \tag{2.92}$$

Then from (2.34), (2.81), (2.88) and (2.91) we have

$$g_{m_k}(x) = f_{m_k}(x) \tag{2.93}$$

for all sufficiently large k satisfying (2.92).

Since, by (2.91), $x \in H$ for the x being considered, it follows from (2.88) and (2.93) that for this x

$$g_{m_k}(x) \to \varphi(x, E), \tag{2.94}$$

as $k \to \infty$ taking on values for which (2.92) holds.

Next suppose that x as before satisfies (2.91) while at the same time

$$F_0(x) < f_{m_k}(x) \tag{2.95}$$

for an infinite set of k. Then from (2.34), (2.81) and (2.91) we shall have, for these k,

$$g_{m_k}(x) = F_0(x) = \varphi(x, E),$$

and hence (2.94) holds also in the case when $k \longrightarrow \infty$ taking on values for which (2.95) holds.

Since x is any point satisfying (2.91) and since (2.94) holds as $k \longrightarrow \infty$ taking on values for which either (2.92) or (2.95) holds (since $x \in H$, by (2.88) the functions $f_{m_k}(x)$ are defined for any $k = 0, 1, 2, \ldots$), it follows that

$$\lim_{k\to\infty} g_{m_k}(x) = \varphi(x, E) \quad (x \in (E_2 \cdot H)). \tag{2.96}$$

Now suppose that

$$x \in (E_3 \cdot H). \tag{2.97}$$

and that

$$G_0(x) \leqslant f_{m_k}(x) \tag{2.98}$$

for an infinite set of k. Then from (2.34), (2.82), (2.88) and (2.97) we have (2.93) for all sufficiently large k satisfying (2.98). Hence (2.94) will hold if $k \longrightarrow \infty$ taking on values for which (2.98) is valid.

Next assume that (2.97) holds as before, and also

$$f_{m_k}(x) < G_0(x) \tag{2.99}$$

for an infinite set of k. Then from (2.34) and (2.82) we have, for these k,

$$g_{m_k}(x) = G_0(x) = \varphi(x, E),$$

and hence (2.94) holds also in the case when $k \longrightarrow \infty$ taking on values for which (2.99) holds.

Since we are now assuming that x is any point satisfying (2.97), and since (2.94) holds as $k \longrightarrow \infty$ taking on values for which either (2.98) or (2.99) holds, it follows that

$$\lim_{k\to\infty} g_{m_k}(x) = \varphi(x, E) \quad (x \in (E_3 \cdot H)). \tag{2.100}$$

Finally, suppose that

$$x \in (E_4 \cdot H). \tag{2.101}$$

Then from (2.34) and (2.83) we have

$$g_{m_k}(x) = F_0(x) = G_0(x) = \varphi(x, E) \quad (k = 0, 1, 2, \ldots)$$

(from (2.34) and (2.88) it follows that $f_{m_k}(x)$ and $g_{m_k}(x)$ are defined at the point x being considered), and hence

$$\lim_{k\to\infty} g_{m_k}(x) = \varphi(x, E) \tag{2.102}$$

for all x satisfying (2.101).

It follows from (2.85), (2.86), (2.89), (2.90), (2.96) and (2.100) that (2.102) holds a.e. on E; and in this case, since the sequence (2.87) is increasing and since conditions (2.79) hold, $\varphi(x, E)$ is a limit function of the sequence (2.4) (see Definition 7). Since we have supposed that $\varphi(x, E)$ is any limit function of the sequence (2.3), we see that any limit function of (2.3) is a limit function of the sequence (2.4).

Now suppose that $\varphi(x, E)$ is any limit function of the sequence (2.4). Then conditions (2.79) hold, and, moreover, there is an increasing sequence

$$\mu_r \quad (r = 0, 1, 2, \ldots), \tag{2.103}$$

such that (2.78) holds and

$$\lim_{r\to\infty} g_{\mu_r}(x) = \varphi(x, E) \tag{2.104}$$

a.e. on E, i.e. this equality holds on some set H' such that

$$\operatorname{meas} H' = \operatorname{meas} E > 0, \qquad H' \subset E \subset [a, b]. \tag{2.105}$$

Let $F^{(1)}(x)$ and $G^{(1)}(x)$ denote the upper and lower limits in measure on $[a, b]$ of the sequence

$$f_{\mu_r}(x) \quad (r = 0, 1, 2, \ldots). \tag{2.106}$$

As we already know from the formulation of Lemma 11, $F(x)$ and $G(x)$ are the upper and lower limits in measure on $[a, b]$ of the sequence (2.3). In this case from Lemma 5 we have

$$G(x) \leqslant G^{(1)}(x) \leqslant F^{(1)}(x) \leqslant F(x) \quad (x \in H_0), \tag{2.107}$$

where

$$\operatorname{meas} H_0 = b - a, \quad H_0 \subset [a, b]. \tag{2.108}$$

Noting the definition of $F^{(1)}(x)$ and $G^{(1)}(x)$ and Lemma 10, in which we take the sequence (2.106) in place of (2.3) and the functions $F^{(1)}(x)$ and $G^{(1)}(x)$ in place of $F(x)$ and $G(x)$, we can assert that there is an increasing sequence of natural numbers

$$n_s \quad (s = 0, 1, 2, \ldots), \tag{2.109}$$

which is contained in (2.103) and for which

$$G^{(1)}(x) \leqslant \varliminf_{s\to\infty} f_{n_s}(x) \leqslant \varlimsup_{s\to\infty} f_{n_s}(x) \leqslant F^{(1)}(x) \quad (x \in H_1), \tag{2.110}$$

where

$$\operatorname{meas} H_1 = b - a, \quad H_1 \subset [a, b]. \tag{2.111}$$

Further, since by the hypothesis of Lemma 11 the inequality (2.33) holds a.e. on $[a, b]$ and also the functions $g_m(x)$ $(m = 0, 1, 2, \ldots)$ are defined from (2.34) a.e. on $[a, b]$, it follows that

$$G_0(x) \leqslant G(x) \leqslant F(x) \leqslant F_0(x) \quad (x \in H_2), \tag{2.112}$$

$$G_0(x) \leqslant g_m(x) \leqslant F_0(x) \quad (x \in H_2, \; m = 0, 1, 2, \ldots), \tag{2.113}$$

where

$$\text{meas } H_2 = b - a, \quad H_2 \subset [a, b]. \tag{2.114}$$

Let us set

$$H'' = (H' \cdot H_0 \cdot H_1 \cdot H_2). \tag{2.115}$$

Then by (2.105), (2.108), (2.111) and (2.114) we have

$$\text{meas } H'' = \text{meas} E > 0, \quad H'' \subset E \subset [a, b], \tag{2.116}$$

and, moreover, by (2.107), (2.110) and (2.112),

$$G_0(x) \leqslant \varliminf_{s\to\infty} f_{n_s}(x) \leqslant \varlimsup_{s\to\infty} f_{n_s}(x) \leqslant F_0(x) \quad (x \in H''). \tag{2.117}$$

As we already know, the increasing sequence (2.109) is contained in the increasing sequence (2.103). In this case from (2.104), (2.115) and from the definition of the set H' (see the remark after (2.104)) it follows that

$$\lim_{s\to\infty} g_{n_s}(x) = \varphi(x, E) \quad (x \in H''). \tag{2.118}$$

For our function $\varphi(x, E)$ consider the sets E_j ($j = 1, 2, 3, 4$), defined from (2.80)–(2.83) and satisfying (2.84)–(2.86). In this case from (2.85), (2.86), (2.115), (2.78) and (2.116) we obtain

$$\sum_{j=1}^{4} (E_j \cdot H'') + e' = H'', \tag{2.119}$$

where

$$\text{meas } e' = 0. \tag{2.120}$$

Suppose at the outset that

$$x \in (E_1 \cdot H''). \tag{2.121}$$

Then, by (2.80) and (2.118), $G_0(x) < g_{n_s}(x) < F_0(x)$ for all sufficiently large s, and hence, by (2.34), $f_{n_s}(x) = g_{n_s}(x)$ for the same s; and in this case

$$\lim_{s\to\infty} f_{ns}(x) = \varphi(x, E) \quad (x \in (E_1 \cdot H'')). \tag{2.122}$$

Suppose further that

$$x \in (E_2 \cdot H'') \tag{2.123}$$

and that for our x we have

$$f_{n_s}(x) \leqslant F_0(x) \tag{2.124}$$

for an uncountable set of s.

From (2.81), (2.113), (2.115), (2.118) and (2.123) we obtain

$$G_0(x) < g_{n_s}(x) \leqslant F_0(x) \tag{2.125}$$

for all sufficiently large s, whence we find that

$$G_0(x) \leqslant f_{n_s}(x) \tag{2.126}$$

for the same s, since were the reverse inequality to hold for any of the values of s being considered, then from (2.34) we should have $g_{n_s}(x) = G_0(x)$.

Combining (2.124) and (2.126), we see that $G_0(x) \leqslant f_{n_s}(x) \leqslant F_0(x)$ holds for an infinite set of s, namely for all sufficiently large s satisfying (2.124). Then from (2.34) we have

$$g_{n_s}(x) = f_{n_s}(x) \tag{2.127}$$

for the same s. Hence from (2.118) and (2.123) we have

$$f_{n_s}(x) \to \varphi(x, E), \tag{2.128}$$

as $s \to \infty$ taking on values for which (2.124) holds.

In the case when as before condition (2.123) holds and, moreover,

$$F_0(x) < f_{n_s}(x) \tag{2.129}$$

for an infinite set of s, let us denote by

$$s_\nu \quad (\nu = 0, 1, 2, \ldots) \tag{2.130}$$

all natural numbers s arranged in increasing order for which (2.129) holds. Then for our point x we have

$$F_0(x) < f_{n'_\nu}(x) \quad (\nu = 0, 1, 2, \ldots), \tag{2.131}$$

where

$$n'_\nu = n_{s_\nu} \cdot (\nu = 0, 1, 2, \ldots). \tag{2.132}$$

Since (2.109) and (2.130) are increasing sequences of natural numbers, it follows from (2.132) that the natural numbers

$$n'_\nu \quad (\nu = 0, 1, 2, \ldots) \tag{2.133}$$

form an increasing sequence which is contained in the sequence (2.109). We see, noting (2.117) and (2.123), that

$$\overline{\lim_{\nu\to\infty}} f_{n'_\nu}(x) \leqslant \overline{\lim_{s\to\infty}} f_{n_s}(x) \leqslant F_0(x) \tag{2.134}$$

and moreover, from (2.131),

$$F_0(x) \leqslant \underline{\lim_{\nu\to\infty}} f_{n'_\nu}(x). \tag{2.135}$$

Combining (2.134) and (2.135) with (2.81) and (2.123), for the x being considered we see that

$$\lim_{\nu\to\infty} f_{n'_\nu}(x) = F_0(x) = \varphi(x, E). \tag{2.136}$$

From the definition of the sequence (2.130) and from (2.132) it follows that the increasing sequence (2.133) consists of all numbers of (2.109) for which (2.129) holds. In this case (2.128) holds as $s \to \infty$ taking on values for which (2.129) holds.

Thus, if $x \in (E_2 \cdot H'')$, then (2.128) holds as $s \to \infty$ taking on values for which either (2.124) or (2.129) holds. Hence

$$\lim_{s\to\infty} f_{n_s}(x) = \varphi(x, E) \quad (x \in (E_2 \cdot H'')). \tag{2.137}$$

Now suppose that

$$x \in (E_3 \cdot H'') \tag{2.138}$$

and that

$$G_0(x) \leqslant f_{n_s}(x) \tag{2.139}$$

for an infinite set of s.

From (2.82), (2.113), (2.115), (2.118) and (2.138) we find that

$$G_0(x) \leqslant g_{n_s}(x) < F_0(x)$$

for all sufficiently large s, and so by (2.34)

$$f_{n_s}(x) \leqslant F_0(x) \tag{2.140}$$

for the same s.

Combining (2.139) with (2.140) and noting (2.34), we see that (2.127) holds for an infinite set of s, namely for all sufficiently large s satisfying (2.139). Hence from (2.118) and (2.138) it follows that (2.128) holds as $s \longrightarrow \infty$ taking on values for which (2.139) holds.

Now if as before condition (2.138) holds and at the same time

$$f_{n_s}(x) < G_0(x) \tag{2.141}$$

for an infinite set of s, then let us denote by

$$t_\nu \quad (\nu = 0, 1, 2, \ldots) \tag{2.142}$$

all natural numbers s arranged in increasing order for which (2.141) holds. Then for the point x we have

$$f_{n''_\nu}(x) < G_0(x) \quad (\nu = 0, 1, 2, \ldots). \tag{2.143}$$

where

$$n''_\nu = n_{t_\nu} \quad (\nu = 0, 1, 2, \ldots). \tag{2.144}$$

Since (2.109) and (2.142) are increasing sequences of natural numbers, it follows from (2.144) that the natural numbers

$$n''_\nu \ (\nu = 0, 1, 2, \ldots) \tag{2.145}$$

form an increasing sequence which is contained in (2.109). In this case from (2.117) and (2.138) we obtain

$$G_0(x) \leqslant \varliminf_{s\to\infty} f_{n_s}(x) \leqslant \varliminf_{\nu\to\infty} f_{n''_\nu}(x) \tag{2.146}$$

and, moreover, by (2.143),

$$\varlimsup_{\nu\to\infty} f_{n''_\nu}(x) \leqslant G_0(x). \tag{2.147}$$

Combining (2.146), (2.147), (2.82) and (2.138), we see that at the point x we have

$$\lim_{\nu\to\infty} f_{n''_\nu}(x) = G_0(x) = \varphi(x, E). \tag{2.148}$$

From the definition of the sequence (2.142) and from (2.144) it follows that the increasing sequence of natural numbers (2.145) consists of all numbers of the sequence (2.109) for which (2.141) is valid. In this case (2.128) holds as $s \longrightarrow \infty$ taking on values for which (2.141) holds.

Thus if $x \in (E_3 \cdot H'')$, then (2.128) holds as $s \longrightarrow \infty$ taking on values for which either (2.139) or (2.141) holds. Hence,

$$\lim_{s\to\infty} f_{n_s}(x) = \varphi(x, E) \quad (x \in (E_3 \cdot H'')). \tag{2.149}$$

Finally, suppose that

$$x \in (E_4 \cdot H''). \tag{2.150}$$

Then from (2.83), (2.117) and (2.150) we obtain

$$\varphi(x, E) = G_0(x) \leqslant \varliminf_{s\to\infty} f_{n_s}(x) \leqslant \varlimsup_{s\to\infty} f_{n_s}(x) \leqslant F_0(x) = \varphi(x, E)$$

(from (2.83), (2.113), (2.115) and (2.150) it follows that at the point x the functions $F_0(x)$, $G_0(x)$, $\varphi(x, E)$ and $g_{n_s}(x)$ $(s = 0, 1, \ldots)$ are defined, and hence by (2.34) the functions $f_{n_s}(x)$ $(s = 0, 1, \ldots)$ are defined). Therefore

$$\lim_{s\to\infty} f_{n_s}(x) = \varphi(x, E) \tag{2.151}$$

for all x satisfying (2.150).

It follows from (2.119), (2.120), (2.116), (2.122), (2.137) and (2.149) that (2.151) holds a.e. on E; and in this case, since the sequence (2.109) is increasing and since the conditions (2.79) hold (see the remark before (2.103)), $\varphi(x, E)$ is a limit function of the sequence (2.3) (see Definition 7). Since we have assumed that $\varphi(x, E)$ is an arbitrary limit function of (2.4), we see that any limit function of (2.4) is a limit function of (2.3).

Earlier we have proved that any limit function of (2.3) is a limit function of (2.4). Hence the set of all limit functions of (2.3) coincides with the set of all limit functions of the sequence (2.4). The proof of Lemma 11 is complete.

Using Lemmas 9 and 11, we shall prove the main lemma needed to prove the theorems formulated in §1.

LEMMA 12. *Consider two measurable functions $F(x)$ and $G(x)$ satisfying the inequality $G(x) \leqslant F(x)$ a.e. on some $[a, b]$, and the set $M = \{\varphi(x, E)\}$ of measurable functions $\varphi(x, E)$ each of which is defined a.e. on the corresponding set E,* meas $E > 0$, $E \subset [a, b]$.

Then, if the set M has property $h(G, F)$ (see Definition 9), there exists a sequence (2.4) of measurable functions $g_m(x)$ $(m = 0, 1, 2, \ldots)$ defined a.e. on $[a, b]$ which has the following properties:

1°. *M is the set of all limit functions of the sequence* (2.4).

2°. *$F(x)$ and $G(x)$ are the upper and lower limits in measure on $[a, b]$ of* (2.4).
3°.

$$G(x) \leqslant g_m(x) \leqslant F(x) \quad (m = 0, 1, 2, \ldots) \tag{2.152}$$

a.e. on $[a, b]$.

PROOF. Since by the hypothesis of Lemma 12 the set M has property $h(G, F)$, by Definition 9 there is a sequence (2.3) of measurable functions $f_m(x)$ defined a.e. on $[a, b]$ for which M is the set of all limit functions and $F(x)$ and $G(x)$ are the upper and lower limits in measure on $[a, b]$.

At the $x \in [a, b]$ at which $F(x)$, $G(x)$ and $f_m(x)$ $(m = 0, 1, 2, \ldots)$ are defined let us define functions $g_m(x)$ of a sequence (2.4) by the equality

$$g_m(x) = \begin{cases} f_m(x), & \text{if} \quad G(x) \leqslant f_m(x) \leqslant F(x), \\ F(x), & \text{if} \quad f_m(x) > F(x) \geqslant G(x), \\ G(x), & \text{if} \quad f_m(x) < G(x) \leqslant F(x) \end{cases} \\ (m = 0, 1, 2, \ldots). \tag{2.153}$$

Then the $g_m(x)$ will be measurable and satisfy (2.152) a.e. on $[a, b]$.

Combining (2.152) and (2.153) with Lemmas 9 and 11, in which we set $F_0(x) = F(x)$ and $G_0(x) = G(x)$, we see that the sequence (2.4) of functions $g_m(x)$ satisfies all the conditions of Lemma 12, as required.

Let us prove a further lemma on equiconvergent sequences.

LEMMA 13. *Suppose that the functions $f_m(x)$ and $g_m(x)$ $(m = 0, 1, 2, \ldots)$ of sequences* (2.3) *and* (2.4) *are measurable and defined a.e. on some $[a, b]$. If the sequences* (2.3) *and* (2.4) *are equiconvergent a.e. on $[a, b]$, then their sets of limit functions coincide.*

PROOF. Since the sequences (2.3) and (2.4) occur symmetrically in the formulation of the lemma, it suffices to show that any limit function of (2.3) is a limit function of (2.4).

Suppose that $\varphi(x, E)$ is a limit function of (2.3). Then, by Definition 7, meas $E > 0$, $E \subset [a, b]$, and there is an increasing sequence of natural numbers m_k such that

$$\lim_{k\to\infty} f_{m_k}(x) = \varphi(x, E)$$

a.e. on E.

From the equiconvergence, from Lemma 1, and from the preceding equality it follows that

$$\lim_{k\to\infty} g_{m_k}(x) = \varphi(x, E)$$

a.e. on E, and this implies that $\varphi(x, E)$ is a limit function of (2.4), as required.

§3. A lemma on defining a sequence of functions from its ordinary limits of indeterminacy and from its limits of indeterminacy in measure

Before formulating the main lemma of this section, let us prove

LEMMA 14. *Let the functions* $f_m(x)$ *and* $g_m(x)$ $(m = 0, 1, \ldots)$ *of sequences* (2.3) *and* (2.4) *be measurable and defined a.e. on some* $[a, b]$. *Suppose that*

$$\lim_{m\to\infty} \text{meas}\, E\,[f_m(x) \neq g_m(x)] = 0, \tag{3.1}$$

let $F(x)$ *and* $G(x)$ *denote the upper and lower limits in measure on* $[a, b]$ *of* (2.3), *and let* $M = \{\varphi(x, E)\}$ *denote the set of all limit functions of* (2.3).

Then $F(x)$ *and* $G(x)$ *will be the upper and lower limits in measure on* $[a, b]$ *of* (2.4), *and* M *will be the set of all limit functions of* (2.4).

PROOF. Let e denote the set of all points in $[a, b]$ at which at least one of $F(x)$, $G(x)$, $f_m(x)$ or $g_m(x)$ $(m = 0, 1, 2, \ldots)$ is not defined. Then, by the conditions of Lemma 14,

$$\text{meas}\, e = 0, \qquad e \subset [a, b]. \tag{3.2}$$

Setting

$$e_m = e + E\,[f_m(x) \neq g_m(x)] \qquad (m = 0, 1, 2, \ldots), \tag{3.3}$$

from (3.1) and (3.2) we get

$$\lim_{m\to\infty} \text{meas}\, e_m = 0, \qquad e_m \subset [a, b] \qquad (m = 0, 1, 2, \ldots). \tag{3.4}$$

If $\varphi(x)$ and $\psi(x)$ are two measurable functions defined a.e. on $[a, b]$, then from the definition of e and (3.3) it follows that

$$\begin{gathered}\{E\,[g_m(x) > \varphi(x)] \cdot E\,[\varphi(x) > F(x)]\} \\ \subset \{E\,[f_m(x) > \varphi(x)] \cdot E\,[\varphi(x) > F(x)]\} + e_m \qquad (m = 0, 1, 2, \ldots),\end{gathered} \tag{3.5}$$

$$\begin{gathered}\{E\,[g_m(x) > \psi(x)] \cdot E\,[F(x) > \psi(x)]\} \\ \supset \{E\,[f_m(x) > \psi(x)] \cdot E\,[F(x) > \psi(x)]\} \setminus e_m \qquad (m = 0, 1, 2, \ldots).\end{gathered} \tag{3.6}$$

Since, by hypothesis, $F(x)$ is the upper limit in measure on $[a, b]$ of (2.3), it follows from Definition 4 that conditions a° and b° therein hold, i.e. (1.9) holds for any measurable $\varphi(x)$ defined a.e. on $[a, b]$ and (1.10) holds for any measurable $\psi(x)$ defined a.e. on $[a, b]$ and satisfying (1.11).

In this case, by (3.4)–(3.6) the same conditions a° and b° will hold if we replace therein $f_m(x)$ and $g_m(x)$ and assume that $\varphi(x)$ and $\psi(x)$ satisfy the previous conditions; and then it follows that $F(x)$ is the upper limit in measure on $[a, b]$ of (2.4).

Further, since, by the hypothesis of Lemma 14, $G(x)$ is the lower limit in measure on $[a, b]$ of (2.3), it follows from Definition 5 that $-G(x)$ is the upper limit in measure on $[a, b]$ of $-f_m(x)$ $(m = 0, 1, \ldots)$. Then, by the part of Lemma 14 already proved, $-G(x)$ is the upper limit in measure on $[a, b]$ of $-g_m(x)$ $(m = 0, 1, \ldots)$, and hence $G(x)$ is the lower limit in measure on $[a, b]$ of (2.4).

Thus to end the proof we need to show that $M = \{\varphi(x, E)\}$ is the set of all limit functions of (2.4). First suppose that $\varphi(x, E)$ is any limit function of (2.3). Then from Definition 7 we have

$$\text{meas } E > 0, \qquad E \subset [a, b] \tag{3.7}$$

and also there is an increasing sequence

$$m_k \ (k = 0, 1, 2, \ldots) \tag{3.8}$$

such that

$$\lim_{k\to\infty} f_{m_k}(x) = \varphi(x, E) \tag{3.9}$$

a.e. on E.

Noting (3.4) and, if need be, replacing (3.8) by a sufficiently rarefied subsequence, we can always assume that besides (3.9) we also have

$$\sum_{k=1}^{\infty} \text{meas } e_{m_k} < +\infty. \tag{3.10}$$

Let us set

$$e' = \lim_{k\to\infty} e_{m_k}. \tag{3.11}$$

Then from (3.10) and the condition $e_m \subset [a, b]$ $(m = 0, 1, \ldots)$ (see (3.4)) we obtain

$$\text{meas } e' = 0, \qquad e' \subset [a, b]. \tag{3.12}$$

By (3.3) and (3.11),

$$g_{m_k}(x) = f_{m_k}(x) \qquad (k > k_x, \ x \in [a, b] \setminus e'), \tag{3.13}$$

where k_x depends in general on x. Noting (3.9) and (3.12), we have

$$\lim_{k\to\infty} g_{m_k}(x) = \varphi(x, E) \tag{3.14}$$

a.e. on E, and in this case, by (3.7), $\varphi(x, E)$ is a limit function of (2.4) on E. Thus, any limit function of (2.3) is also one of (2.4).

Now let $\varphi(x, E)$ be any limit function of (2.4). Then (3.7) will hold and there will exist an increasing sequence (3.8) which will satisfy (3.14) a.e. on E. Hence we can take the sequence (3.8) to be increasing rapidly enough so that (3.10) holds.

Taking e' to be defined from (3.11), we obtain (3.12) and (3.13), and in this case (3.9) will hold a.e. on E, i.e., $\varphi(x, E)$ is a limit function of (2.3) on E. We have thus proved that any limit function of (2.4) is also one of (2.3). Combining this with what was already proved, we find that the sets of limits functions of (2.3) and (2.4) coincide; and since, by the hypothesis of Lemma 14, $M = \{\varphi(x, E)\}$ is the set of limit functions of (2.3), it is also that of (2.4). The proof of Lemma 14 is complete.

Now we shall prove the main lemma of this section.

LEMMA 15. *Suppose that* $\varphi_1(x)$, $\varphi_2(x)$, $\psi_1(x)$ *and* $\psi_2(x)$ *are measurable and satisfy*

$$\varphi_1(x) \leqslant \psi_1(x) \leqslant \psi_2(x) \leqslant \varphi_2(x) \tag{3.15}$$

a.e. on some $[a, b]$ *(these functions can take on the values* $\pm\infty$ *on sets of positive measure). Consider the set* $M = \{\varphi(x, E)\}$ *of functions* $\varphi(x, E)$ *each of which is defined a.e. on the corresponding set* E *of positive measure lying in* $[a, b]$, *and suppose that* M *has property* $h(\psi_1, \psi_2)$.

In this case there is a sequence of measurable functions

$$G_m(x) \quad (m = 0, 1, 2, \ldots), \tag{3.16}$$

defined a.e. on $[a, b]$ *and having the following properties:*

1) $\psi_2(x)$ *and* $\psi_1(x)$ *are the upper and lower limits in measure on* $[a, b]$ *of the sequence* (3.16).

2)

$$\underline{\lim}_{m\to\infty} G_m(x) = \varphi_1(x), \quad \overline{\lim}_{m\to\infty} G_m(x) = \varphi_2(x) \tag{3.17}$$

a.e. on $[a, b]$.

3) M *is the set of all limit functions of* (3.16).

4)

$$\varphi_1(x) \leqslant G_m(x) \leqslant \varphi_2(x) \quad (m = 0, 1, 2, \ldots) \tag{3.18}$$

a.e. on $[a, b]$.

PROOF. Since, by hypothesis of Lemma 15, $\psi_1(x) \leqslant \psi_2(x)$ a.e. on $[a, b]$ and M has property $h(\psi_1, \psi_2)$, by Lemma 12 one can find a sequence (2.3) of measurable functions $f_m(x)$ defined a.e. on $[a, b]$ which have the following properties:

A″. M is the set of all limit functions of (2.3).

B″. $\psi_1(x)$ and $\psi_2(x)$ are the upper and lower limits in measure on $[a, b]$ of (2.3).

C″.

$$\psi_1(x) \leqslant f_m(x) \leqslant \psi_2(x) \quad (m = 0, 1, 2, \ldots) \tag{3.19}$$

a.e. on $[a, b]$.

Now we define a sequence of intervals δ_ν ($\nu = 0, 1, \ldots$) which have the following properties:

a°. Any x in $[a, b]$, except for perhaps a countable set e, lies in an infinite set of different intervals δ_ν.

b°.

$$\lim_{\nu\to\infty} \operatorname{meas} \delta_\nu = 0. \tag{3.20}$$

These intervals δ_ν can be defined, in particular, as follows. If $2^k - 1 \leqslant \nu < 2^{k+1} - 1$ ($k = 0, 1, \ldots$), then we set

$$\delta_\nu = \left(a + \frac{\nu - 2^k + 1}{2^k}(b - a),\ a + \frac{\nu - 2^k + 2}{2^k}(b - a)\right).$$

These intervals will be defined for all $\nu = 0, 1, \ldots$ and will have properties a° and b°, where the countable set e will consist of all points of the form

$$a + \frac{s}{2^k}(b-a) \quad (0 \leqslant s \leqslant 2^k,\ k = 0,\ 1,\ 2, \ldots).$$

Further, let e' denote the set of all x in $[a, b]$ at which at least one of the following functions is not defined:

$$\varphi_1(x),\ \varphi_2(x),\ \psi_1(x),\ \psi_2(x),\ f_m(x) \quad (m = 0,\ 1,\ 2, \ldots) \tag{3.21}$$

or at which one of the inequalities (3.15) or (3.19) does not hold. Then

$$\operatorname{meas} e = 0,\ \operatorname{meas} e' = 0,\ e \subset [a, b],\ e' \subset [a, b], \tag{3.22}$$

and if

$$x \in [a, b] \setminus e \setminus e', \tag{3.23}$$

then at this point x all the functions (3.21) are defined, and (3.15) and (3.19) hold. Moreover, for this x there is an increasing sequence

$$\nu_s \ (s = 0,\ 1,\ 2, \ldots), \tag{3.24}$$

such that

$$x \in \delta_{\nu_s} \ (s = 0,\ 1,\ 2, \ldots). \tag{3.25}$$

Now we define the sequence (3.16) of functions $G_m(x)$ as follows.

a′. If $m = 2\nu$ ($\nu = 0, 1, \ldots$), we set

$$G_m(x) = \varphi_1(x) \quad (x \in \delta_\nu \setminus e \setminus e'). \tag{3.26}$$

b′. If $m = 2\nu + 1$, we set

$$G_m(x) = \varphi_2(x) \quad (x \in \delta_\nu \setminus e \setminus e'). \tag{3.27}$$

c′.

$$G_m(x) = f_m(x) \quad (x \in [a, b] \setminus \delta_\nu \setminus e \setminus e'), \tag{3.28}$$

if $m = 2\nu$ or $m = 2\nu + 1$.

Now, since the functions (3.21) are measurable on $[a, b]$, by (3.22) the functions $G_m(x)$ will be measurable and defined a.e. on $[a, b]$.

Since (3.15) and (3.19) hold for all x satisfying (3.23), by (3.26), (3.27) and (3.28) we have

$$\varphi_1(x) \leqslant G_m(x) \leqslant \varphi_2(x) \quad (x \in [a, b] \setminus e \setminus e',\ m = 0,\ 1,\ 2, \ldots), \tag{3.29}$$

whence

$$\varphi_1(x) \leqslant \varliminf_{m\to\infty} G_m(x) \leqslant \varlimsup_{m\to\infty} G_m(x) \leqslant \varphi_2(x) \quad (x \in [a, b] \setminus e \setminus e'). \tag{3.30}$$

On the other hand, if x satisfies (3.23), then, as we already know, for this x there is an increasing sequence (3.24) for which (3.25) holds, and hence, by a′ and b′ (see (3.26) and (3.27)),

$$G_{2\nu_s}(x) = \varphi_1(x) \quad (s = 0, 1, 2, \ldots), \tag{3.31}$$

$$G_{2\nu_s+1}(x) = \varphi_2(x) \quad (s = 0, 1, 2, \ldots), \tag{3.32}$$

and in this case

$$\varliminf_{m\to\infty} G_m(x) \leqslant \varphi_1(x), \quad \varlimsup_{m\to\infty} G_m(x) \geqslant \varphi_2(x) \quad (x \in [a, b] \setminus e \setminus e').$$

Combining this with (3.30), we have

$$\varliminf_{m\to\infty} G_m(x) = \varphi_1(x), \quad \varlimsup_{m\to\infty} G_m(x) = \varphi_2(x) \quad (x \in [a, b] \setminus e \setminus e'). \tag{3.33}$$

From (3.22), (3.29) and (3.33) it follows that the functions of the sequence (3.16) satisfy conditions 2) and 4) in Lemma 15. Hence to end the proof we need to show that they also satisfy 1) and 3).

If $m = 2\nu$ or $m = 2\nu + 1$, we set

$$e_m = \delta_\nu + e + e'. \tag{3.34}$$

Then the set e_m will be defined for any $m = 0, 1, \ldots,$ and in (3.34) ν is a function of m satisfying the condition

$$\lim_{m\to\infty} \nu = +\infty. \tag{3.35}$$

Since (3.22) implies

$$\operatorname{meas} e_m = \operatorname{meas} \delta_\nu, \tag{3.36}$$

it follows from (3.20) and (3.35) that

$$\lim_{m\to\infty} \operatorname{meas} e_m = 0. \tag{3.37}$$

Moreover, from c′ (see (3.28)) and (3.34) it follows that

$$E\,[G_m(x) \neq f_m(x)] \subset e_m \quad (m = 0, 1, 2, \ldots),$$

whence, from (3.37),

$$\lim_{m\to\infty} \operatorname{meas} E\,[G_m(x) \neq f_m(x)] = 0. \tag{3.38}$$

Combining this with properties A″ and B″ of the $f_m(x)$ and applying Lemma 14, in which we take $\psi_1(x)$, $\psi_2(x)$, $f_m(x)$ and $G_m(x)$ $(m = 0, 1, 2, \ldots)$ in place of $G(x)$, $F(x)$, $f_m(x)$ and $g_m(x)$, we see that (3.16) has properties 1) and 3), which completes the proof.

Next let us recall the definition of a uniformly convergent sequence in the case when the limit function can take on the values $\pm\infty$ on a set of positive measure.

DEFINITION 15. We shall say that a sequence (2.3) of functions $f_m(x)$ defined on some set E *converges uniformly* to a function $f(x)$ on E if for any positive ϵ one can find a natural number N such that for any $x \in E$ and for any $m > N$ one of the following conditions holds:

$\alpha°$. $-\infty < f_m(x) < +\infty$, $|f_m(x) - f(x)| < \varepsilon$, if $f(x)$ is finite;

$\beta°$. $f_m(x) > \frac{1}{\varepsilon}$, if $f(x) = +\infty$;

$\gamma°$. $f_m(x) < -\frac{1}{\varepsilon}$, if $f(x) = -\infty$.

(See [1], §1, Definition 3. In this definition the $f_m(x)$ can take on the values $\pm\infty$ on sets of positive measure.)

For this definition one can obtain a generalization of Egorov's theorem (see [1], §1, Lemma K). Namely, we have

LEMMA 16. *If the sequence* (2.3) *of measurable functions* $f_m(x)$ *defined a.e. on a measurable set* E *converges to* $f(x)$ *a.e. on* E, *then for any positive* σ *one can find a measurable set* M *satisfying*

$$M \subset E, \quad \text{meas}\, M > \text{meas}\, E - \sigma \tag{3.39}$$

and such that (2.3) *converges to* $f(x)$ *uniformly on* M.

§4. Proof of Theorem 5

Before going to the proof of Theorem 5 (see §1) we shall prove a lemma on universal sequences of functions.

LEMMA 17. *Suppose that the sequence* (2.3) *of measurable functions* $f_m(x)$ *defined a.e. on some* $[a, b]$ *is a sequence of type* $A(F_1, F_2)$, *where the measurable functions* $F_1(x)$ *and* $F_2(x)$ *are defined and satisfy*

$$F_1(x) \leqslant F_2(x) \tag{4.1}$$

a.e. on $[a, b]$. *Further, suppose that the measurable functions*

$$\tau_k(x) \quad (k=0, 1, 2, \ldots) \tag{4.2}$$

satisfy

$$F_1(x) \leqslant \tau_k(x) \leqslant F_2(x) \tag{4.3}$$

a.e. on $[a, b]$. *Then there is an increasing sequence*

$$m_k \quad (k = 0, 1, 2, \ldots), \tag{4.4}$$

such that the sequence

$$f_{m_k}(x) \quad (k = 0, 1, 2, \ldots) \tag{4.5}$$

and the sequence (4.2) *will be equiconvergent a.e. on* $[a, b]$.

PROOF. Since, by the hypothesis of Lemma 17, each of the measurable functions $\tau_k(x)$ satisfies (4.3) a.e. on $[a, b]$ and (2.3) is a sequence of type $A(F_1, F_2)$, by Definition 3 and the remark thereafter (see §1) for any $k = 0, 1, \ldots$ one can find an increasing sequence

$$\lambda_{ks} \quad (s = 0, 1, 2, \ldots), \tag{4.6}$$

for which

$$\lim_{s\to\infty} f_{\lambda_{ks}}(x) = \tau_k(x) \quad (k = 0, 1, 2, \ldots) \tag{4.7}$$

a.e. on $[a, b]$. Then, by Lemma 16, for each $k = 0, 1, \ldots$ there is a measurable set E_k which satisfies

$$\text{meas } E_k > b - a - \frac{1}{k^2}, \quad E_k \subset [a, b] \quad (k = 0, 1, 2, \ldots) \tag{4.8}$$

and on which the sequence

$$f_{\lambda_{ks}}(x) \quad (s = 0, 1, 2, \ldots) \tag{4.9}$$

converges uniformly to $\tau_k(x)$, and $\tau_k(x)$ and the functions (4.9) are defined for all $x \in E_k$.

Now we define the sequence (4.4) as follows. Set $m_0 = 1$ and assume that m_{k-1}, where k is any natural number, has already been defined. Then, since (4.6) is an increasing sequence for any k, by Definition 15 and the property of (4.9) just established we can find a natural number s_k such that on setting $\lambda_{k,s_k} = m_k$ we shall have

$$m_k > m_{k-1} \tag{4.10}$$

and for any $x \in E_k$ either

a.
$$-\infty < f_{m_k}(x) < +\infty, \quad |f_{m_k}(x) - \tau_k(x)| < \frac{1}{2^k}, \tag{4.11}$$

if $\tau_k(x)$ is finite, or

b.
$$f_{m_k}(x) > 2^k, \tag{4.12}$$

if $\tau_k(x) = +\infty$, or

c.
$$f_{m_k}(x) < -2^k, \tag{4.13}$$

if $\tau_k(x) = -\infty$.

Thus we define the natural number m_k knowing the value of m_{k-1}, and conditions (4.10)–(4.13) will hold for any $k = 1, 2, \ldots$. It follows, in particular, that (4.4) is increasing.

Set

$$E = \varliminf_{k\to\infty} E_k. \tag{4.14}$$

Then from (4.8) we obtain

$$\text{meas } E = b - a, \quad E \subset [a, b], \tag{4.15}$$

and, moreover, for any $x \in E$ we can find a natural number $k(x)$ such that

$$x \in E_k \quad (k > k(x), \ x \in E). \tag{4.16}$$

Now suppose that $x \in E$ and $k \to \infty$ taking on values for which $f_{m_k}(x)$ and $\tau_k(x)$ are finite. Then by conditions a, b, c (see (4.11)–(4.13)) and (4.16) the inequality (4.11) will hold if k is sufficiently large, and hence $f_{m_k}(x) - \tau_k(x) \longrightarrow 0$.

Suppose, further, that $x \in E$ and $k \longrightarrow \infty$ taking on values for which at least one of $f_{m_k}(x)$ or $\tau_k(x)$ is equal to $+\infty$. Then from conditions a, b, and c (4.16) it follows that $\tau_k(x) = +\infty$ and $f_{m_k}(x) > 2^k$ if k is sufficiently large, and so $f_{m_k}(x) \longrightarrow +\infty$ and $\tau_k(x) \longrightarrow +\infty$.

Finally, suppose that $x \in E$ and $k \longrightarrow \infty$ taking on values for which at least one of $f_{m_k}(x)$ or $\tau_k(x)$ is equal to $-\infty$. Then from conditions a, b and c and (4.16) we have $\tau_k(x) = -\infty$, $f_{m_k}(x) < -2^k$ if k is sufficiently large, and hence $f_{m_k}(x) \longrightarrow -\infty$ and $\tau_k(x) \longrightarrow -\infty$.

It follows from Definition 14 that (4.2) and the sequence (4.5) we have defined are equiconvergent for all $x \in E$, and so by (4.15) these sequences satisfy the conditions of Lemma 17, as required.

PROOF OF THEOREM 5. Consider arbitrary $F_1(x)$ and $F_2(x)$ satisfying the conditions of Theorem 1, i.e. suppose that $F_1(x)$ and $F_2(x)$ are measurable, are defined a.e. on some $[a, b]$, and $F_1(x) \leqslant F_2(x)$ (see (1.7)) a.e. on $[a, b]$. Suppose, further, that the sequence (2.3) (or, what is the same, (1.4)) of measurable functions $f_m(x)$ defined a.e. on $[a, b]$ is a sequence of type $A(F_1, F_2)$ (see Definition 3 in §1, and the remark thereafter). We shall show that this is a sequence of type $C(F_1, F_2)$ (see Definition 10).

Consider any four measurable functions $\varphi_1(x)$, $\varphi_2(x)$, $\psi_1(x)$ and $\psi_2(x)$, defined a.e. on $[a, b]$ and satisfying (1.12) a.e. on $[a, b]$. Suppose that $M = \{\varphi(x, E)\}$ is any set of functions possessing property $h(\psi_1, \psi_2)$, and each of the $\varphi(x, E) \in M$ is defined a.e. on the corresponding set E satisfying meas $E > 0$ and $E \subset [a, b]$.

Since the four functions mentioned above satisfy (1.12) and hence also (3.15) a.e. on $[a, b]$, and since M has property $h(\psi_1, \psi_2)$, where the sets E corresponding to all possible $\varphi(x, E) \in M$ satisfy (3.7), by Lemma 15 we can find a sequence (4.2) of measurable functions $\tau_k(x)$ satisfying conditions 1)–4) of Lemma 15, in which in place of $G_m(x)$ we take the $\tau_k(x)$. Hence the $\tau_k(x)$ satisfy the following conditions:

1°. $\psi_2(x)$ and $\psi_1(x)$ are the upper and lower limits in measure on $[a, b]$ of (4.2).

2°.

$$\underline{\lim_{k\to\infty}}\, \tau_k(x) = \varphi_1(x), \quad \overline{\lim_{k\to\infty}}\, \tau_k(x) = \varphi_2(x) \tag{4.17}$$

a.e. on $[a, b]$.

3°. M is the set of all limit functions of (4.2).

4°.

$$\varphi_1(x) \leqslant \tau_k(x) \leqslant \varphi_2(x) \quad k = 0,1,2,\ldots) \tag{4.18}$$

a.e. on $[a, b]$.

Combining (1.12) with (4.18), we see that

$$F_1(x) \leqslant \tau_k(x) \leqslant F_2(x) \quad k = 0,1,2,\ldots) \tag{4.19}$$

a.e. on $[a, b]$, and since (2.3) is of type $A(F_1, F_2)$, by Lemma 17 there is an increasing sequence (4.4) of natural numbers m_k such that the sequences (4.2) and (4.5) of functions $\tau_k(x)$ and $f_{m_k}(x)$ are equiconvergent a.e. on $[a, b]$. In this case, by condition

1° and Lemma 4 the $f_{m_k}(x)$ of (4.5) satisfy condition a° in Definition 10. Further, from condition 2° and Lemma 3 it follows that the functions of the sequence (4.5) satisfy condition b° in Definition 10. Finally, by condition 3° and Lemma 13 the functions (4.5) satisfy condition c° in Definition 10.

Thus we have taken arbitrary measurable $\varphi_1(x)$, $\varphi_2(x)$, $\psi_1(x)$ and $\psi_2(x)$ defined a.e. on $[a, b]$ and satisfying (1.12) a.e. on $[a, b]$, and proved that one can find an increasing sequence m_k such that the sequence $f_{m_k}(x)$ satisfies conditions a°, b°, and c° in Definition 10. Hence the sequence (2.3) of functions $f_m(x)$ is of type $C(F_1, F_2)$.

Since, by assumption, a sequence (2.3) of measurable functions $f_m(x)$ defined a.e. on $[a, b]$ is an arbitrary sequence of type $A(F_1, F_2)$, Theorem 5 is proved.

Thus, as noted at the end of §1, Theorems 1, 2, 3 and 4 are also proved.

§5. Proof of Theorem 6

First note that the following assertion is valid.

If the linear method T defined by the matrix $\|a_{mk}\|$ $(m, k = 0, 1, \dots)$ is normal, then for any sequence of finite numbers T_m $(m = 0, 1, \dots)$ one can find a sequence of finite numbers S_m such that

$$T_m = \sum_{k=0}^{m} a_{mk} S_k \quad (m = 0, 1, 2, \dots). \tag{5.1}$$

Here

$$S_m = \sum_{k=0}^{m} b_{mk} T_k \quad (m = 0, 1, 2, \dots), \tag{5.2}$$

where $\|b_{mk}\|$ is the inverse of $\|a_{mk}\|$ (and hence $\|b_{mk}\|$ depends only on $\|a_{mk}\|$ and is defined uniquely therefrom).

Further, note that S_m and b_{mk} are real if T_m and a_{mk} are real.

The preceding assertion follows from the fact that $a_{mm} \neq 0$ $(m = 0, 1, \dots)$ in the case of a normal T, and hence from the infinite system of linear equations (5.1) with unknowns S_k one can sequentially find these unknowns.

Now for the proof of Theorem 6. Let the sequence (1.4) (or, what is the same, the sequence (2.3)) of measurable functions $f_m(x)$ finite a.e. on $[a, b]$ be universal (Definition 2), and let T defined by $\|a_{mk}\|$ be any pure, normal linear method (see Definitions 12 and 13).

Consider any measurable $f(x)$ defined a.e. on $[a, b]$. Then for any $m = 0, 1, \dots$, setting

$$\begin{aligned} g_m(x) &= f(x), && \text{if} \quad -\infty < f(x) < +\infty, \\ g_m(x) &= m, && \text{if} \quad f(x) = +\infty, \\ g_m(x) &= -m. && \text{if} \quad f(x) = -\infty, \end{aligned}$$

we see that

$$\lim_{m\to\infty} g_m(x) = f(x) \tag{5.3}$$

a.e. on $[a, b]$, where the $g_m(x)$ are measurable and finite a.e. on $[a, b]$.

Since T is normal, by the assertion at the beginning of this section we can find a sequence

$$\lambda_k(x) \quad (k = 0, 1, 2, \ldots), \tag{5.4}$$

for which

$$g_m(x) = \sum_{k=0}^{m} a_{mk} \lambda_k(x) \quad (m = 0, 1, 2, \ldots) \tag{5.5}$$

a.e. on $[a, b]$ and

$$\lambda_m(x) = \sum_{k=0}^{m} b_{mk} g_k(x) \quad (m = 0, 1, 2, \ldots) \tag{5.6}$$

a.e. on $[a, b]$, where the b_{mk} depend only on $\|a_{mk}\|$ and hence are constant.

Since the $g_m(x)$ are measurable and finite a.e. on $[a, b]$, by (5.6) the functions $\lambda_k(x)$ have the same property. Further, since by the hypothesis of Theorem 6 the sequence (1.4) is universal, i.e. is a sequence of type $A(F_1, F_2)$ and $F_1(x) = -\infty$ and $F_2(x) = +\infty$ (see the remark after Definition 3), it follows from Lemma 17, in which we take $\tau_k(x) = \lambda_k(x)$ $(k = 0, 1, \ldots)$, $F_1(x) = -\infty$ and $F_2(x) = +\infty$, that we can find an increasing sequence m_k such that the sequence (5.4) of functions $\lambda_k(x)$ and the sequence (4.5) of functions $f_{m_k}(x)$ will be equiconvergent a.e. on $[a, b]$.

By the hypothesis of Theorem 6, the functions $f_m(x)$ of the sequence (1.4) are finite a.e. on $[a, b]$. Hence the $f_{m_k}(x)$ of the sequence (4.5) have the same property. Thus, since the functions of (5.4) are also finite a.e. on $[a, b]$, from the definition of equiconvergent sequence (see Definition 14) we have

$$\lim_{k\to\infty} [\lambda_k(x) - f_{m_k}(x)] = 0 \tag{5.7}$$

a.e. on $[a, b]$.

Since by the hypothesis of Theorem 6 the method T defined by $\|a_{mk}\|$ is pure, from Definition 12 and (5.7) we have

$$\lim_{m\to\infty} \sum_{k=0}^{m} a_{mk} \cdot [\lambda_k(x) - f_{m_k}(x)] = 0$$

a.e. on $[a, b]$, and hence, by (5.5),

$$\lim_{m\to\infty} \left[g_m(x) - \sum_{k=0}^{m} a_{mk} \cdot f_{mk}(x) \right] = 0 \tag{5.8}$$

a.e. on $[a, b]$.

From (5.8) it follows by (5.3) that

$$\lim_{m\to\infty} \sum_{k=0}^{m} a_{mk} f_{m_k}(x) = f(x)$$

a.e. on $[a, b]$, and this means that the sequence (4.5) of functions $f_{m_k}(x)$ is summable by T to $f(x)$ a.e. on $[a, b]$.

Thus we have taken an arbitrary measurable function $f(x)$ defined a.e. on $[a, b]$ and defined an increasing sequence m_k such that the sequence (4.5) sums by T a.e. on $[a, b]$ to $f(x)$. It follows from Definition 11 that the sequence (1.4) is universal relative to T.

Since T is an arbitrary pure normal linear method and since (1.4) is an arbitrary sequence of measurable functions finite a.e. on $[a, b]$, Theorem 6 is proved.

BIBLIOGRAPHY

1. D. E. Men'šov, *On limits of indeterminacy of partial sums of universal trigonometric series,* Moskov. Gos. Univ. Učen. Zap. **165**, Mat. 7 (1954), 3–33; English transl., Amer. Math. Soc. Transl. (2) **111** (1978), 1–33. MR **16**, 467.

2. ———, *On limit functions of trigonometric series,* Trudy Moskov. Mat. Obšč. **7** (1958), 291–334; English transl., Amer. Math. Soc. Transl. (2) **111** (1978), 35–79. MR **21** #2152.

3. ———, *On convergence in measure of trigonometric series,* Trudy Mat. Inst. Steklov. **32** (1950); English transl., Amer. Math. Soc. Transl. (1) **3** (1962), 196–270. MR **12**, 254.

4. G. H. Hardy, *Divergent series,* Clarendon Press, Oxford, 1949. MR **11**, 25.

Translated by J. ELLIS

Amer. Math. Soc. Transl.
(2) Vol. 111, 1978

ON THE QUOTIENT NORM OF SINGULAR INTEGRAL OPERATORS*

I. C. GOHBERG AND N. Ja. KRUPNIK

§1. Singular integral operator

Let Γ be a set of simple nonintersecting closed contours of Ljapunov type, and let S_Γ be the singular integral operator along Γ:

$$(S_\Gamma\varphi)(t) = \frac{1}{\pi i}\int_\Gamma \frac{\varphi(\tau)}{\tau - t}\,d\tau \qquad (t \in \Gamma);$$

the contour Γ is considered oriented in some manner.

As is known, the operator S_Γ is bounded in each of the spaces $L_p(\Gamma)$ $(1 < p < \infty)$. Let $\|S_\Gamma\|_p$ be the norm of S_Γ in $L_p(\Gamma)$ and let $|S_\Gamma|_p$ be its quotient norm:

$$|S_\Gamma|_p = \inf_{T \in \mathfrak{P}(L_p(\Gamma))} \|S_\Gamma + T\|_p,$$

where $\mathfrak{P}(L_p(\Gamma))$ is the two-sided ideal of all completely continuous linear operators in the space $L_p(\Gamma)$. It is easy to show that for a fixed p $(1 < p < \infty)$, the norm $\|S_\Gamma\|$ depends on the contour Γ.

THEOREM 1. *The equality*

$$|S_\Gamma|_p = |S_C|_p, \tag{1}$$

holds, where $C = \{\zeta\colon |\zeta| = 1\}$ *is the unit circle and* p *is an arbitrary number from the interval* $(1 < p < \infty)$.

In particular, if $p = 2^n$ $(p = 2^n(2^n - 1)^{-1})$, *where* $n = 1, 2, \ldots,$ *then*

$$|S_\Gamma|_p = \cot(\pi/2p) \qquad (|S_\Gamma|_p = \tan(\pi/2p)). \tag{2}$$

PROOF. First we consider the case when Γ consists of one closed curve. Let $t = \beta(z)$ be the function which conformally maps the circle $|z| < 1$ onto the set bounded by Γ. Since Γ is a contour of Ljapunov type, the derivative $\beta'(z)$ satisfies a Hölder condition on C. We denote by B the bounded linear operator from $L_p(\Gamma)$ into $L_p(C)$ defined by $(B\varphi)(z) = |\beta'(z)|^{1/p}\varphi(\beta(z))$.

It is easy to see that $BS_\Gamma - S_CB = T_1 + T_2$, where the operators T_1 and T_2 are defined by

$$(T_1\varphi)(z) = |\beta'(z)|^{1/p}\int_C \left(\frac{\beta'(\zeta)}{\beta(\zeta) - \beta(z)} - \frac{i}{\zeta - z}\right)\varphi(\beta(\zeta))\,d\zeta,$$

$$(T_2\varphi)(z) = \int_C \frac{|\beta'(z)|^{1/p} - |\beta'(\zeta)|^{1/p}}{\zeta - z}\,\varphi(\beta(\zeta))\,d\zeta$$

and act from $L_p(\Gamma)$ into $L_p(C)$.

AMS (MOS) subject classifications (1970). Primary 47G05.

*Translation of Mat. Issled. 4 (1969), no. 3 (13), 136–139. MR 41 #4306.

Since the function $\beta'(z)$ satisfies a Hölder condition on C, it is easy to show that the kernel

$$\frac{\beta'(\zeta)}{\beta(z)-\beta(\zeta)}-\frac{1}{z-\zeta}$$

of the operator T_1 has a weak singularity, and consequently T_1 is a completely continuous operator acting from $L_p(\Gamma)$ into $L_p(C)$. The operator T_2 is also completely continuous.

Indeed, T_2 can be represented in the form

$$T_2=(\alpha(z)S_C-S_C\alpha(z))Y,$$

where $(Y\varphi)(\zeta)=\varphi(\beta(\zeta))$ and $\alpha(z)=|\beta'(z)|^{1/p}$. The function $\alpha(z)$ ($|z|=1$) is continuous; consequently, by virtue of a well-known proposition in the theory of singular integral equations, the operator $\alpha(z)S_C-S_C\alpha(z)$ is completely continuous in $L_p(C)$.

The operator B maps $L_p(\Gamma)$ isometrically onto $L_p(C)$, and, by virtue of what we have proved, $S_\Gamma=B^{-1}S_CB+T$, where $T\in\mathfrak{P}(L_p(\Gamma))$. Hence it follows that

$$|S_\Gamma|_p=|S_C|_p. \tag{3}$$

We proceed to the proof of the theorem in the general case. We assume that Γ consists of a finite number of closed contours $\Gamma_1,\dots,\Gamma_m$ and let P_k ($k=1,\dots,m$) be the projections defined in $L_p(\Gamma)$ by the equality

$$(P_k\varphi)(t)=\begin{cases}\varphi(t), & \text{if} \quad t\in\Gamma_k\\ 0, & \text{if} \quad t\in\Gamma\setminus\Gamma_k.\end{cases}$$

For the operator $R=S_\Gamma+T$, where T is an arbitrary operator from $\mathfrak{P}(L_p(\Gamma))$, we have

$$\|R\|_{L_p(\Gamma)}\geqslant\|P_1RP_1\|_{L_p(\Gamma)}\geqslant|S_{\Gamma_1}|_p.$$

By virtue of what we have proved, we have $|S_{\Gamma_1}|_p=|S_C|_p$. Thus, $|S_\Gamma|_p\geqslant|S_C|_p$.

Let us prove the reverse inequality. From (3) it follows that for each $k=1,\dots,m$ and $\epsilon>0$ there exists an operator $T_k\in\mathfrak{P}(L_p(\Gamma_k))$ such that

$$|S_C|_p+\epsilon>\|S_{\Gamma_k}+T_k\|_{L_p(\Gamma_k)}.$$

Denoting by T the operator from $\mathfrak{P}(L_p(\Gamma))$ defined by $T=\Sigma_1^m P_kT_kP_k$, we have

$$|S_C|_p+\epsilon>\max\|S_{\Gamma_k}+T_k\|_{L_p(\Gamma_k)}=\left\|\sum_{k=1}^m P_k(S_\Gamma+T)P_k\right\|_{L_p(\Gamma)}. \tag{4}$$

Since the operators $P_kS_\Gamma P_n$ ($k\neq n$) are completely continuous in $L_p(\Gamma)$, we have

$$\left\|\sum_{k=1}^m P_k(S_\Gamma+T)P_k\right\|_{L_p(\Gamma)}=\|S_\Gamma+\widetilde{T}\|_{L_p(\Gamma)},$$

where $\widetilde{T}$ is some operator from $\mathfrak{P}(L_p(\Gamma))$. From this and (4) it follows that $|S_C|_p\geqslant|S_\Gamma|_p$.

Relation (1) is proved, while (2) follows from (1) and from the results of the note [1] (see also [2]).

§2. The general singular integral operator

Let A be the singular integral operator defined in $L_p(\Gamma)$ by

$$(A\varphi)(t) = a(t)\varphi(t) + b(t)(S_\Gamma\varphi)(t), \tag{5}$$

where $a(t)$ and $b(t)$ are bounded measurable functions.

We denote by $\beta_n(z)$ $(n = 1, \ldots, m)$ the function which maps the circle $|z| < 1$ into the domain bounded by Γ_n, and by A_n the operator acting in $L_p(C)$ and defined by

$$(A_n\psi)(z) = a(\beta_n(z))\psi(z) + b(\beta_n(z))(S_C\psi)(z).$$

THEOREM 2. *For every operator A of the form* (5),

$$|A|_{L_p(\Gamma)} = \max_n |A_n|_{L_p(C)} \qquad (1 < p < \infty).$$

PROOF. If $m = 1$, the assertion of the theorem follows from the easily verifiable equality $A = B^{-1}AB + T$, where B is the isometric operator defined in the proof of Theorem 1, while T is a completely continuous operator. In the general case the proof of Theorem 2 is similar to the proof of Theorem 1.

We note that if $a(t)$ and $b(t)$ are constants, then $|A|_{L_p(\Gamma)} = |A|_{L_p(C)}$. For example, for the operators $P_\Gamma = ½(I + S_\Gamma)$ and $Q_\Gamma = ½(I - S_\Gamma)$ we have the equalities $|P_\Gamma|_p = |P_C|_p$ and $|Q_\Gamma|_p = |Q_C|_p$.

In conclusion, the authors wish to thank A. S. Markus for his valuable remarks.

BIBLIOGRAPHY

1. I. C. Gohberg and N. Ja. Krupnik, *On the norm of the Hilbert transform in the space* L_p, Funkcional. Anal. i Priložen. **2** (1968), no. 2, 91–92 = Functional Anal. Appl. **2** (1968), 180–181.

2. ———, *The spectrum of singular integral operators in* L_p *spaces*, Studia Math. **31** (1968), 347–362. (Russian) MR **38** #5068.

Translated by E. DEUTSCH

Amer. Math. Soc. Transl.
(2) Vol. 111, 1978

ON COMPOSITE LINEAR SINGULAR INTEGRAL OPERATORS*

I. C. GOHBERG AND N. Ja. KRUPNIK

Usually, by a one-dimensional linear singular integral operator we mean an operator A of the form

$$(A\varphi)(t) = c(t)\varphi(t) + \frac{d(t)}{\pi i}\int_\Gamma \frac{\varphi(\tau)}{\tau - t}d\tau, \tag{0.1}$$

where $c(t)$ and $d(t)$ are given functions, called the coefficients of the operator, and Γ is a sufficiently smooth oriented (closed or open) contour.

Methods of inversion and qualitative investigations have been worked out for operators of this form and also for matrices consisting of such operators in the case of various constraints on the coefficients and on the space.

In the present paper we show that the investigation and the inversion of more complicated singular integral operators of the form

$$A = \sum_{j=1}^{k} \prod_{l=1}^{m} A_{jl}, \tag{0.2}$$

where A_{jl} are operators of the form (0.1), reduce to the same problems for matrix operators of the form (0.1).

In the first section we present the abstract foundation of this method. In the second section some applications to composite singular integral operators are given. In the third section we determine conditions of applicability of a special approximation method for solving composite singular equations. The proofs are based on the results of the first section.

§1. The linear extension of the elements of an algebra

Let $\mathfrak{A}$ be an algebra with identity e. By $\mathfrak{A}_{r\times r}$ we denote the algebra of square matrices of order r with elements from $\mathfrak{A}$. We denote the identity of the algebra $\mathfrak{A}_{r\times r}$ $(r > 1)$ by E. Let

$$x = \sum_{j=1}^{k} \prod_{l=1}^{m} x_{jl},$$

where $x_{jl} \in \mathfrak{A}$; the matrix X $(\in \mathfrak{A}_{r\times r})$ is called a *linear extension* (*of order* r) of the element x, if X has the following properties: 1) the elements of X are linear combinations of the elements x_{jl} and of the identity e; 2) there exist triangular matrices $Y, Z \in \mathfrak{A}_{r\times r}$ with identities on the diagonal such that

---

AMS (MOS) subject classifications (1970). Primary 47G05; Secondary 47A50, 45E05, 45L05.

*Translation of Mat. Issled. 4 (1969), no. 4 (14), 20–32. MR 42 #996.

$$X = Y \begin{pmatrix} E & 0 \\ 0 & x \end{pmatrix} Z. \tag{1.1}$$

Obviously the matrices Y and Z are invertible in $\mathfrak{A}_{r \times r}$, and their inverses are constructed by means of the well-known simple rule.

The element x and its linear extension X have a series of common properties. For example, they simultaneously are or are not left (right) invertible in their algebras. Moreover, if x^{-1} is a one-sided inverse of the element x, then the matrix

$$Z^{-1} \begin{pmatrix} E & U \\ 0 & x^{-1} \end{pmatrix} Y^{-1}$$

is the inverse of X on the same side. If X^{-1} is a one-sided inverse of X, then the element $[ZX^{-1}Y]_{rr}$ of the matrix $ZX^{-1}Y$ (situated in the lower right corner) is the inverse of the element x on the same side.

If $\mathfrak{A}\,(\mathfrak{A}_{r \times r})$ is the algebra of bounded linear operators of some Banach space $\mathfrak{B}\ (\mathfrak{B}^r)$, then, simultaneously, the operator $x \in \mathfrak{A}$ and its linear extension $X \in \mathfrak{A}_{r \times r}$ are or are not normally solvable operators, and moreover $\dim \ker x = \dim \ker X$ and $\dim \operatorname{coker} x = \dim \operatorname{coker} X$.

THEOREM 1.1. *Every element $x \in \mathfrak{A}$ of the form*

$$x = \sum_{j=1}^{k} \prod_{l=1}^{m} x_{jl} \qquad (x_{jl} \in \mathfrak{A})$$

admits a linear extension (of order $r \leqslant mk - k + 1$).

PROOF. We form the square matrix of order $k(m-1)$ $(m > 2)$

$$M = \left\| \begin{array}{cccccccccc} e & & -x_{12} & & & & & & & \\ & e & & \ddots & & & & & & \\ & & \ddots & & -x_{k2} & & & & & \\ & & & \ddots & & -x_{13} & & & & \\ & & & & e & & \ddots & & & \\ & & & & & \ddots & & -x_{k3} & & \\ & & & & & & \ddots & & \ddots & \\ & & & & & & & \ddots & & -x_{1,m-1} \\ & & & & & & & & \ddots & \\ & & & & & & & & & -x_{k,m-1} \\ & & & & & & & & e & \\ & & & & & & & & & e \end{array} \right\| \begin{array}{l} \\ \\ \\ \\ \\ \\ \\ \\ \\ \left.\begin{array}{l} \\ \\ \\ \end{array}\right\} k+1 \\ \end{array}$$

in which all elements except those on the two indicated diagonals are equal to zero.

If $m = 2$, then for M we choose the identity matrix of order k. By F, G, and H we denote respectively a matrix of one column and two matrices of one row, given by

$$F = \left\| \begin{matrix} 0 \\ \vdots \\ 0 \\ -x_{1m} \\ \vdots \\ -x_{km} \end{matrix} \right\| \left.\vphantom{\begin{matrix} 0 \\ \vdots \\ 0 \end{matrix}}\right\} k(m-2), \qquad G = \|x_{11}, \dots, x_{n1}, \underbrace{0, \dots, 0}_{k(m-2)}\|,$$

$$H = \|x_{11}, \dots, x_{k1}, x_{11}x_{12}, \dots, x_{k1}x_{k2}, \dots, x_{11} \cdots x_{1,m-1}, \dots, x_{k1} \cdots x_{k,m-1}\|.$$

It is easy to verify that

$$\begin{pmatrix} M & F \\ G & 0 \end{pmatrix} = \begin{pmatrix} E & 0 \\ H & e \end{pmatrix} \begin{pmatrix} E & 0 \\ 0 & x \end{pmatrix} \begin{pmatrix} M & F \\ 0 & e \end{pmatrix} \tag{1.2}$$

From this it follows that the matrix

$$X = \begin{pmatrix} M & F \\ G & 0 \end{pmatrix} \tag{1.3}$$

is a linear extension (of order $mk - k + 1$) of the element x. The theorem is proved.

REMARK. If $x_0 \in A$, then from the equality

$$\begin{pmatrix} M & F \\ G & x_0 \end{pmatrix} = \begin{pmatrix} E & 0 \\ H & e \end{pmatrix} \begin{pmatrix} E & 0 \\ 0 & x + x_0 \end{pmatrix} \begin{pmatrix} M & F \\ 0 & e \end{pmatrix} \tag{1.4}$$

it follows that the matrix

$$\begin{pmatrix} M & F \\ G & x_0 \end{pmatrix}$$

is a linear extension of the element $x + x_0$.

In the sequel, the linear extension X of the element

$$\sum_{j=1}^{k} \prod_{l=1}^{m} x_{jl},$$

defined by (1.3), will be denoted by $\Xi(x_{jl})$. By $\Xi'(x_{jl})$ we denote the transpose of $\Xi(x_{jl})$, and by $\Xi(x_{jl}, x_0)$ we denote the linear extension of the element $x + x_0$, defined by (1.4).

Let F', M' and G' be the transposes of the matrices F, M and G, and let H' be the matrix obtained from H by taking its transpose and by replacing the elements $x_{\alpha 1}x_{\alpha 2} \cdot \ldots \cdot x_{\alpha\beta}$ by the elements $x_{\alpha\beta} \cdot \ldots \cdot x_{\alpha 2}x_{\alpha 1}$ respectively. Then it is easy to verify that

$$\Xi'(x_{jl}) = \begin{pmatrix} M' & 0 \\ F' & e \end{pmatrix} \begin{pmatrix} E & 0 \\ 0 & x' \end{pmatrix} \begin{pmatrix} E & H' \\ 0 & e \end{pmatrix}, \tag{1.5}$$

where

$$x' = \sum_{j=1}^{k} \prod_{l=1}^{m} x_{j,m-l+1}.$$

It follows from (1.5) that the matrix $\Xi'(x_{jl})$ is a linear extension of the element x'.

§2. Applications to singular integral equations

Let Γ be a contour consisting of a finite number of closed curves of Ljapunov type, bounding a connected set which contains the point $z = 0$; let S be the singular integral operator along Γ

$$(S\varphi)(t) = \frac{1}{\pi i} \int_\Gamma \frac{\varphi(\tau)}{\tau - t} d\tau \qquad (t \in \Gamma),$$

acting in the space $L_p(\Gamma)$ $(1 < p < \infty)$, and let $P = (I + S)/2$ and $Q = (I - S)/2$. The operator $\|\delta_{jk}P\|_{j,k=1}^r$ $(\|\delta_{jk}Q\|_{j,k=1}^r)$, acting in the space $L_p^r(\Gamma)$ of vector-valued functions $f = \{f_j\}_1^r$ with components in $L_p(\Gamma)$, will be also denoted by the letter P (Q).

We consider in $L_p(\Gamma)$ $(1 < p < \infty)$ the operator

$$A = \sum_{j=1}^{k} \prod_{l=1}^{m} (a_{jl}P + b_{jl}Q), \tag{2.1}$$

where a_{jl} and b_{jl} are bounded measurable functions on Γ.

The linear extension $\Xi(a_{jl}P + b_{jl}Q)$ of the operator A represents a singular integral operator with matrix coefficients

$$\Xi(a_{jl}P + b_{jl}Q) = \Xi(a_{jl})P + \Xi(b_{jl})Q,$$

acting in $L_p^r(\Gamma)$. This allows us to carry over different propositions from the theory of singular integral equations with matrix coefficients to operators of the form (2.1). We mention some of them.

1°. *Let $a_{jl}(t)$ and $b_{jl}(t)$ be bounded measurable functions on Γ. If the operator A, defined by* (2.1), *is a Φ-operator* in $L_p(\Gamma)$, then*

$$\operatorname{ess\,inf} \left| \sum_{j=1}^{k} \prod_{l=1}^{m} a_{jl}(t) \right| \neq 0 \quad \textit{and} \quad \operatorname{ess\,inf} \left| \sum_{j=1}^{k} \prod_{l=1}^{m} b_{jl}(t) \right| \neq 0. \tag{2.2}$$

PROOF. If A is a Φ-operator in $L_p(\Gamma)$, then $\Xi(a_{jl})P + \Xi(b_{jl})Q$ is a Φ-operator in $L_p^r(\Gamma)$. We denote by γ the set of the points of Γ at which $|\det \Xi(a_{jl})| > \delta/2$, and by $a(t)$ we denote the function

$$a(t) = \begin{cases} 0, & \text{if } t \in \gamma, \\ \delta, & \text{if } t \in \Gamma \backslash \gamma, \end{cases}$$

where the number $\delta > 0$ is selected in such a way that $\Xi(a_{jl}, a)P + \Xi(b_{jl})Q$ is a

**Editor's note.* This is an abbreviation for "Fredholm operator".

Φ-operator in $L_p^r(\Gamma)$. Since $\det \Xi(a_{jl}, a) = \det \Xi(a_{jl}) + a$, we have $|\det \Xi(a_{jl}, a)| \geqslant \delta/2$. From this it follows that $P + |\Xi(a_{jl}, a)|^{-1}\Xi(b_{jl})Q$ is a Φ-operator in $L_p^r(\Gamma)$, and therefore (see [1], Lemma 2)

$$\operatorname{ess\,inf} |\det \Xi(b_{jl}(t))| \neq 0.$$

The last condition means that

$$\operatorname{ess\,inf} \left| \sum_{j=1}^{k} \prod_{l=1}^{m} b_{jl}(t) \right| \neq 0.$$

The second of the relations (2.2) is proved in a similar way. The proposition is proved.

If the functions a_{jl} and b_{jl} are continuous on Γ, then, as is known (see [2]), the operator A defined by (2.1) differs from the operator $B = a(t)P + b(t)Q$, where

$$a(t) = \sum_{j=1}^{k} \prod_{l=1}^{m} a_{jl}(t) \quad \text{and} \quad b(t) = \sum_{j=1}^{k} \prod_{l=1}^{m} b_{jl}(t),$$

by a completely continuous term. From this it follows that A is a Φ $(\Phi_\pm)$-operator in $L_p(\Gamma)$ if and only if

$$\sum_{j=1}^{k} \prod_{l=1}^{m} a_{jl}(t) \neq 0, \qquad \sum_{j=1}^{k} \prod_{l=1}^{m} b_{jl}(t) \neq 0. \tag{2.3}$$

In addition to this we have the following proposition.

2°. *Assume that $a_{jl}(t)$ and $b_{jl}(t)$ are continuous on Γ and that conditions* (2.3) *hold. Then*

$$\dim \ker A = -\sum_{\kappa_j<0} \kappa_j \quad \textit{and} \quad \dim \operatorname{coker} A = \sum_{\kappa_j>0} \kappa_j,$$

where $\kappa_1, \ldots, \kappa_r$ $(r = mk - k + 1)$ are the right partial indices of the matrix-valued function $[\Xi(b_{j1}(t))]^{-1}\Xi(a_{j1}(t))$.

PROOF. As is known (see, for example, [3] or [4]),

$$\dim \ker \Xi(a_{jl}P + b_{jl}Q) = -\sum \kappa_j \quad (\kappa_j < 0).$$

and

$$\dim \operatorname{coker} \Xi(a_{jl}P + b_{jl}Q) = \sum \kappa_j \quad (\kappa_j > 0).$$

Since, in addition,

$$\dim \ker A = \dim \ker \Xi(a_{jl}P + b_{jl}Q)$$

and

$$\dim \operatorname{coker} A = \dim \operatorname{coker} \Xi(a_{jl}P + b_{jl}Q),$$

the proposition is proved.

Similarly we obtain the following proposition from the results of the previous section and of [5].

3°. *Let $a_{jl}(t)$ and $b_{jl}(t)$ be piecewise-continuous and left-continuous functions on the contour Γ. In order that the operator A defined in $L_p(\Gamma)$ $(1 < p < \infty)$ by* (2.1) *be a Φ-operator in $L_p(\Gamma)$, it is necessary and sufficient that the following two conditions hold*:

$$\sum_{j=1}^{k} \prod_{l=1}^{m} b_{jl}(t \pm 0) \neq 0 \qquad (t \in \Gamma) \tag{2.4}$$

and

$$\det F_p(t, \mu) \neq 0 \qquad (t \in \Gamma, 0 \leqslant \mu \leqslant 1), \tag{2.5}$$

where

$$F_p(t, \mu) = \frac{e^{i\mu\theta}}{\sin \theta} [\sin(1 - \mu)\theta F(t) + e^{-i\theta} \sin \mu\theta F(t + 0)]$$

$$(F(t) = [\Xi(b_{jl}(t))]^{-1}\Xi(a_{jl}(t)), \qquad \theta = \pi - 2\pi/p)$$

if $p \neq 2$, and $F_2(t, \mu) = (1 - \mu)F(t) + \mu F(t + 0)$.

If (2.4) *holds, then the index of A is obtained from the formula* $\operatorname{ind} A = -\operatorname{ind} \det F_p(t, \mu)$.

We note that if (2.4) holds, then $P + F^{-1}(t)Q$ is a Φ-operator in $L_p^r(\Gamma)$. From this it follows that the matrix-valued function $F^{-1}(t)$ admits the factorization defined in [1]. Let κ_j' be the partial indices of $F^{-1}(t)$ corresponding to this factorization. Then

$$\dim \ker A = \sum_{\kappa_j'>0} \kappa_j', \qquad \dim \operatorname{coker} A = -\sum_{\kappa_j'<0} \kappa_j'.$$

We consider one more example. Let $a(t)$, $b(t)$, $c(t)$, and $d(t)$ be bounded measurable functions on Γ, and let $A = PaP + PbQ + QcP + QdQ$. The operator

$$\begin{pmatrix} I & (c - a)P + (b - d)Q \\ P & cP + dQ \end{pmatrix} = \begin{pmatrix} 1 & 1 \\ 0 & 1 \end{pmatrix} \left[\begin{pmatrix} 0 & -a \\ 1 & c \end{pmatrix} P + \begin{pmatrix} 1 & -b \\ 0 & d \end{pmatrix} Q \right] \tag{2.6}$$

is a linear extension of the second order of the operator A. From this one can derive a series of consequences.

For example, if $A = PaP + PbQ + QcP + QdQ$ is a Φ-operator in $L_p(\Gamma)$, then $\operatorname{ess\,inf} |a(t)| \neq 0$, $\operatorname{ess\,inf} |d(t)| \neq 0$,

$$\dim \ker A = \sum \kappa_j' \quad (\kappa_j' > 0) \quad \text{and} \quad \operatorname{dimcoker} A = -\sum \kappa_j' \quad (\kappa_j' < 0),$$

where κ_1' and κ_2' are the partial indices of the function

$$\begin{pmatrix} c/a & (ad - bc)/a \\ -1/a & b/a \end{pmatrix}.$$

We consider more explicitly the special case of this operator when $A = cS - Sc - \lambda I$, where $c(t)$ is a piecewise-continuous function and λ is a complex number. We prove the following proposition.

4°. *Let $c(t)$ be a piecewise-continuous and left-continuous function on the contour Γ; let $t_1, \ldots, t_n$ be its points of discontinuity and let $\alpha_k = c(t_k + 0) - c(t_k)$ $(k = 1, \ldots, n)$. The complement of the set of the Φ-points of the operator $A = cS - Sc$ in $L_2(\Gamma)$ consists of n segments, each of them joining the two points $i\alpha_k$ and $-i\alpha_k$ of the plane; in $L_p(\Gamma)$ $(p \neq 2)$ it consists of $2n$ loops, defined by the equations*

$$\lambda = \pm \frac{2i\alpha_k}{\sin \theta} e^{i(\mu - \theta/2)} \sqrt{\sin \mu\theta \sin(1 - \mu)\theta} \qquad (\theta = \pi - 2\pi/p, 0 \leqslant \mu \leqslant 1).$$

Proof. Let $A_\lambda = cS - Sc - \lambda I$; for this operator the condition (2.4) means $\lambda \neq 0$. We consider the linear extension of A_λ $(\lambda \neq 0)$ defined by (2.6):

$$\begin{pmatrix} I & c(t)I \\ S & c(t)S - \lambda I \end{pmatrix} = \begin{pmatrix} 1 & c(t) \\ -1 & -\lambda - c(t) \end{pmatrix} \left[\begin{pmatrix} 2c/\lambda + 1 & 2c^2/\lambda \\ -2/\lambda & 1 - 2c/\lambda \end{pmatrix} P + Q \right].$$

The conditions (2.4) and (2.5) mean that

$$\lambda \neq \pm 2i(c(t + 0) - c(t))\sqrt{\mu(1 - \mu)}$$

if $p = 2$, and

$$\lambda \neq 2i(c(t + 0) - c(t))e^{i(\mu - 1/2)\theta}\sqrt{\sin \mu\theta \sin(1 - \mu)\theta}/\sin \theta$$

if $p \neq 2$ $(1 < p < \infty)$. The proposition is proved.

3. Approximation solution of composite equations

In this section we justify the application of generalized projection methods for the inversion of operators of the form

$$A = \sum_{j=1}^{k} \prod_{l=1}^{m} A_{jl}, \tag{3.1}$$

where A_{jl} are Wiener-Hopf operators or singular integral operators.

1. Let $\mathfrak{B}$ be a Banach space and let $R = R(\mathfrak{B})$ be the algebra of bounded linear operators acting in $\mathfrak{B}$. By Λ we denote some unbounded set of positive numbers and by $\{P_\tau\}_{\tau \in \Lambda}$ the set of the projections from R. We will assume that $\|P_\tau x - x\| \longrightarrow 0$ as $\tau \longrightarrow \infty$ for any vector $x \in \mathfrak{B}$.

Following [6], we will say that the projection method (P_τ) is applicable to the operator A if, starting with some τ_0, for every $y \in \mathfrak{B}$ the equation $P_\tau A P_\tau x = P_\tau y$

has in the subspace $P_\tau \mathfrak{B}$ a unique solution x_τ which tends as $\tau \longrightarrow \infty$ to the solution of the equation $Ax = y$.

Let A be a bounded linear operator represented in the form (3.1), where $A_{jl} \in R(\mathfrak{B})$. For the approximate inversion of this operator we will consider the method which consists in the determination of the solution $x \in P_\tau \mathfrak{B}$ of the equation

$$A_\tau x = P_\tau y, \tag{3.2}$$

where

$$A_\tau = \sum_{j=1}^{k} \prod_{l=1}^{m} P_\tau A_{jl} P_\tau.$$

If, starting with some τ_0, for every $y \in \mathfrak{B}$ there exists in the subspace $p_\tau \mathfrak{B}$ a unique solution x_τ of (3.2) and for $\tau \longrightarrow \infty$ the vectors x_τ tend to the solution of the equation $Ax = y$, then we will say that the *generalized projection method* (P_τ, A_{jl}) is applicable to the operator (3.1).

Let $\mathfrak{B}^r$ be the space of vectors $(x_1, \ldots, x_r)$ with components $x_j \in \mathfrak{B}$ $(j = 1, \ldots, r; r = mk - k + 1)$, and let P_τ be the operator defined in $\mathfrak{B}^r$ by the matrix $P_\tau = \|\delta_{jk} P_\tau\|_{j,k=1}^r$. From (1.2) and from Pol'skiĭ's theorem [6] one can easily prove the following proposition.

The generalized projection method (P_τ, A_{jl}) *is applicable to the operator* (3.1) *if and only if the projection method* (P_τ) *is applicable to the operator* $\Xi(A_{jl})$.

This simple proposition is the origin of a series of applications. We mention some of them.

2. To each function $a(\zeta)$ $(|\zeta| = 1)$, continuous on the unit circle, we associate two operators $T(a)$ and $\widetilde{T}(a)$, defined by the matrices $\|a_{j-k}\|_{j,k=1}^{\infty}$ and $\|a_{j-k}\|_{j,k=-\infty}^{+\infty}$, in the corresponding spaces l_2 and $\widetilde{l}_2$,(1) where the a_j are the Fourier coefficients of the function $a(\zeta)$. By $T_n(a)$ we denote the operator defined in finite-dimensional space by the matrix $\|a_{j-k}\|_{j,k=1}^n$.

THEOREM 3.1. *Let* $a_{jl}(\zeta)$ *be functions continuous on the unit circle, and let*

$$A = \sum_{j=1}^{k} \prod_{l=1}^{m} T(a_{jl}).$$

In order that, starting with some n, the system of equations

$$\sum_{l=1}^{n} b_{jl}^{(n)} \xi_l = \eta_j \qquad (j = 1, \ldots, n),$$

where $\|b_{jl}^{(n)}\|_{j,l=1}^n$ *is the matrix of the operator*

$$\sum_{j=1}^{k} \prod_{l=1}^{m} T_n(a_{jl}),$$

should have a unique solution $\{\xi_j^{(n)}\}_{j=1}^n$ *and for* $n \longrightarrow \infty$ *the vectors*

(1) By l_2 we denote the space of sequence $\{\xi_j\}_1^{\infty}$, and by $\widetilde{l}_2$ the space of sequences $\{\xi_j\}_{-\infty}^{+\infty}$.

$\xi^{(n)} = (\xi_1^{(n)}, \ldots, \xi_n^{(n)}, 0, \ldots)$ *should converge in the norm of the space* l_2 *to the solution of the equation* $A\xi = \eta$, *where* $\eta = (\eta_1, \eta_2, \ldots)$, *it is necessary and sufficient that each of the operators*

$$A = \sum_{j=1}^{k} \prod_{l=1}^{m} T(a_{jl}) \quad \text{and} \quad B = \sum_{j=1}^{k} \prod_{l=1}^{m} T(a_{j,m-l+1}) \tag{3.3}$$

be invertible in l_2.

The theorem means that the projection method $(P_n, T(a_{jl}))$, where P_n $(n = 1, 2, \ldots)$ are projections defined by the equalities $P_n\{\xi_j\}_1^\infty = (\xi_1, \ldots, \xi_n, 0, \ldots)$, is applicable to the operator A if and only if the operators A and B are invertible in l_2. Let $\Xi(T(a_{jl}))$ and $\Xi'(T(a_{jl}))$ be linear extensions of the operators A and B, respectively. The theorem we have proved is equivalent to the following proposition: *The projection method* (P_n) *is applicable to the operator* $\Xi(T(a_{jl}))$ *if and only if the operators* $\Xi(T(a_{jl}))$ *and* $\Xi'(T(a_{jl}))$ *are invertible in* l_2^r. Since $\Xi'(T(a_{jl}))$ is the transpose of $\Xi(T(a_{jl}))$, the last proposition follows from Theorem 2.1 in Chapter IV of [7] (1967 ed.).

We denote by $\widetilde{P}_n$ the projection defined in $\widetilde{l}_2$ by the equality $\widetilde{P}_n\{\xi_j\}_{-\infty}^{+\infty} = (\ldots, 0, \xi_{-n}, \ldots, \xi_n, 0, \ldots)$.

THEOREM 3.2. *Let*

$$\widetilde{A} = \sum_{j=1}^{k} \prod_{l=1}^{m} \widetilde{T}(a_{jl}),$$

where the $a_{jl}(\zeta)$ *are continuous functions on the unit circle* $|\zeta| = 1$. *In order that the generalized projection method* $(\widetilde{P}_n, T(a_{jl}))$ *be applicable to the operator* $\widetilde{A}$, *it is necessary and sufficient that the operators* A *and* B, *defined by* (3.3), *be invertible in* l_2.

The proof of this theorem is carried out in the same way as that of the previous one. It is only necessary to replace the reference to [7] by a reference to Theorem 4.4 of Chapter IV of the same book.

3. Let $a(\zeta)$ and $b(\zeta)$ be functions continuous on $|\zeta| = 1$, let a_j and b_j $(j = 0, \pm 1, \ldots)$ be their corresponding Fourier coefficients, and let $T(a, b)$ be the operator defined in l_2 by the matrix $T(a, b) = \|c_{jl}\|_{j,l=-\infty}^{+\infty}$, where

$$c_{jl} = \begin{cases} a_{j-l}, & \text{if } l \geqslant 0, \\ b_{j-l}, & \text{if } l < 0. \end{cases}$$

THEOREM 3.3. *If the operator*

$$A = \sum_{j=1}^{k} \prod_{l=1}^{m} T(a_{jl}, b_{jl})$$

is invertible in $\widetilde{l}_2$, *and the operators*

$$\sum_{j=1}^{k} \prod_{l=1}^{m} T(a_{jl}), \quad \sum_{j=1}^{k} \prod_{l=1}^{m} T(a_{j,m-l+1}), \quad \sum_{j=1}^{k} \prod_{l=1}^{m} T(b_{jl}), \quad \sum_{j=1}^{k} \prod_{l=1}^{m} T(b_{j,m-l+1})$$

are invertible in l_2, *then the generalized projection method* $(\widetilde{P}_n, T(a_{jl}, b_{jl}))$ *is applicable to the operator* A.

PROOF. From the conditions of the theorem it follows that the operator $\Xi(T(a_{jl}, b_{jl}))$ is invertible in the space $\widetilde{l}_2$, and the operators $\Xi(T(a_{jl}))$, $\Xi(T(b_{jl}))$, $\Xi'(T(a_{jl}))$ and $\Xi'(T(b_{jl}))$ are invertible in l_2^r. From this, with the help of Theorem 4.3 from Chapter IV of [7] (1967 ed.), it can be easily established that the projection method $(\widetilde{P}_n)$ is applicable to the operator $\Xi(T(a_{jl}, b_{jl}))$, and consequently the generalized projection method $(\widetilde{P}_n, T(a_{jl}, b_{jl}))$ is applicable to the operator A. The theorem is proved.

The results presented above hold also in the spaces l_p and $\widetilde{l}_p$ $(1 \leqslant p \leqslant \infty)$ respectively, provided we restrict in a natural manner the class of functions $a(\zeta)$ and $b(\zeta)$ which define the operators $T(a)$ and $T(a, b)$ (see [7]). For example, in the case $p = 1$, one considers only the functions which can be expanded in absolutely convergent Fourier series.

We also note that our results can be carried over to other operators and projection methods, investigated in [7].

4. We return to singular integral equations. Let Γ be the unit circle and let S be the singular integral operator along Γ. We consider in $L_p(\Gamma)$ $(1 < p < \infty)$ the bounded linear operator

$$A = \sum_{j=1}^{k} \prod_{l=1}^{m} A_{jl}, \tag{3.4}$$

where $A_{jl} = a_{jl}P + b_{jl}Q$ are singular integral operators with continuous coefficients a_{jl} and b_{jl} $(P = (I + S)/2;\ Q = I - P)$.

By $\hat{P}_n$ we denote the projections acting in $L_p(\Gamma)$, defined by

$$(\hat{P}_n f)(\zeta) = \sum_{j=-n}^{n} f_j \zeta^j \qquad (|\zeta| = 1),$$

where f_j $(j = 0, \pm 1, \dots)$ are the Fourier coefficients of the function $f(\zeta)$.

THEOREM 3.4. *Assume that the operator* A *defined by* (3.4) *is invertible in* $L_p(\Gamma)$. *If each of the operators*

$$\sum_{j=1}^{k} \prod_{l=1}^{m} T(a_{jl}), \quad \sum_{j=1}^{k} \prod_{l=1}^{m} T(a_{j,m-l+1}), \quad \sum_{j=1}^{k} \prod_{l=1}^{m} T(b_{jl}), \quad \sum_{j=1}^{k} \prod_{l=1}^{m} T(b_{j,m-l+1})$$

is invertible in l_2, *then the generalized projection method* $(\hat{P}_n, A_{jl})$ *is applicable to the operator* A.

The proof is similar to the proof of Theorem 3.3.

BIBLIOGRAPHY

1. I. B. Simonenko, *Certain general questions of the theory of the Riemann boundary value problem*, Izv. Akad. Nauk SSSR Ser. Mat. **32** (1968), 1138–1146 = Math. USSR Izv. **2** (1968), 1091–1099. MR **38** #3447.

2. S. G. Mihlin, *Singular integral equations*, Uspehi Mat. Nauk **3** (1948), no. 3 (25), 29–112; English transl., Amer. Math. Soc. Transl. (1) **10** (1962), 84–198. MR **10**, 305.

3. G. F. Mandžavidze and B. V. Hvedelidze, *On the Riemann-Privalov problem with continuous coefficients*, Dokl. Akad. Nauk SSSR **123** (1958), 791–794. (Russian) MR **22** #3806.

4. I. B. Simonenko, *The Riemann boundary value problem for n pairs of functions with continuous coefficients*, Izv. Vysš. Učebn. Zaved. Matematika **1961**, no. 1 (20), 140–145. (Russian) MR **24** #A838.

5. I. C. Gohberg and N. Ja. Krupnik, *Systems of singular integral equations in L_p spaces with weight*, Dokl. Akad. Nauk SSSR **186** (1969), 998–1001 = Soviet Math. Dokl. **10** (1969), 688–691. MR **40** #1818.

6. N. I. Pol'skiĭ, *Projection methods for the solution of linear problems*, Uspehi Mat. Nauk **18** (1963), no. 2 (110), 179–180. (Russian)

7. I. C. Gohberg and I. A. Fel'dman, *Projection methods for solving Wiener-Hopf equations*, Acad. Sci. Moldav. SSR Inst. Math. Computing Center, Kishinev, 1967; rev. ed., *Convolution equations and projection methods for their solution*, "Nauka", Moscow, 1971; English transl., Amer. Math. Soc., Providence, R. I., 1974. MR **37** #1915; **50** #8148; **50** #8150.

Translated by E. DEUTSCH

Amer. Math. Soc. Transl.
(2) Vol. 111, 1978

Dedicated to Acadamician
M. V. Keldyš
On His Sixtieth Birthday

INFINITE HANKEL BLOCK MATRICES AND RELATED EXTENSION PROBLEMS*

V. M. ADAMJAN, D. Z. AROV AND M. G. KREĬN

In this article we continue our investigations [1] of infinite Hankel matrices and their applications. Here we generalize the results obtained earlier to the case when the elements of an infinite Hankel matrix are not numbers but blocks, i.e. operators, acting from one Hilbert space N_1 into another N_2.

All of the Hilbert spaces considered are assumed to be separable. The space of all bounded linear operators acting from one Hilbert space H_1 into another H_2 is denoted by $[H_1, H_2]$.

If N is a fixed Hilbert space, $l_2^+(N)$ denotes the Hilbert space of one-sided sequences $\xi = \{\xi_k\}_1^\infty$ of vectors $\xi_k \in N$ for which

$$\|\xi\|^2 = \sum_{k=1}^{\infty} \|\xi_k\|^2 < \infty.$$

It is natural to consider N as the subspace in $l_2^+(N)$ of vectors ξ whose "coordinates" $\xi_k = 0$ for $k > 1$.

An operator $\Gamma \in [l_2^+(N_1), l_2^+(N_2)]$ is called a *Hankel operator* if it is defined by a Hankel block matrix $(\gamma_{j+k-1})_1^\infty$, where the blocks $\gamma_k \in [N_1, N_2]$ $(k = 1, 2, \ldots)$:

$$\Gamma\{\xi_j\}_1^\infty = \{\eta_j\}_1^\infty, \qquad \eta_j = \sum_{k=1}^{\infty} \gamma_{j+k-1}\xi_k, \; j = 1, 2, \ldots. \tag{0.1}$$

The set of all members of $[l_2^+(N_1), l_2^+(N_2)]$ that are Hankel operators is denoted below by $H^+(N_1, N_2)$.

Let $\mathbf{L}_\infty(N_1, N_2)$ be the space of weakly measurable(1) operator-valued functions on the unit circle $C = \{\zeta: |\zeta| = 1\}$ with values in $[N_1, N_2]$ and with norm

$$\|F\|_\infty = \text{ess sup}\|F(\zeta)\| < \infty.$$

The class of operators $H^+(N_1, N_2)$ plays an important role in the harmonic analysis of the functions in $\mathbf{L}_\infty(N_1, N_2)$.

THEOREM 0.1. *Suppose given a sequence $\{\gamma_k\}_1^\infty$ of operators from $[N_1, N_2]$ and a number $\rho > 0$. In order for there to exist an operator-valued function $F \in \mathbf{L}_\infty(N_1, N_2)$ such that*(2)

AMS (MOS) subject classifications (1970). Primary 15A57, 47B10.

*Translation of Izv. Akad. Nauk Armjan. SSR Ser. Mat. 6 (1971), 87–112. MR 45 #7506.

(1) All of the weakly measurable abstract functions encountered in this paper are also strongly measurable by virtue of the separability of the Hilbert spaces considered.

(2) The integral in the right side of (0.2) is understood in the weak sense.

$$c_{-k}(F) \overset{\text{def}}{=} \frac{1}{2\pi} \int_C \zeta^k F(\zeta)|d\zeta| = \gamma_k \quad (k = 1, 2, \ldots), \tag{0.2}$$

and

$$\|F\|_\infty \leqslant \rho, \tag{0.3}$$

it is necessary and sufficient that the matrix $(\gamma_{j+k-1})_1^\infty$ *define an operator* $\Gamma \in H^+(N_1, N_2)$ *by formula* (0.1), *and that* $\rho \geqslant \|\Gamma\|$.

In the scalar case ($\dim N_1 = \dim N_2 = 1$) this proposition was first established by Nehari [2] as a generalization of a classical result of I. Schur [3].

A function $F(\zeta) \in \mathbf{L}_\infty(N_1, N_2)$ satisfying conditions (0.2) and (0.3) for fixed γ_k ($k = 1, 2, \ldots$) and $\rho > 0$ is called a $(\Gamma; \rho)$*-function.* When $\rho = \|\Gamma\|$ it is also called a Γ*-minifunction.*

The problem of describing all of the $(\Gamma; \rho)$-functions and, in particular, of establishing the uniqueness of the Γ-minifunction remained open after [2] even for the scalar case.

The authors obtained a solution of these problems in [1b] on the basis of the theory of extension of isometric operators. At the same time a new proof which immediately carries over to the general case ($1 \leqslant \dim N_k \leqslant \infty$, $k = 1, 2$) was obtained for Nehari's theorem.

For the case $\dim N_1 = \dim N_2$ ($\leqslant \infty$) the proof of Theorem 0.1, which is a direct generalization of the scalar version of the proof presented in [2] and [1a], is essentially contained in [4]. For the same case ($\dim N_1 = \dim N_2 \leqslant \infty$) it is shown in [5] that Theorem 0.1 is a simple corollary of a comparatively recent result of Sz.-Nagy and Foiaş (see [16], Theorem II.2.3). A by-product of the present paper is another proof of Theorem 0.1 for the most general case.

In §1 we show that for fixed $\Gamma \in H^+(N_1, N_2)$ and $\rho \geqslant \|\Gamma\|$ the "zeroth" Fourier coefficients $c_0(F)$ of the various $(\Gamma; \rho)$-functions $F(\zeta)$ fill out an operator quasi-ball $K(\Gamma; \rho)$ in $[N_1, N_2]$, the center $\gamma_0^{(c)}(\Gamma; \rho)$ and half-radii $\mathbf{R}_l(\Gamma; \rho)$ and $\mathbf{R}_r(\Gamma; \rho)$ (left and right) of which are found from the operator Γ and the number ρ by means of the following procedure.

Let

$$R_{\rho^2} = (\rho^2 I_1 - \Gamma^*\Gamma)^{-1}, \qquad \overset{*}{R}_{\rho^2} = (\rho^2 I_2 - \Gamma\Gamma^*)^{-1},$$

where I_k ($k = 1, 2$) is the identity operator in $l_2^+(N_k)$, and let

$$G(\Gamma; \rho) = [P_1 R_{\rho^2} | N_1]^{-1/2} (\gg 0), \qquad \overset{*}{G}(\Gamma; \rho) = [P_2 \overset{*}{R}_{\rho^2} | N_2]^{-1/2} (\gg 0),$$

where P_k ($k = 1, 2$) is the orthogonal projection in $l_2^+(N_k)$ onto N_k.

The right shift operator in $l_2^+(N_k)$ ($k = 1, 2$) is denoted by T_k:

$$T_k\{\xi_1, \xi_2, \ldots\} = \{0, \xi_1, \xi_2, \ldots\}, \ \{\xi_j\}_1^\infty \in l_2^+(N_k) \qquad (k = 1, 2).$$

We emphasize that the equality $\Gamma T_1 = T_2^* \Gamma$ characterizes the membership of an operator $\Gamma \in [l_2^+(N_1), l_2^+(N_2)]$ in the class $H^+(N_1, N_2)$.

When $\rho > \|\Gamma\|$ the "parametric" equation for a quasi-ball $K(\Gamma; \rho)$ is written in the form

$$c_0 = \gamma_0^{(c)}(\Gamma; \rho) + \mathbf{R}_{l}(\Gamma; \rho)\mathbf{E}\mathbf{R}_{r}(\Gamma; \rho), \qquad \mathbf{E} \in B(N_1, N_2),$$

where

$$\mathbf{R}_{l}(\Gamma; \rho) = \rho^{-1/2}\overset{*}{G}(\Gamma; \rho), \qquad \mathbf{R}_{r}(\Gamma; \rho) = \rho^{-1/2}G(\Gamma; \rho),$$

$$\gamma_0^{(c)}(\Gamma; \rho) = \begin{cases} -P_2\Gamma T_1^* R_{\rho^2} G^2(\Gamma; \rho), \\ -\overset{*}{G}{}^2(\Gamma; \rho)P_2\overset{*}{R}_{\rho^2}T_2\Gamma|N_1, \end{cases}$$

and $B(N_1, N_2)$ is the set of all contractions $\mathbf{E} \in [N_1, N_2]$ ($\|\mathbf{E}\| \leqslant 1$).

When $\rho = \|\Gamma\|$ the corresponding quasi-ball $K(\Gamma)$ ($= K(\Gamma; \|\Gamma\|)$ is obtained as the limit for $p' \uparrow \|\Gamma\|$ of a family of quasi-balls $K(\Gamma; \rho)$ that are successively imbedded in each other.

For the sake of definiteness we assume everywhere that $\dim N_1 \leqslant \dim N_2$. The identity operator in N_k ($k = 1, 2$) is denoted by E_k.

In §§2 and 3 we show that when $\rho > \|\Gamma\|$ there corresponds to a coefficient $c_0(F)$ lying on the "skeleton" of a quasi-ball $K(\Gamma; \rho)$ ($\mathbf{E}^*\mathbf{E} = E_1$) one and only one $(\Gamma; \rho)$-function $F(\zeta)$ ($c_0 = c_0(F)$); the latter is called a *canonical* $(\Gamma; \rho)$*-function.*

The following operators play a fundamental role in the study of $(\Gamma; \rho)$-functions with $\rho > \|\Gamma\|$:

$$\mathbf{P} = \rho R_{\rho^2}G(\Gamma; \rho) \ (\in [N_1, l_2^+(N_1)]), \qquad \mathbf{Q} = T_2\Gamma R_{\rho^2}G(\Gamma; \rho) \ (\in [N_1, l_2^+(N_2)]),$$

$$\overset{*}{\mathbf{P}} = \rho \overset{*}{R}_{\rho^2}\overset{*}{G}(\Gamma; \rho) \ (\in [N_2, l_2^+(N_2)]), \qquad \overset{*}{\mathbf{Q}} = T_1\Gamma^*\overset{*}{R}_{\rho^2}\overset{*}{G}(\Gamma; \rho) \ (\in [N_2, l_2^+(N_1)]).$$

The "positive" Fourier coefficients $c_j(F)$ ($j = 1, 2, \ldots$) of a canonical $(\Gamma; \rho)$-function $F(\zeta)$ can be found with the use of the recursion relations

$$c_j(F) = \rho^{-1}\left[\sum_{k=0}^{j-1} c_k(F)X_{j-k+1} + \sum_{k=1}^{\infty} \gamma_k X_{j+k+1}\right] G(\Gamma; \rho),$$

where $X_k = P_1 T_1^{*k-1}(\mathbf{P} + \overset{*}{\mathbf{Q}}\mathbf{E})$ ($k = 1, 2, \ldots$).

In §6, under the assumption that $\dim N_k < \infty$ ($k = 1, 2$), we give a description of all of the $(\Gamma; \rho)$-functions for $\rho > \|\Gamma\|$ that is analogous to the description formulated in [1b] (Theorem 4.2) and proved in [1d] by the authors in the scalar case. Also under this assumption we show (§5) that $\det G(\Gamma; \rho) = \det \overset{*}{G}(\Gamma; \rho)$ and thereby establish that if one of the *limit* half-radii $\mathbf{R}_{l}(\Gamma)$ ($= \lim_{\rho \downarrow \|\Gamma\|}\mathbf{R}_{l}(\Gamma; \rho)$) and $\mathbf{R}_{r}(\Gamma)$ ($= \lim \mathbf{R}_{r}(\Gamma; \rho)$) is nondegenerate, so is the other. The latter result can be generalized to the case when only the following condition is satisfied: $P_1\Gamma$ (or, equivalently, $P_2\Gamma^*$) is a Hilbert-Schmidt operator, i.e.

$$\sum_{k=1}^{\infty} \operatorname{Tr}(\gamma_k^*\gamma_k) < \infty. \tag{0.4}$$

This condition is satisfied in particular if at least one of the numbers $\dim N_k$ ($k = 1,2$) is finite.

The important class of Hankel operators Γ with property (0.4) will be investigated in another paper. For the operators of this class (and hence, a fortiori, for the operators

$\Gamma \in H^+(N_1, N_2)$ with $\dim N_k < \infty$, $k = 1, 2$) it is possible to give a perspicuous description not only of the set of all $(\Gamma; \rho)$-functions for any $\rho > \|\Gamma\|$ but also of the set of all Γ-minifunctions in the *completely indeterminate case* (i.e. when the limit half-radii $\mathbf{R}_l(\Gamma)$ and $\mathbf{R}_r(\Gamma)$ are nondegenerate).

We will also consider in another paper applications of the results to extrapolation questions for matrix-valued and operator-valued functions of various classes. In particular, we will use new approaches to obtain (i) a number of known results on the Nevanlinna-Pick matrix problem [6]–[8] and the more general matrix problem of Sarason [4], which, as will be shown, includes the problem of extending positive definite matrices from a segment (see [9]–[11]), as well as (ii) a number of additions to them.

Some interesting details arise in the case when the operator Γ is completely continuous. It should be noted that a description of the $(\Gamma; \rho)$-functions for $\rho > \|\Gamma\|$ has been obtained for an important special class of these operators (with $\dim N_1 = \dim N_2 < \infty$) in [12], in which the relationship between the circle of questions under consideration and the theory of S-matrices for canonical differential operators is simultaneously elucidated.

We consider it our pleasant duty to mention that a corresponding result of V. P. Potapov and I. V. Kovališina on the Nevanlinna-Pick matrix problem ([13], [14]; personal communication in the spring of 1970) motivated the establishment of our theorem on the coincidence of the determinants of the left and right half-radii $\mathbf{R}_l(\Gamma)$ and $\mathbf{R}_r(\Gamma)$.

§1. One-step $(\Gamma; \rho)$-extensions

1. Let $\Gamma \in H^+(N_1, N_2)$ be defined by a Hankel block matrix (γ_{j-k+1}). Then every operator $\Gamma^{(1)} \in H^+(N_1, N_2)$ with a Hankel block matrix $(\gamma^{(1)}_{j-k+1})$, where $\gamma^{(1)}_1 = \gamma_0$ is an arbitrary operator from $[N_1, N_2]$ and $\gamma^{(1)}_k = \gamma_{k-1}$ for $k = 1, 2, \dots$, will be called a *one-step Γ-extension.* For such an operator $\Gamma^{(1)}$ we have $\|\Gamma\| \leqslant \|\Gamma^{(1)}\|$ and

$$\Gamma^{(1)} = \gamma_0 P_1 + T_2\Gamma + P_2\Gamma T_1^*. \tag{1.1}$$

If in conjunction with (1.1) we have $\|\Gamma^{(1)}\| \leqslant \rho$ for a fixed number $\rho \geqslant \|\Gamma\|$ ($\|\Gamma^{(1)}\| = \|\Gamma\|$), the operator $\Gamma^{(1)}$ is called a *one-step $(\Gamma; \rho)$-extension* (*one-step miniextension*).

It turns out that the set of all one-step $(\Gamma; \rho)$-extensions for $\rho > \|\Gamma\|$ admits a simple description:

THEOREM 1.1. *For $\Gamma \in H^+(N_1, N_2)$ and $\rho > \|\Gamma\|$ the one-step $(\Gamma; \rho)$-extensions (1.1) are defined by those and only those $\gamma_0 \in [N_1, N_2]$ which belong to the operator quasi-ball*

$$\gamma_0 = -P_2\Gamma T_1^* R_{\rho^2} G^2(\Gamma; \rho) + \rho^{-1}\overset{*}{G}(\Gamma; \rho)\mathbf{E}G(\Gamma; \rho), \qquad \mathbf{E} \in B(N_1, N_2). \tag{1.2}$$

PROOF. The condition $\|\Gamma^{(1)}\| \leqslant \rho$ means that

$$\rho^2 I_2 - \Gamma^{(1)}\Gamma^{(1)*} = \rho^2 I_2 - \Gamma\Gamma^* - (\gamma_0 + T_2\Gamma)P_1(\gamma_0^* P_2 + \Gamma^* T_2^*) \geqslant 0,$$

or, what is the same thing, that $I_2 - A^*A \geqslant 0$, where

$$A = P_1(\gamma_0^* P_2 + \Gamma^* T_2^*) R_{\rho 2}^{1/2}, \qquad \overset{*}{R}{}_{\rho 2}^{1/2} = (\rho^2 I_2 - \Gamma\Gamma^*)^{-1/2} (\gg 0).$$

Since the condition $I_2 - A^*A \geqslant 0$ is equivalent to the condition $I_1 - AA^* \geqslant 0$, we arrive at the inequality

$$P_1 - P_1(\gamma_0^* P_2 + \Gamma^* T_2^*)\overset{*}{R}_{\rho 2}(\gamma_0 + T_2\Gamma)P_1 \geqslant 0. \tag{1.3}$$

We now note that an arbitrary $\gamma_0 \in [N_1, N_2]$ is uniquely representable in the form (1.2) for some $\mathbf{E} \in [N_1, N_2]$. We substitute this expression into (1.3). Since in this connection

$$\begin{aligned}
(\gamma_0^* P_2 + P_1\Gamma^* T_2^*)\overset{*}{R}_{\rho 2}\gamma_0 &= - P_1\Gamma^* T_2^* \overset{*}{R}_{\rho 2} \overset{*}{G}{}^2 P_2 \overset{*}{R}_{\rho 2}\gamma_0 \\
&+ \rho^{-1} G\mathbf{E}^* \overset{*}{G} P_2 \overset{*}{R}_{\rho 2}\gamma_0 + P_1\Gamma^* T_2^* \overset{*}{R}_{\rho 2}\gamma_0 = - P_1\Gamma^* T_2^* \overset{*}{R}_{\rho 2} \overset{*}{G}{}^2 \overset{*}{G}{}^{-2}\gamma_0 \\
&+ \rho^{-1} G\mathbf{E}^* \overset{*}{G}\overset{*}{G}{}^{-2}\gamma_0 + P_1\Gamma^* T_2^* \overset{*}{R}_{\rho 2}\gamma_0 = \rho^{-1} G\mathbf{E}^* \overset{*}{G}{}^{-1}\gamma_0 \\
&= - \rho^{-1} G\mathbf{E}^* \overset{*}{G} P_2 \overset{*}{R}_{\rho 2} T_2\Gamma P_1 + \rho^{-2} G\mathbf{E}^* \overset{*}{G}{}^{-1} \overset{*}{G}\mathbf{E}G \\
&= - \rho^{-1} G\mathbf{E}^* \overset{*}{G} P_2 \overset{*}{R}_{\rho 2} T_2\Gamma P_1 + \rho^{-2} G\mathbf{E}^*\mathbf{E}G,
\end{aligned}$$

where $G = G(\Gamma; \rho)$, $\overset{*}{G} = \overset{*}{G}(\Gamma; \rho)$ and

$$\gamma_0^* P_2 \overset{*}{R}_{\rho 2} T_2\Gamma P_1 = - P_1\Gamma^* T_2^* \overset{*}{R}_{\rho 2} \overset{*}{G}{}^2 P_2 \overset{*}{R}_{\rho 2} T_2\Gamma P_1 + \rho^{-1} G\mathbf{E}^* \overset{*}{G} P_2 \overset{*}{R}_{\rho 2} T_2\Gamma P_1,$$

inequality (1.3) is equivalent to the following condition on $\mathbf{E}$ ($\in [N_1, N_2]$):

$$P_1 - \rho^{-2} G\mathbf{E}^*\mathbf{E}GP_1 + P_1\Gamma^* T_2^* \overset{*}{R}_{\rho 2} \overset{*}{G}{}^2 P_2 \overset{*}{R}_{\rho 2} T_2\Gamma P_1 - P_1\Gamma^* T_2^* \overset{*}{R}_{\rho 2} T_2\Gamma P_1 \geqslant 0. \tag{1.4}$$

Let us now verify the identity

$$P_1 + P_1\Gamma^* T_2^* \overset{*}{R}_{\rho 2} \overset{*}{G}{}^2 P_2 \overset{*}{R}_{\rho 2} T_2\Gamma P_1 - P_1\Gamma^* T_2^* \overset{*}{R}_{\rho 2} T_2\Gamma P_1 = \rho^{-2} G^2 P_1, \tag{1.5}$$

by virtue of which (1.4) is equivalent to the inequality

$$G(E_1 - \mathbf{E}^*\mathbf{E})G \geqslant 0.$$

The proof of the theorem will then be complete.

For $\Gamma \in H^+(N_1, N_2)$ it is easily seen that

$$\Gamma T_1^* = T_2\Gamma + P_2\Gamma T_1^* - T_2\Gamma P_1,$$

which by virtue of the basic relation $\Gamma T_1 = T_2^*\Gamma$ gives

$$(\rho^2 I_2 - \Gamma\Gamma^*)T_2 = T_2(\rho^2 I_2 - \Gamma\Gamma^*) + T_2\Gamma P_1\Gamma^* - P_2\Gamma T_1^*\Gamma^*,$$

or

$$T_2\overset{*}{R}_{\rho 2} = \overset{*}{R}_{\rho 2} T_2 + \overset{*}{R}_{\rho 2} T_2\Gamma P_1\Gamma^* \overset{*}{R}_{\rho 2} - \overset{*}{R}_{\rho 2} P_2\Gamma T_1^*\Gamma^* \overset{*}{R}_{\rho 2}, \tag{1.6}$$

whence

$$\overset{*}{R}_{\rho 2} = T_2^* \overset{*}{R}_{\rho 2} T_2 + T_2^* \overset{*}{R}_{\rho 2} T_2\Gamma P_1\Gamma^* \overset{*}{R}_{\rho 2} - T_2^* \overset{*}{R}_{\rho 2} P_2\Gamma T_1^*\Gamma^* \overset{*}{R}_{\rho 2}.$$

Therefore

$$
\begin{aligned}
P_1\Gamma^*T_2^*\overset{*}{R}_{\rho^2}T_2\Gamma P_1 &= P_1\Gamma^*\overset{*}{R}_{\rho^2}\Gamma P_1 - P_1\Gamma^*T_2^*\overset{*}{R}_{\rho^2}T_2\Gamma P_1\Gamma^*\overset{*}{R}_{\rho^2}\Gamma P_1 \\
&+ P_1\Gamma^*T_2^*\overset{*}{R}_{\rho^2}P_2\Gamma T_1^*\Gamma^*\overset{*}{R}_{\rho^2}\Gamma P_1 = -P_1 + \rho^2G^{-2}P_1 + P_1\Gamma^*T_2^*\overset{*}{R}_{\rho^2}T_2\Gamma P_1 \\
&- \rho^2P_1\Gamma^*T_2^*\overset{*}{R}_{\rho^2}T_2\Gamma G^{-2}P_1 + \rho^2P_1\Gamma^*T_2^*\overset{*}{R}_{\rho^2}P_2\Gamma T_2^*\overset{*}{R}_{\rho^2}P_1.
\end{aligned}
$$

We get that

$$\rho^2P_1\Gamma^*T_2^*\overset{*}{R}_{\rho^2}T_2\Gamma P_1 = -G^2P_1 + \rho^2P_1 + \rho^2P_1\Gamma^*T_2^*\overset{*}{R}_{\rho^2}P_2\Gamma T_1^*R_{\rho^2}G^2P_1,$$

i.e.

$$P_1 - P_1\Gamma^*T_2^*\overset{*}{R}_{\rho^2}T_2\Gamma P_1 + P_1\Gamma^*T_2^*\overset{*}{R}_{\rho^2}P_2\Gamma T_1^*R_{\rho^2}G^2P_1 = \rho^{-2}G^2P_1.$$

For the proof of (1.5) it remains to verify that

$$P_2\Gamma T_1^*R_{\rho^2}G^2P_1 = G^2P_2\overset{*}{R}_{\rho^2}T_2\Gamma P_1. \tag{1.7}$$

On the basis of (1.6) we have

$$
\begin{aligned}
P_1\Gamma^*T_2^*\overset{*}{R}_{\rho^2}P_2 &= P_1\Gamma^*\overset{*}{R}_{\rho^2}T_2^*P_2 - P_1\Gamma^*\overset{*}{R}_{\rho^2}\Gamma P_1\Gamma^*T_2^*\overset{*}{R}_{\rho^2}P_2 \\
&+ P_1\Gamma^*\overset{*}{R}_{\rho^2}\Gamma T_1\Gamma^*P_2\overset{*}{R}_{\rho^2}P_2 = P_1\Gamma^*T_2^*\overset{*}{R}_{\rho^2}P_2 - \rho^2P_1R_{\rho^2}P_1\Gamma^*T_2^*\overset{*}{R}_{\rho^2}P_2 \\
&+ \rho^2P_1R_{\rho^2}T_1\Gamma^*P_2\overset{*}{R}_{\rho^2}P_2,
\end{aligned}
$$

which implies that

$$P_1R_{\rho^2}P_1\Gamma^*T_2^*\overset{*}{R}_{\rho^2}P_2 = P_1R_{\rho^2}T_1\Gamma^*\overset{*}{R}_{\rho^2}P_2,$$

or

$$G^{-2}P_1\Gamma^*T_2^*\overset{*}{R}_{\rho^2}P_2 = P_1R_{\rho^2}T_1\Gamma^*\overset{*}{G}^{-2}P_2,$$

which gives (1.7). The theorem is proved.

2. As can be seen from (1.2), the operators

$$\mathbf{R}_{\mathrm{l}}(\Gamma;\rho) = \rho^{-1/2}\overset{*}{G}(\Gamma;\rho), \qquad \mathbf{R}_{\mathrm{r}}(\Gamma;\rho) = \rho^{-1/2}G(\Gamma;\rho)$$

are the left and right half-radii of the operator quasi-ball (1.2). Two expressions have been obtained for the center of this quasi-ball, viz.

$$\gamma_0^{(c)}(\Gamma;\rho) = \begin{cases} -P_2\Gamma T_1^*R_{\rho^2}G^2(\Gamma;\rho), \\ -\overset{*}{G}{}^2(\Gamma;\rho)P_2\overset{*}{R}_{\rho^2}T_2\Gamma|N_1. \end{cases}$$

It can be seen from the calculations that $\|\Gamma^{(1)}\| = \rho$ if and only if we have

$\|\mathbf{E}\|$ ($= \|\mathbf{E}^*\|$) $= 1$ in (1.2) for $\mathbf{E} \in B(N_1, N_2)$, the norms of the operators $\Gamma^{(1)}$ and $\mathbf{E}$ in this case being attained together if they are attained at all.

3. We have called a set $K(\Gamma; \rho)$ of operators $\gamma_0 \in [N_1, N_2]$ that are representable in the form (1.2) an operator quasi-ball since, by definition, a set $K \subset [N_1, N_2]$ is called a *quasi-ball* with *center* $\gamma^{(c)} \in [N_1, N_2]$ if it consists of all the operators γ of the form

$$\gamma = \gamma^{(c)} + \mathbf{R}_{\mathfrak{l}}\mathbf{E}\mathbf{R}_{\mathfrak{r}} \qquad (\mathbf{E} \in B(N_1, N_2)),$$

where $\mathbf{R}_{\mathfrak{r}}$ and $\mathbf{R}_{\mathfrak{l}}$ ($\in [N_1, N_2]$) are fixed nonnegative operators called the *right* and *left half-radii* respectively of the quasi-ball K.

The center $\gamma^{(c)}$, being the center of symmetry of K, is uniquely defined by K.

A quasi-ball K degenerates into a point if and only if at least one of the half-radii $\mathbf{R}_{\mathfrak{l}}$ or $\mathbf{R}_{\mathfrak{r}}$ is equal to zero. In this case the ball consists of only its center and is said to be *trivial.*

If a quasi-ball K is nontrivial, its half-radii $\mathbf{R}_{\mathfrak{r}}$ and $\mathbf{R}_{\mathfrak{l}}$ are defined to within multiplication and division by a positive number r: $\mathbf{R}_{\mathfrak{r}} \longrightarrow r\mathbf{R}_{\mathfrak{r}}$ and $\mathbf{R}_{\mathfrak{l}} \longrightarrow r^{-1}\mathbf{R}_{\mathfrak{l}}$.

If $K^{(1)} \supset K^{(2)} \supset \cdots$ is a sequence of quasi-balls imbedded in each other, its intersection is again a quasi-ball whose center is the weak limit of the centers of the quasi-balls $K^{(n)}$. In addition, the right and left half-radii $\mathbf{R}_{\mathfrak{r}}^{(n)}$ and $\mathbf{R}_{\mathfrak{l}}^{(n)}$ of the quasi-balls $K^{(n)}$ can be chosen so that they constitute a nonincreasing sequence $\mathbf{R}^{(n)} \geqslant \mathbf{R}^{(n+1)} (n = 1, 2, \ldots)$. Their strong limits $\mathbf{R}_{\mathfrak{r}} = s\text{-}\lim_{n\to\infty}\mathbf{R}_{\mathfrak{r}}^{(n)}$ and $\mathbf{R}_{\mathfrak{l}} = s\text{-}\lim_{n\to\infty}\mathbf{R}_{\mathfrak{l}}^{(n)}$ will then be the right and left half-radii respectively of the intersection K.

All of the assertions formulated above concerning operator quasi-balls were established in [15].(3)

The quasi-balls $K(\Gamma; \rho)$ ($\rho > \|\Gamma\|$) introduced in Theorem 1.1 obviously form a nested one-parameter family of quasi-balls: $K(\Gamma; \rho) \supset K(\Gamma; \rho')$ for $\rho > \rho'$ ($> \|\Gamma\|$), with nonincreasing half-radii. Consequently, according to what has been said above the intersection

$$K(\Gamma) = \bigcap_{\rho > \|\Gamma\|} K(\Gamma; \rho)$$

is a nonempty quasi-ball with center

$$\gamma_0^{(c)}(\Gamma) = w\text{-}\lim_{\rho \downarrow \|\Gamma\|} \gamma_0^{(c)}(\Gamma; \rho)$$

and half-radii

$$\mathbf{R}_{\mathfrak{r}}(\Gamma) = s\text{-}\lim_{\rho \downarrow \|\Gamma\|} \|\Gamma\|^{-1/2}[P_1(\rho^2 I_1 - \Gamma^*\Gamma)^{-1}|N_1]^{-1/2},$$

$$\mathbf{R}_{\mathfrak{l}}(\Gamma) = s\text{-}\lim_{\rho \downarrow \|\Gamma\|} \|\Gamma\|^{-1/2}[P_2(\rho^2 I_2 - \Gamma\Gamma^*)^{-1}|N_2]^{-1/2}.$$

Thus, on the basis of Theorem 1.1 we get

(3) In [15] a *quasi-ball* (*half-radius*) is simply called a *ball* (*radius*).

THEOREM 1.2. *The one-step Γ-miniextensions are defined by those and only those* $\gamma_0 \in [N_1, N_2]$ *which belong to the quasi-ball*

$$K(\Gamma):\ \gamma_0 = \gamma_0^{(c)}(\Gamma) + \mathbf{R}_l(\Gamma)\mathbf{E}\mathbf{R}_\tau(\Gamma)\ (\mathbf{E} \in B(N_1, N_2)).$$

The first assertion of the following proposition is obtained as a direct corollary of Theorem 1.2.

THEOREM 1.3. *For* $\Gamma \in H^+(N_1, N_2)$ *there exists a unique one-step Γ-miniextension if and only if at least one of the following two conditions is satisfied*:

$$\begin{aligned} &\text{a)}\ \ \underset{\rho\downarrow\|\Gamma\|}{s\text{-lim}}\ [P_1(\rho^2 I_1 - \Gamma^*\Gamma)^{-1}|N_1]^{-1/2} = 0, \\ &\text{a*)}\ \ \underset{\rho\downarrow\|\Gamma\|}{s\text{-lim}}\ [P_2(\rho^2 I_2 - \Gamma\Gamma^*)^{-1}|N_2]^{-1/2} = 0. \end{aligned} \tag{1.8}$$

Conditions a) *and* a*) *are respectively equivalent to the conditions*

$$\text{b)}\ \ N_1 \cap (\|\Gamma\|^2 I_1 - \Gamma^*\Gamma)^{1/2} l_2^+(N_1) = \{0\},$$

$$\text{b*)}\ \ N_2 \cap (\|\Gamma\|^2 I_2 - \Gamma\Gamma^*)^{1/2} l_2^+(N_2) = \{0\}.$$

Let us prove, for example, the equivalence of conditions a) and b). Suppose condition a) is satisfied and

$$g \in N_1 \cap (\|\Gamma\|^2 I_1 - \Gamma^*\Gamma)^{1/2} l_2^+(N_1).$$

Then

$$\lim_{\rho\downarrow\|\Gamma\|} ((\rho^2 I_1 - \Gamma^*\Gamma)^{-1} g, g) = \lim_{\rho\downarrow\|\Gamma\|} (G^{-2}(\Gamma;\rho)g, g) = \lim_{\rho\downarrow\|\Gamma\|} \|G^{-1}(\Gamma;\rho)g\|^2 < \infty. \tag{1.9}$$

Inasmuch as

$$\underset{\rho\downarrow\|\Gamma\|}{s\text{-lim}}\ G(\Gamma;\rho) = 0$$

and the family of vectors $h_\rho = G^{-1}(\Gamma;\rho)g$ $(\rho > \|\Gamma\|)$ is bounded, we have

$$g = G(\Gamma;\rho)h_\rho = \underset{\rho\downarrow\|\Gamma\|}{w\text{-lim}}\ G(\Gamma;\rho)h_\rho = 0.$$

Thus a) implies b). Now suppose condition a) is not satisfied, and

$$g = \lim_{\rho\downarrow\|\Gamma\|} G(\Gamma;\rho)h \neq 0$$

for some $h \in N_1$. Then the vector g satisfies (1.9), and hence

$$g \in N_1 \cap (\|\Gamma\|^2 I_1 - \Gamma^*\Gamma)^{1/2} l_2^+(N_1).$$

We will need the following sufficient condition for the uniqueness of a Γ-miniextension.

COROLLARY 1.1. *If* ρ^2 *is an eigenvalue of the operator* $\Gamma^*\Gamma$ ($\Gamma\Gamma^*$) *and the eigenspace* $L \subset l_2^+(N_1)$ ($\overset{*}{L} \subset l_2^+(N_2)$) *corresponding to it is such that* $\overline{P_1 L} = N_1$ ($\overline{P_2 \overset{*}{L}} = N_2$), *then the one-step* Γ*-miniextension is unique.*

For suppose, for example, $\overline{P_1 L} = N_1$. Then $g \perp L$ for $g = (\rho^2 I_1 - \Gamma^*\Gamma)^{1/2} f$ ($f \in l_2^+(N_1)$). Therefore, if, in addition, $g \in N_1$, then $g = P_1 g \perp \overline{P_1 L} = N_1$, whence $g = 0$. Thus condition b) of Theorem 1.3 is satisfied.

§2. One-step canonical (Γ; ρ)-extensions

It will be assumed without loss of generality that $\dim N_1$ $(= n_1) \leqslant \dim N_2$ $(= n_2)$. Otherwise it would be possible to consider $\Gamma^* \in H^+(N_2, N_1)$ in place of $\Gamma \in H^+(N_1, N_2)$. In this section we will dwell in more detail on those one-step (Γ; ρ)-extensions $\Gamma^{(1)}$ ($\rho > \|\Gamma\|$) to which there correspond in (1.2) mappings $\mathbf{E}$ of N_1 into N_2 that are isometric, i.e. for which $\mathbf{E}^*\mathbf{E} = E_1$. Such ($\Gamma$; ρ)-extensions are said to be *canonical.* A fundamental role is played in the sequel by

THEOREM 2.1. *For a canonical one-step* (Γ; ρ)*-extension* $\Gamma^{(1)}$ *defined by an operator* $\mathbf{E} \in B(N_1, N_2)$

$$\Gamma^{(1)}(\mathbf{P} + \overset{*}{\mathbf{Q}}\mathbf{E}) = \rho(\mathbf{Q} + \overset{*}{\mathbf{P}}\mathbf{E}); \qquad \Gamma^{(1)*}(\mathbf{Q} + \overset{*}{\mathbf{P}}\mathbf{E}) = \rho(\mathbf{P} + \overset{*}{\mathbf{Q}}\mathbf{E}), \tag{2.1}$$

and the linear manifolds

$$L_\rho = (\mathbf{P} + \overset{*}{\mathbf{Q}}\mathbf{E})N_1, \qquad \overset{*}{L}_\rho = (\mathbf{Q} + \overset{*}{\mathbf{P}}\mathbf{E})N_1 \tag{2.2}$$

are the eigenspaces corresponding to the eigenvalue ρ^2 *of the operators* $\Gamma^{(1)*}\Gamma^{(1)}$ *and* $\Gamma^{(1)}\Gamma^{(1)*}$, *respectively.*

All of the points of the interval $(\|\Gamma\|^2, \rho^2)$ *are regular points of the operator* $\Gamma^{(1)*}\Gamma^{(1)}$ *or, equivalently, of the operator* $\Gamma^{(1)}\Gamma^{(1)*}$.

REMARK 2.1. Formula (2.1) implies

$$P_1 L_\rho = N_1, \qquad P_2 \overset{*}{L}_\rho = N_2. \tag{2.3}$$

In fact,

$$P_1 L_\rho = P_1 \mathbf{P} N_1 = \rho G^{-2}(\Gamma; \rho) N_1 = N_1; \quad P_2 \overset{*}{L}_\rho = P_2 \overset{*}{\mathbf{P}} \mathbf{E} N_1 = \rho \overset{*}{G}{}^{-2}(\Gamma; \rho) N_2 = N_2.$$

PROOF OF THE THEOREM. We can verify (2.1) if we take into account the fact that for any $\mathbf{E} \in [N_1, N_2]$

$$\Gamma(\mathbf{P} + \overset{*}{\mathbf{Q}}\mathbf{E}) = \rho T_2^*(\mathbf{Q} + \overset{*}{\mathbf{P}}\mathbf{E}), \qquad \Gamma^*(\mathbf{Q} + \overset{*}{\mathbf{P}}\mathbf{E}) = \rho T_2^*(\mathbf{P} + \overset{*}{\mathbf{Q}}\mathbf{E})$$

and make use of (1.1), (1.2), (1.7) and the condition $\mathbf{E}^*\mathbf{E} = E_1$. It can be seen from (2.1) that the linear manifolds L_ρ and $\overset{*}{L}_\rho$ defined by (2.2) are eigenspaces of the operators $\Gamma^{(1)*}\Gamma^{(1)}$ and $\Gamma^{(1)}\Gamma^{(1)*}$ respectively. Their maximality follows from Remark 2.1. In fact, it follows from (2.3) that if there were to exist an eigenvector of the operator $\Gamma^{(1)*}\Gamma^{(1)}$ corresponding to the eigenvalue ρ^2 and not belonging to L_ρ, there would exist an eigenvector $\xi \notin L_\rho$ with $P_1 \xi = 0$, i.e. of the form $\xi = T_1 \mathbf{h}$, where $\mathbf{h}$ $(\neq 0) \in l_2(N_1)$. But then we would have

$$\|\Gamma^{(1)}\xi\| = \|\Gamma^{(1)}\| \, \|\xi\| = \rho\|\mathbf{h}\|.$$

On the other hand, on the basis of (1.1)

$$\|\Gamma^{(1)}\xi\| = \|\Gamma^{(1)}T_1\mathbf{h}\| = \|T_2^*\Gamma^{(1)}\mathbf{h}\| = \|\Gamma\mathbf{h}\|,$$

so that $\|\Gamma\mathbf{h}\| \geqslant \rho\|\mathbf{h}\|$, which contradicts the condition $\|\Gamma\| < \rho$. The maximality of $\overset{*}{L}_\rho$ is verified analogously.

It remains to prove the last assertion of the theorem. From arguments analogous to those developed at the beginning of the proof of Theorem 1.1 it follows that a point $\lambda > \|\Gamma\|^2$ is a regular point of the operator $\Gamma^{(1)}\Gamma^{(1)*}$ precisely when the following operator has an inverse in N_2:

$$\Phi(\lambda) = E_1 - P_1(\gamma_0^* P_2 + \Gamma^* T_2^*)\overset{*}{R}_\lambda(\gamma_0 + T_2\Gamma)P_1|N_1, \qquad \overset{*}{R}_\lambda = (\lambda I_2 - \Gamma\Gamma^*)^{-1}.$$

On the other hand, $\Phi(\rho^2)$ coincides with the restriction to N_1 of the left side of (1.3), which was shown to be equal to $\overset{*}{\rho}{}^{-2}G(E_1 - \mathbf{E}^*\mathbf{E})G = 0$. The operator-valued function $\Phi(\lambda)$ increases with increasing λ for $\lambda > \|\Gamma\|^2$; in particular, the operator $-\Phi(\lambda) = \Phi(\rho^2) - \Phi(\lambda)$ is uniformly positive for $\rho^2 > \lambda > \|\Gamma\|^2$. In fact, since the spectral resolution of the resolvent R_λ implies

$$((\overset{*}{R}_\lambda - \overset{*}{R}_{\rho^2})g, g) \geqslant c(\lambda)\|g\|^2 \qquad (g \in l_2^+(N_2)),$$

where $c(\lambda) = (\rho^2 - \lambda)/\lambda\rho^2$, we have

$$-(\Phi(\lambda)f, f) \geqslant c(\lambda)\|g\|^2 \qquad (\rho^2 > \lambda > \|\Gamma\|^2),$$

where

$$\|g\| = \|(\gamma_0 + T_2\Gamma)f\| = \|\overset{*}{R}{}_{\rho^2}^{-1/2}\overset{*}{R}{}_{\rho^2}^{1/2}(\gamma_0 + T_2\Gamma)f\| \geqslant \rho^{-1}\|\overset{*}{R}{}_{\rho^2}^{1/2}(\gamma_0 + T_2\Gamma)f\|$$
$$= \rho^{-1}\|f\|, \qquad f \in N_1.$$

In the last equality we have once more taken into account the fact that $\Phi(\rho^2) = 0$. Thus

$$-(\Phi(\lambda)f, f) \geqslant (\rho^2 - \lambda)[(\rho^2 - \lambda)/\lambda\rho^4]\|f\|^2 \qquad (f \in N_1),$$

and hence the operator $\Phi(\lambda)$ has a bounded inverse for $\|\Gamma\|^2 < \lambda < \rho^2$. The theorem is proved.

It follows from this theorem that ρ^2 is an isolated eigenvalue of the operator $\Gamma^{(1)*}\Gamma^{(1)}$ with a maximal eigenspace L_ρ $(= (\mathbf{P} + \overset{*}{\mathbf{Q}}\mathbf{E})N_1)$ such that $P_1L_\rho = N_1$. Since $\|\Gamma^{(1)}\| = \rho$, a one-step $(\Gamma^{(1)}; \rho)$-extension will be a $\Gamma^{(1)}$-miniextension. To this miniextension we can apply Corollary 1.1, from which we get

THEOREM 2.2. *An arbitrary canonical one-step* $(\Gamma; \rho)$*-extension* $(\rho > \|\Gamma\|)$ $\Gamma^{(1)}$ *has a unique one-step* $\Gamma^{(1)}$*-miniextension* $\Gamma^{(2)}$.

§3. Full $(\Gamma; \rho)$-extensions

1. If N is an arbitrary Hilbert space, $l_2^+(N)$ can be regarded as a subspace of the space $l_2(H)$ of two-sided sequences $h = \{h_k\}_{-\infty}^{\infty}$ of vectors $h_k \in N$ $(k = 0, \pm 1, \pm 2, \ldots)$ for which

$$\|h\|^2 = \sum_{-\infty}^{\infty} \|h_k\|^2 < \infty.$$

The orthogonal projection of $l_2(N)$ onto $l_2^+(N)$ will be denoted by P_+. The minimal unitary dilation of the right shift operator T in $l_2^+(N)$ is the right shift operator U in $l_2(N)$: $U\{h_k\}_{-\infty}^{\infty} = \{h'_k\}_{-\infty}^{\infty}$, where $h'_k = h_{k-1}$ $(k = 0, \pm 1, \pm 2, \ldots)$. Let N_1 and N_2 be arbitrary Hilbert spaces and let U_1 and U_2 be the right shift operators in $l_2(N_1)$ and $l_2(N_2)$ respectively. An operator $\widetilde{\Gamma}$ from $[l_2(N_1), l_2(N_2)]$ is called a *full Hankel operator* if it is defined by a Hankel block matrix $(\widetilde{\gamma}_{j+k-1})_{-\infty}^{\infty}$:

$$\widetilde{\Gamma}\{h_k\}_{-\infty}^{\infty} = \left\{ \sum_{k=-\infty}^{\infty} \widetilde{\gamma}_{j+k-1} h_k \right\}_{-\infty}^{\infty}, \qquad \{h_k\}_{-\infty}^{\infty} \in l_2(N_1),\ \widetilde{\gamma}_k \in [N_1, N_2],$$

$$k = 0, \pm 1, \pm 2, \ldots .$$

For such (and only such) an operator, we have

$$\widetilde{\Gamma} U_1 = U_2^* \widetilde{\Gamma}.$$

The subspace of all full Hankel operators of $[l_2(N_1), l_2(N_2)]$ is denoted by $H(N_1, N_2)$. Let Γ be a fixed operator from $H^+(N_1, N_2)$. An operator $\widetilde{\Gamma} \in H(N_1, N_2)$ is called a *full Γ-extension* if

$$\Gamma = P_+ \widetilde{\Gamma} | l_2^+(N_1).$$

If an operator $\Gamma \in H^+(N_1, N_2)$ is defined by a Hankel block matrix $(\gamma_{j+k-1})_1^{\infty}$, an operator $\widetilde{\Gamma} \in H(N_1, N_2)$ is a full Γ-extension if and only if it is defined by a Hankel block matrix $(\widetilde{\gamma}_{j+k-1})_{-\infty}^{\infty}$ in which $\widetilde{\gamma}_k = \gamma_k$ for $k = 1, 2, \ldots$.

Clearly, for any full Γ-extension $\widetilde{\Gamma}$, $\|\widetilde{\Gamma}\| \geqslant \|\Gamma\|$. If a full Γ-extension $\widetilde{\Gamma}$ is such that $\|\widetilde{\Gamma}\| \leqslant \rho$ for a fixed number $\rho \geqslant \|\Gamma\|$, we will call it a *full* $(\Gamma; \rho)$*-extension,* and if $\rho = \|\Gamma\|$, we will call it a *full Γ-miniextension* and denote it by $\widetilde{\Gamma}_\mu$. We have

THEOREM 3.1. *For any $\Gamma \in H^+(N_1, N_2)$ there exists a full Γ-miniextension $\widetilde{\Gamma}_\mu$. It is unique if and only if the one-step Γ-miniextension is unique, i.e. at least one of the conditions* (1.8) *is satisfied.*

PROOF. According to Theorem 1.2 the operator Γ has at least one one-step Γ-miniextension. Suppose $\{\Gamma^{(p)}\}_1^{\infty} \subset H^+(N_1, N_2)$, where $\Gamma^{(p)}$ is a one-step $\Gamma^{(p-1)}$-miniextension $(p = 1, 2, \ldots; \Gamma^{(0)} = \Gamma)$. Then $\|\Gamma^{(p)}\| = \|\Gamma\|$, and if the operator Γ is defined by a Hankel block matrix $(\gamma_{j+k-1-p})_1^{\infty}$, where the blocks $\gamma_0, \gamma_{-1}, \ldots, \gamma_{-p+1}$ $(p = 1, 2, \ldots)$ are operators from $[N_1, N_2]$. We consider the matrix $(\gamma_{j+k-1})_{-\infty}^{\infty}$ composed from the two-sided sequence of blocks $\{\gamma_k\}_{-\infty}^{\infty}$ $(\gamma_k \in [N_1, N_2])$ obtained in this way.

Clearly, this matrix defines a full Γ-miniextension, i.e. the formula

$$\widetilde{\Gamma}\{h_k\}_{-\infty}^{\infty} = \left\{ \sum_{k=-\infty}^{\infty} \gamma_{j+k-1} h_k \right\}_{j=-\infty}^{\infty}$$

defines an operator $\widetilde{\Gamma} \in H(N_1, N_2)$ such that $\|\widetilde{\Gamma}\| = \|\Gamma\|$.

Thus the first assertion of the theorem is proved. It follows that an operator $\Gamma \in H^+(N_1, N_2)$ has a unique one-step Γ-miniextension $\Gamma^{(1)}$ if it has a unique full Γ-miniextension $\widetilde{\Gamma}_\mu$. The converse assertion can be seen to hold by verifying that $\Gamma^{(2)}$ is uniquely defined whenever $\Gamma^{(1)}$ is uniquely defined.

Since $\Gamma = T_2^*\Gamma^{(1)}$, we have

$$\Gamma^*\Gamma = \Gamma^{(1)*}T_2T_2^*\Gamma^{(1)} = \Gamma^{(1)*}\Gamma^{(1)} - \Gamma^{(1)*}P_2\Gamma^{(1)} \geqslant \Gamma^{(1)*}\Gamma^{(1)},$$

and hence $(\rho^2 I_1 - \Gamma^*\Gamma)^{-1} \leqslant (\rho^2 I_1 - \Gamma^{(1)*}\Gamma^{(1)})^{-1}$. Therefore

$$G^{-2}(\Gamma; \rho) = P_1(\rho^2 I_1 - \Gamma^*\Gamma)^{-1}|N_1 \leqslant P_1(\rho^2 I_1 - \Gamma^{(1)*}\Gamma^{(1)})^{-1}|N_1 = G^{-2}(\Gamma^{(1)}; \rho),$$

so that $G(\Gamma^{(1)}; \rho) \leqslant G(\Gamma; \rho)$. Analogously, we get

$$\overset{*}{G}(\Gamma^{(1)}; \rho) \leqslant \overset{*}{G}(\Gamma; \rho).$$

It now follows directly from these inequalities and Theorem 2.2 that the uniquely defined operator $\Gamma^{(1)}$ has a uniquely defined $\Gamma^{(1)}$-miniextension $\Gamma^{(2)}$.

The theorem is proved.

COROLLARY 3.1. *An operator $\Gamma \in H^+(N_1, N_2)$ has a unique full Γ-miniextension $\widetilde{\Gamma}_\mu$ if the number $\rho^2 = \|\Gamma\|^2$ is an eigenvalue of the operator $\Gamma^*\Gamma$ $(\Gamma\Gamma^*)$ and the corresponding eigenspace L $(\overset{*}{L})$ is such that $\overline{P_1 L} = N_1$ $(\overline{P_2\overset{*}{L}} = N_2)$.*

2. Let us dwell in more detail on the case, important for what follows, when a stronger form of the conditions of Corollary 3.1 is satisfied. Suppose, for example, that in place of the condition $\overline{P_1 L} = N_1$ the following condition is satisfied: the operator P_1 is a one-to-one projection of L onto all of N_1.

By Banach's theorem the inverse, which we denote by X, of the restriction $P_1|L$ $(\in [L, N_1])$ is bounded. Let $Y = \rho^{-1}\Gamma X$ Then, inasmuch as L is the eigenspace corresponding to the eigenvalue ρ^2 of the operator $\Gamma^*\Gamma$, we will have

$$\Gamma X = \rho Y, \qquad \Gamma^* Y = \rho X. \tag{3.1}$$

For the (unique) full Γ-miniextension $\widetilde{\Gamma}$ we get

$$\|P_+\widetilde{\Gamma}Xh\| = \|\Gamma Xh\| = \rho\|Xh\| = \|\widetilde{\Gamma}\| \cdot \|Xh\| \qquad (h \in N_1),$$

from which it can be seen that $\Gamma X = P_+\widetilde{\Gamma}X = \widetilde{\Gamma}X$. Analogously, we get that $\Gamma^*Y = P_+\widetilde{\Gamma}^*Y = \widetilde{\Gamma}^*Y$. Thus we have

$$\widetilde{\Gamma}X = \rho Y, \qquad \widetilde{\Gamma}^*Y = \rho X. \tag{3.2}$$

We put

$$X_k = P_1T_1^{*k-1}X\,(\in [N_1, N_1]), \quad Y_k = P_2T_2^{*k-1}Y\,(\in [N_1, N_2]) \quad (k = 1, 2, \ldots)$$

so that

$$Xh = \sum_{k=1}^{\infty} T_1^{k-1} X_k h = \{X_k h\}_1^\infty \ (\in l_2^+ (N_1)),$$

$$Yh = \sum_{k=1}^{\infty} T_2^{k-1} Y_k h = \{Y_k h\}_1^\infty \ (\in l_2^+ (N_2)) \ (h \in N_1).$$

If $(\gamma_{j+k-1})_{-\infty}^{\infty}$ is the matrix corresponding to the operator $\widetilde{\Gamma}$, relations (3.1) are equivalent to the system of equalities (here the convergence is understood in the strong sense)

$$\sum_{k=1}^{\infty} \gamma_{j+k-1} X_k = \rho Y_j, \qquad \sum_{k=1}^{\infty} \gamma_{j+k-1}^* Y_k = \rho X_j \qquad (j = 1, 2, \ldots),$$

while relations (3.2) give in addition the system of equalities

$$\sum_{k=1}^{\infty} \gamma_{j+k-1} X_k = 0, \qquad \sum_{k=1}^{\infty} \gamma_{j+k-1}^* Y_k = 0 \qquad (j = 0, -1, -2, \ldots). \tag{3.3}$$

The definitions of the operator X and the sequence $\{X_k\}_1^\infty$ imply that $X_1 = E_1$. Using (3.3), we can find all of the unknown blocks γ_k $(k = 0, -1, -2, \ldots)$ of the matrix $(\gamma_{j+k-1})_{-\infty}^{\infty}$ from the matrix $(\gamma_{j+k-1})_1^\infty$ corresponding to the operator Γ and from the sequence $\{X_k\}_1^\infty$. In fact, from (3.3) we successively obtain: when $j = 0$

$$\gamma_0 = -\sum_{k=2}^{\infty} \gamma_{k-1} X_k = -\sum_{k=1}^{\infty} \gamma_k X_{k+1},$$

when $j = -1$

$$\gamma_{-1} = -\sum_{k=2}^{\infty} \gamma_{k-2} X_k = \sum_{k=0}^{\infty} \gamma_k X_{k+2},$$

etc.

For the operators $X \in [N_1, l_2^+(N_1)]$ and $Y \in [N_1, l_2^+(N_2)]$ satisfying (3.1) when $\rho \neq 0$ we have

$$X^* T_1^k X = Y^* T_2^{*k} Y \quad (k = 0, 1, 2, \ldots). \tag{3.4}$$

In fact, when $k \geqslant 0$

$$Y^* T_2^{*k} Y = \rho^{-1} Y^* T_2^{*k} \Gamma X = \rho^{-1} Y^* \Gamma T_1^k X = X^* T_1^k X.$$

The relations obtained in this subsection assume a particularly perspicuous form in a function theoretic interpretation.

Let us consider the following isomorphisms of the spaces $l_2(N_1)$ and $l_2(N_2)$ onto the spaces $\mathbf{L}_2(N_1)$ and $\mathbf{L}_2(N_2)$ respectively:

$$\xi_+(\zeta) = \sum_{-\infty}^{\infty} \xi_k \zeta^{k-1}, \quad \text{if } \{\xi_k\}_{-\infty}^{\infty} \in l_2(N_1),$$

and

$$\eta_-(\zeta) = \sum_{-\infty}^{\infty} \eta_k \zeta^{-k}, \quad \text{if } \{\eta_k\}_{-\infty}^{\infty} \in l_2(N_2).$$

The equalities here are understood in the sense of convergence in the mean. Under these isomorphisms the subspaces $l_2^+(N_1)$ and $l_2^+(N_2)$ go over into the Hardy subspaces $\mathbf{H}_2(N_1)$ and $\mathbf{H}_2^-(N_2) = \mathbf{L}_2(N_2) \ominus \mathbf{H}_2(N_2)$. The orthogonal projections onto these subspaces are denoted by π_+ and π_-. If $\widetilde{\Gamma} \in H(N_1, N_2)$ is an arbitrary full Hankel operator defined by the matrix $(\gamma_{j+k-1})_{-\infty}^{\infty}$, it goes over under the above isomorphisms into an operator of "multiplication" by the operator-valued function $F(\zeta)$ (with values in $[N_1, N_2]$) defined by

$$F(\zeta)h = \sum_{-\infty}^{\infty} \gamma_j \zeta^{-j} h, \quad h \in N_1,$$

i.e. for a given function

$$\xi(\zeta) = \sum_{-\infty}^{\infty} \xi_k \zeta^{k-1} \in \mathbf{L}_2(N_1)$$

we will have

$$F(\zeta)\zeta(\xi) = \sum_{-\infty}^{\infty} \eta_j \zeta^{-j} \in \mathbf{L}_2(N_2),$$

where

$$\eta_j = \sum_{k=-\infty}^{\infty} \gamma_{j+k-1} \xi_k \qquad (j = 0, \pm 1, \pm 2, \ldots),$$

with

$$\|\widetilde{\Gamma}\| = \|F\|_\infty = \operatorname*{ess\,sup}_{\zeta \in C} \|F(\zeta)\|.$$

We define a pair of operator-valued functions $X_+(\zeta)$ and $Y_-(\zeta)$ corresponding respectively to the operators X and Y defined above:

$$X_+(\zeta)h = [Xh]_+(\zeta) = \sum_{k=1}^{\infty} X_k \zeta^{k-1} h,$$

$$Y_-(\zeta)h = [Yh]_-(\zeta) = \sum_{k=1}^{\infty} Y_k \zeta^{-k} h, \quad h \in N_1.$$

Relations (3.1) and (3.2) can then be rewritten in the form

$$(\pi_- F X_+)(\zeta) = \rho Y_-(\zeta), \quad (\pi_+ F^* Y_-)(\zeta) = \rho X_+(\zeta), \tag{3.5}$$

$$F(\zeta) X_+(\zeta) = \rho Y_-(\zeta), \quad F^*(\zeta) Y_-(\zeta) = \rho X_+(\zeta). \tag{3.6}$$

The system (3.4) means that the functions

$$(X_+(\zeta)h', X_+(\zeta)h'') \quad \text{and} \quad (Y_-(\zeta)h', Y_-(\zeta)h'') \quad (h', h'' \in N_1)$$

have the same Fourier coefficients, i.e. coincide almost everywhere. *The following equality holds in this (weak) sense*:

$$X_+^*(\zeta) X_+(\zeta) = Y_-^*(\zeta) Y_-(\zeta). \tag{3.7}$$

The functional considerations are simplified and reduce to interesting results if it is assumed that the spaces N_1 and N_2 are finite-dimensional, i.e. that $n_k = \dim N_k < \infty$ $(k = 1, 2)$. The mean square convergence of the Fourier series of an operator-valued function with values in $[N_j, N_k]$ $(j, k = 1, 2)$ in the strong sense then implies its uniform mean square convergence. For example, in the case of the functions $X_+(\zeta)$ and $Y_-(\zeta)$ introduced above we will have

$$\lim_{n\to\infty} \int_C \left\| X_+(\zeta) - \sum_{k=1}^{n} X_k \zeta^{k-1} \right\|^2 |d\zeta| = 0, \qquad \int_C \|X_+(\zeta)\|^2 \, |d\zeta| < \infty;$$

$$\lim_{n\to\infty} \int_C \left\| Y_-(\zeta) - \sum_{k=1}^{n} Y_k \zeta^{-k} \right\|^2 |d\zeta| = 0, \qquad \int_C \|Y_-(\zeta)\|^2 |d\zeta| < \infty.$$

The determinant $\det X_+(\zeta)$ is a function of the Hardy class $\mathbf{H}_{2/n_1}$. Since $\det X_1$ $(= 1) \neq 0$, we have $\det X_+(\zeta) \neq 0$ almost everywhere on the unit circle. Therefore (3.6) can be rewritten in the form

$$F(\zeta) = \rho Y_-(\zeta) X_+^{-1}(\zeta). \tag{3.8}$$

Relations (3.8) and (3.7) show that the operator-valued function $F(\zeta)$ isometrically maps all of N_1 into N_2 for almost all ζ. It can be shown that $X_+(\zeta)$ is an *outer* operator-valued function, i.e.

$$(0 =) \ln \det X_1 = \ln \det X_+(0) = \frac{1}{2\pi} \int_C \ln |\det X_+(\zeta)| \, |d\zeta|.$$

$\zeta^{-1} Y_-(\zeta^{-1})$ is also an outer operator-valued function in the sense of the general definition given, for example, in [16].

3. Suppose now that $\Gamma \in H^+(N_1, N_2)$, $\rho > \|\Gamma\|$ and $\Gamma^{(1)}$ is an arbitrary canonical one-step $(\Gamma; \rho)$-extension defined by an isometric parameter $\mathbf{E}$ $(\mathbf{E}^*\mathbf{E} = E_1)$ through formulas (1.1) and (1.2). It follows from Theorem 2.1 that the constructions of subsection 2 are applicable for $\Gamma^{(1)}$ $(\in H^+(N_1, N_2))$ and that the operators $\mathbf{P} + \overset{*}{\mathbf{Q}}\mathbf{E}$ and $\mathbf{Q} + \overset{*}{\mathbf{P}}\mathbf{E}$ will play the roles of the operators X and Y respectively (in these considerations the normalization $X_1 = E_1$ is not essential; it is only important that the operator X_1 have a bounded inverse). Thus, if $\widetilde{\Gamma}^{(1)}$ is the (unique) full $\Gamma^{(1)}$-miniextension, it can be determined from the relations

$$\widetilde{\Gamma}^{(1)}(\mathbf{P} + \overset{*}{\mathbf{Q}}\mathbf{E}) = \rho(\mathbf{Q} + \overset{*}{\mathbf{P}}\mathbf{E}), \qquad \widetilde{\Gamma}^{(1)*}(\mathbf{Q} + \overset{*}{\mathbf{P}}\mathbf{E}) = \rho(\mathbf{P} + \overset{*}{\mathbf{Q}}\mathbf{E}).$$

Since $\Gamma^{(1)} T_1 = T_2^* \Gamma^{(1)} = \Gamma$, it follows that $\widetilde{\Gamma} = \widetilde{\Gamma}^{(1)} U_1 = U_2^* \widetilde{\Gamma}^{(1)}$ is a full $(\Gamma; \rho)$-extension. For it we have

$$\widetilde{\Gamma}(\mathbf{P} + \overset{*}{\mathbf{Q}}\mathbf{E}) = \rho U_2(\mathbf{Q} + \overset{*}{\mathbf{P}}\mathbf{E}), \qquad \widetilde{\Gamma}^+ U_2(\mathbf{Q} + \overset{*}{\mathbf{P}}\mathbf{E}) = \rho(\mathbf{P} + \overset{*}{\mathbf{Q}}\mathbf{E}). \tag{3.9}$$

We define a set of operator-valued functions $P_\pm(\zeta)$ and $Q_\pm(\zeta)$ by means of the formulas

$$\mathbf{P}_+(\zeta)h = [\mathbf{P}h]_+(\zeta), \qquad \mathbf{Q}_+(\zeta)h = [\overset{*}{\mathbf{Q}}h]_+(\zeta) \qquad (h \in N_1),$$

$$\mathbf{P}_-(\zeta)g = [\overset{*}{\mathbf{P}}g]_+(\bar{\zeta}), \qquad \mathbf{Q}_-(\zeta)g = [\mathbf{Q}g]_+(\bar{\zeta}) \qquad (g \in N_2).$$

If the full $(\Gamma; \rho)$-extension $\widetilde{\Gamma}$ is defined by the matrix $(\gamma_{j+k-1})_{-\infty}^{\infty}$, relations (3.9) can then be rewritten for the function $F(\zeta)$ defined by (3.8) in the form

$$F(\zeta)[\mathbf{P}_+(\zeta) + \mathbf{Q}_+(\zeta)\mathbf{E}] = \rho[\mathbf{Q}_-(\zeta) + \mathbf{P}_-(\zeta)\mathbf{E}],$$
$$F^*(\zeta)[\mathbf{Q}_-(\zeta) + \mathbf{P}_-(\zeta)\mathbf{E}] = \rho[\mathbf{P}_+(\zeta) + \mathbf{Q}_+(\zeta)\mathbf{E}]. \tag{3.10}$$

Thus the blocks γ_k $(k = 0, \pm 1, \pm 2, \ldots)$ of the matrix $(\gamma_{j+k-1})_{-\infty}^{\infty}$ corresponding to the full $(\Gamma; \rho)$-extension $\widetilde{\Gamma}$ are the Fourier coefficients of the operator function $F(\zeta)$, while the unknown blocks γ_k $(k = 0, -1, -2, \ldots)$ can be uniquely determined from the known blocks and the functions

$$[\mathbf{P}_+(\zeta) + \mathbf{Q}_+(\zeta)\mathbf{E}] \quad \text{and} \quad [\mathbf{Q}_-(\zeta) + \mathbf{P}_-(\zeta)\mathbf{E}]$$

by means of the functional relations (3.10). In this sense

$$F(\zeta) = \rho[\mathbf{Q}_-(\zeta) + \mathbf{P}_-(\zeta)\mathbf{E}][\mathbf{P}_+(\zeta) + \mathbf{Q}_+(\zeta)\mathbf{E}]^{-1}. \tag{3.11}$$

A full $(\Gamma; \rho)$-extension obtained in this way will be called *canonical.*

When n_k $(= \dim N_k,\ k = 1, 2) < \infty$, formula (3.11) is understood in the usual sense, i.e. as an equality holding almost everywhere for $|\zeta| = 1$.

§4. Basic identities

Here we establish a series of identities, which play a fundamental role in the sequel, for the four operators $\mathbf{P}, \overset{*}{\mathbf{P}}, \overset{*}{\mathbf{Q}}, \mathbf{Q}$. We recall that these operators define four operator-valued functions $\mathbf{P}_+(\zeta), \mathbf{P}_-(\zeta), \mathbf{Q}_+(\zeta), \mathbf{Q}_-(\zeta)$ by means of the formulas

$$\mathbf{P}_+(\zeta)h = [\mathbf{P}h]_+(\zeta), \qquad \mathbf{Q}_-(\zeta)h = [\mathbf{Q}h]_+(\bar{\zeta}) \qquad (h \in N_1),$$
$$\mathbf{P}_-(\zeta)g = [\overset{*}{\mathbf{P}}g]_+(\bar{\zeta}), \qquad \mathbf{Q}_+(\zeta)g = [\overset{*}{\mathbf{Q}}g]_+(\zeta) \qquad (g \in N_2), \tag{4.1}$$

where, for a given vector $\xi = \{\xi_k\}_1^{\infty} \in l_2^+(N)$, the symbol $\xi_+(\zeta)$ denotes the vector-valued function $\xi_+(\zeta) = \Sigma_1^{\infty} \xi_k \zeta^{k-1}$ of the Hardy class $\mathbf{H}_2(N)$. The product $\mathbf{P}_+^*(\zeta)\mathbf{P}_+(\zeta)$, which will be encountered below, can be understood in the weak sense:

$$(\mathbf{P}_+^*(\zeta)\mathbf{P}_+(\zeta)h', h'') \overset{\text{def}}{=\!=} (\mathbf{P}_+(\zeta)h', \mathbf{P}_+(\zeta)h'') \qquad (h', h'' \in N_1).$$

The right side of this relation is defined for almost all ζ, since the factors of the scalar product are vector-valued functions of the Hardy class $\mathbf{H}_2(N_1)$. The other products of this kind for the four functions in question are understood analogously.

THEOREM 4.1. *The following identities hold for the operators* $\mathbf{P}, \overset{*}{\mathbf{P}}, \overset{*}{\mathbf{Q}}$ *and* $\mathbf{Q}$ *constructed with respect to* Γ $(\in H^+(N_1, N_2))$ *and* ρ $(> \|\Gamma\|)$:

$$\mathbf{P}_+^*(\zeta)\mathbf{P}_+(\zeta) - \mathbf{Q}_-^*(\zeta)\mathbf{Q}_-(\zeta) = E_1,$$
$$\mathbf{P}_-^*(\zeta)\mathbf{P}_-(\zeta) - \mathbf{Q}_+^*(\zeta)\mathbf{Q}_+(\zeta) = E_2, \qquad \textit{(almost everywhere)} \tag{4.2}$$
$$\mathbf{P}_-^*(\zeta)\mathbf{Q}_-(\zeta) - \mathbf{Q}_+^*(\zeta)\mathbf{P}_+(\zeta) = 0.$$

PROOF. We will show that

$$\begin{gathered}\mathbf{P}^*T_1^k\mathbf{P} - \mathbf{Q}^*T_2^{*k}\mathbf{Q} = \delta_{0k}E_1, \qquad \overset{*}{\mathbf{P}}{}^*T_2^k\overset{*}{\mathbf{P}} - \overset{*}{\mathbf{Q}}{}^*T_1^{*k}\overset{*}{\mathbf{Q}} = \delta_{0k}E_2, \\ \overset{*}{\mathbf{P}}{}^*T_2^k\mathbf{Q} = \overset{*}{\mathbf{Q}}{}^*T_1^{*k}\mathbf{P}, \qquad \overset{*}{\mathbf{P}}{}^*T_2^{*k}\mathbf{Q} = \overset{*}{\mathbf{Q}}{}^*T_1^k\mathbf{P},\end{gathered} \tag{4.3}$$

where $k = 0, 1, 2, \ldots$ and $\delta_{00} = 1$, while $\delta_{0k} = 0$ for $k > 0$.

To this end we note that

$$T_2\Gamma\mathbf{P} = \rho\mathbf{Q}, \qquad \Gamma^*T_2^*\mathbf{Q} = \rho\mathbf{P} - G.$$

From the second equality we get that $\mathbf{P}^* = \rho^{-1}(G + \mathbf{Q}^*T_2\Gamma)$. Therefore

$$\mathbf{P}^*T_1^k\mathbf{P} = \rho^{-1}(GP_1 + \mathbf{Q}^*T_2\Gamma)T_1^k\mathbf{P} = \rho^{-1}\mathbf{Q}^*T_2^{*k}T_2\Gamma\mathbf{P} = \mathbf{Q}^*T_2^{*k}\mathbf{Q}$$

for $k > 0$ and

$$\mathbf{P}^*\mathbf{P} - \mathbf{Q}^*\mathbf{Q} = \mathbf{P}^*\mathbf{P} - \rho^{-2}\mathbf{P}^*\Gamma^*\Gamma\mathbf{P} = \rho^{-2}\mathbf{P}^*(\rho^2 I_1 - \Gamma^*\Gamma)\mathbf{P} = GP_1R_{\rho^2}G = E_1.$$

Thus the first system of equalities in (4.3) is proved. The second is deduced analogously from the relations

$$T_1\Gamma^*\overset{*}{\mathbf{P}} = \rho\overset{*}{\mathbf{Q}}, \qquad \Gamma T_1^*\overset{*}{\mathbf{Q}} = \rho\overset{*}{\mathbf{P}} - \overset{*}{G}.$$

Further, we have for $k \geqslant 0$

$$\begin{aligned}\overset{*}{\mathbf{P}}{}^*T_2^k\mathbf{Q} &= \rho^{-1}(\overset{*}{G}P_2 + \overset{*}{\mathbf{Q}}{}^*T_1\Gamma^*)T_2^k\mathbf{Q} = \rho^{-1}\overset{*}{\mathbf{Q}}{}^*T_1^{*k}\Gamma^*\Gamma R_{\rho^2}G \\ &= \rho\overset{*}{\mathbf{Q}}{}^*T_1^{*k}R_{\rho^2}G = \overset{*}{\mathbf{Q}}{}^*T_1^{*k}\mathbf{P}\end{aligned}$$

and for $k > 0$

$$\overset{*}{\mathbf{Q}}{}^*T_1^k\mathbf{P} = \rho^{-1}\overset{*}{\mathbf{P}}{}^*\Gamma T_1^*T_1^k\mathbf{P} = \rho^{-1}\overset{*}{\mathbf{P}}{}^*T_2^{*k}T_2\Gamma\mathbf{P} = \overset{*}{\mathbf{P}}{}^*T_2^{*k}\mathbf{Q}.$$

Thus the last system of equalities in (4.3) is proved.

From the first system of equalities in (4.3) we get for $h', h'' \in N_1$

$$(T_1^k\mathbf{P}h', \mathbf{P}h'') - (\mathbf{Q}h', T_2^k\mathbf{Q}h'') = \delta_{0k}(h', h''),$$

i.e.

$$\frac{1}{2\pi}\int_C \zeta^k(\mathbf{P}_+(\zeta)h', \mathbf{P}_+(\zeta)h'')|d\zeta| - \frac{1}{2\pi}\int_C \zeta^k(\mathbf{Q}_-(\zeta)h', \mathbf{Q}_-(\zeta)h'')|d\zeta| = \delta_{0k}(h', h'')$$

for $k \geqslant 0$. By virtue of the arbitrariness of h' and h'' $(\in N_1)$ the latter system of equalities holds for all integral k. Thus the first identity in (4.2) is proved. The remaining two identities in (4.2) are obtained analogously from (4.3). The theorem is proved.

§5. Analytic properties of the functions $\mathbf{P}_\pm(\zeta)$ and $\mathbf{Q}_\pm(\zeta)$ in the finite-dimensional case ($\dim N_k < \infty$, $k = 1, 2$)

1. In this and the final section we will assume that $n_1 = \dim N_1 \leqslant n_2 = \dim N_2 < \infty$. The operator-valued functions $\mathbf{P}_+(\zeta)$ and $\mathbf{P}_-(\zeta)$ introduced above can

then be regarded as square-matrix-valued functions of orders n_1 and n_2 respectively, while $\mathbf{Q}_+(\zeta)$ and $\mathbf{Q}_-(\zeta)$ can be regarded as $n_1 \times n_2$ and $n_2 \times n_1$ matrix-valued functions respectively.

Consider the square-matrix-valued function

$$\mathbf{U}(\zeta) = \begin{pmatrix} \mathbf{P}_+(\zeta) & \mathbf{Q}_+(\zeta) \\ \mathbf{Q}_-(\zeta) & \mathbf{P}_-(\zeta) \end{pmatrix},$$

of order $n_1 + n_2$, and the matrix

$$\mathbf{J} = \begin{pmatrix} E_1 & 0 \\ 0 & -E_2 \end{pmatrix}.$$

It is easily seen that the identities (4.2) express the fact that $\mathbf{U}(\zeta)$ satisfies the relation

$$\mathbf{U}^*(\zeta)\mathbf{J}\mathbf{U}(\zeta) = \mathbf{J} \qquad \text{(almost everywhere)},$$

i.e. $\mathbf{J}$ is unitary (a.e.). Since $n_1 + n_2 < \infty$, the function $\mathbf{U}^*(\zeta)$ is also $\mathbf{J}$-unitary, i.e. $\mathbf{U}(\zeta)\mathbf{J}\mathbf{U}^*(\zeta) = \mathbf{J}$, and hence the following identities hold in conjunction with (4.2):

$$\begin{aligned} &\mathbf{P}_+(\zeta)\mathbf{P}_+^*(\zeta) - \mathbf{Q}_+(\zeta)\mathbf{Q}_+^*(\zeta) = E_1, \\ &\mathbf{P}_-(\zeta)\mathbf{P}_-^*(\zeta) - \mathbf{Q}_-(\zeta)\mathbf{Q}_-^*(\zeta) = E_2, \qquad \text{(almost everywhere)}, \\ &\mathbf{P}_+(\zeta)\mathbf{Q}_-^*(\zeta) - \mathbf{Q}_+(\zeta)\mathbf{P}_-^*(\zeta) = 0. \end{aligned} \tag{5.1}$$

2. The identities (4.2) (or (5.1)) imply that the matrices $\mathbf{P}_\pm(\zeta)$ are invertible and that $\|\mathbf{P}_\pm^{-1}(\zeta)\| \leqslant 1$ (a.e.). Consider the function

$$\chi(\zeta) = \mathbf{P}_+^{-1}(\zeta)\mathbf{Q}_+(\zeta) \ (= \mathbf{Q}_-^*(\zeta)\mathbf{P}_-^{*-1}(\zeta)). \tag{5.2}$$

On the basis of (5.1) we have

$$\|\chi(\zeta)\|^2 = \|\chi(\zeta)\chi^*(\zeta)\| = \|E_1 - \mathbf{P}_+^{*-1}(\zeta)\mathbf{P}_+^{*-1}(\zeta)\| = 1 - \|\mathbf{P}_+(\zeta)\|^{-2} < 1 \qquad \text{(a.e.)}.$$

Since all of the elements of the matrices $\mathbf{P}_\pm(\zeta)$ are functions from the space $\mathbf{L}_2$, we have

$$\frac{1}{2\pi}\int_C \|\mathbf{P}_\pm(\zeta)\|^2 |d\zeta| \leqslant \frac{1}{2\pi}\int_C \operatorname{Tr}(\mathbf{P}_\pm^*(\zeta)\mathbf{P}_\pm(\zeta))|d\zeta| < \infty.$$

Therefore

$$\frac{1}{2\pi}\int_C (1 - \|\chi(\zeta)\|)^{-1}|d\zeta| \leqslant \frac{1}{\pi}\int_C (1 - \|\chi(\zeta)\|)^2 |d\zeta| = \frac{1}{\pi}\int_C \|\mathbf{P}_+(\zeta)\|^2 |d\zeta| < \infty,$$

so that

$$(1 - \|\chi(\zeta)\|)^{-1} \in \mathbf{L}_1. \tag{5.3}$$

As was shown at the end of §3, for every isometric matrix $\mathbf{E}$ ($\mathbf{E}^*\mathbf{E} = E_1$) the elements of the matrix-valued function

$$F_{\mathbf{E}}(\zeta) = \rho[\mathbf{Q}_-(\zeta) + \mathbf{P}_-(\zeta)\mathbf{E}][\mathbf{P}_+(\zeta) + \mathbf{Q}_+(\zeta)\mathbf{E}]^{-1} \tag{5.4}$$

are bounded functions, and its Fourier coefficients $\{\widetilde{\gamma}_j(\mathbf{E})\}_{-\infty}^{\infty}$ are the blocks of a canonical full $(\Gamma; \rho)$-extension $\widetilde{\Gamma}$, so that $\widetilde{\gamma}_j(\mathbf{E}) = \gamma_j$ for $j > 0$.

By virtue of (4.2), definition (5.4) is equivalent to

$$F_{\mathrm{E}}(\zeta) = \rho[\mathbf{P}_-^*(\zeta) + \mathbf{E}\mathbf{Q}_-^*(\zeta)]^{-1}[\mathbf{Q}_+^*(\zeta) + \mathbf{E}\mathbf{P}_+^*(\zeta)]. \tag{5.5}$$

From the same identities it follows that for any contractions $\mathbf{E}_1$ and $\mathbf{E}_2$

$$F_{\mathrm{E}_2}(\zeta) - F_{\mathrm{E}_1}(\zeta) = \rho[\mathbf{P}_-^*(\zeta) + \mathbf{E}_2\mathbf{Q}_-^*(\zeta)]^{-1}(\mathbf{E}_2 - \mathbf{E}_1)[\mathbf{Q}_+(\zeta)\mathbf{E}_1 + \mathbf{P}_+(\zeta)]^{-1}; \tag{5.6}$$

here (5.5) is used for F_{E_2}, while (5.4) is used for F_{E_1}.

We will say that a matrix-valued function belongs to the Hardy class $\mathbf{H}_p$ $(0 < p \leq \infty)$ if all of its elements are functions of this class. According to what has been said, for isometric $\mathbf{E}_1$ and $\mathbf{E}_2$ the matrix-valued functions F_{E_1} and F_{E_2} have the same "negative" Fourier coefficients, and hence their difference belongs to $\mathbf{H}_\infty$. By construction, the matrix-valued functions

$$[\mathbf{P}_+(\zeta) + \mathbf{Q}_+(\zeta)\mathbf{E}_1] \qquad \text{and} \qquad [\mathbf{P}_-^*(\zeta) + \mathbf{E}_2\mathbf{Q}_-^*(\zeta)]$$

belong to $\mathbf{H}_2$. It follows on the basis of (5.6) that

$$(\mathbf{E}_2 - \mathbf{E}_1)[\mathbf{P}_+(\zeta) + \mathbf{Q}_+(\zeta)\mathbf{E}_1]^{-1} \qquad \text{and} \qquad [\mathbf{P}_-^*(\zeta) + \mathbf{E}_2\mathbf{Q}_-^*(\zeta)]^{-1}(\mathbf{E}_2 - \mathbf{E}_1)$$

also belong to $\mathbf{H}_2$.

We now note that if $\mathbf{E}_1$ is a fixed isometric matrix, an isometric matrix $\mathbf{E}_2$ can be chosen so that the product $(\mathbf{E}_2^* - \mathbf{E}_1^*)(\mathbf{E}_2 - \mathbf{E}_1)$ is a nondegenerate matrix (for example, by putting $\mathbf{E}_2 = -\mathbf{E}_1$), while if the value of $\mathbf{E}_2$ is fixed, the matrix $\mathbf{E}_1$ can be suitably chosen so that the range of the mapping $\mathbf{E}_2 - \mathbf{E}_1$ includes a preassigned vector $h \in N_2$. It follows by virtue of these considerations that the matrix-valued functions

$$[\mathbf{P}_+(\zeta) + \mathbf{Q}_+(\zeta)\mathbf{E}_1]^{-1} \qquad \text{and} \qquad [\mathbf{P}_-^*(\zeta) + \mathbf{E}_2\mathbf{Q}_-^*(\zeta)]^{-1}$$

will belong to the class $\mathbf{H}_2$.

3. Using an arbitrary $n_2 \times n_1$ contraction matrix $\mathbf{E}$, we form a function of class $\mathbf{H}_1$:

$$\Phi_{\mathrm{E}}(z) = [\mathbf{P}_+(z) + \mathbf{Q}_+(z)\mathbf{E}]^{-1}[\mathbf{P}_+(z) - \mathbf{Q}_+(z)\mathbf{E}] \quad (|z| < 1).$$

For its boundary values we get from (5.2) that

$$\Phi_{\mathrm{E}}(\zeta) = [E_1 + \chi(\zeta)\mathbf{E}]^{-1}[E_1 - \chi(\zeta)\mathbf{E}],$$

and since $\|\chi(\zeta)\| < 1$ (a.e.),

$$\Phi_{\mathrm{E}}^*(\zeta) + \Phi_{\mathrm{E}}(\zeta) > 0 \quad \text{(a.e.)}.$$

Using the representation of $\Phi_{\mathrm{E}}(z)$ on the disk (a.e.) $|z| < 1$ in the form of a Poisson integral of its boundary values, we immediately see that the inequality $\Phi_{\mathrm{E}}^*(z) + \Phi_{\mathrm{E}}(z) > 0$ is satisfied. Therefore $\Phi_{\mathrm{E}}(z)$ is representable on the disk $|z| < 1$ in the form

$$\Phi_{\mathrm{E}}(z) = [E_1 + \widetilde{\chi}(z)]^{-1}[E_1 - \widetilde{\chi}(z)],$$

where $\widetilde{\chi}(z)$ is a holomorphic matrix-valued function such that $\|\widetilde{\chi}(z)\| < 1$. The boundary values $\widetilde{\chi}(\zeta)$ of $\widetilde{\chi}(z)$ must coincide with $\chi(\zeta)\mathbf{E}$ at almost every point ζ. Consequently, $\chi(\zeta)\mathbf{E}$ is a matrix-valued function of class $\mathbf{H}_\infty$.

The fact that this assertion is valid for an arbitrary choice of the isometric mapping $\mathbf{E}$ implies that $\chi(\zeta)$ also belongs to $\mathbf{H}_\infty$.

Finally, multiplying $[\mathbf{P}_+(\zeta) + \mathbf{Q}_+(\zeta)\mathbf{E}]^{-1}$ from the left by $[E_1 + \chi(\zeta)\mathbf{E}]$, we arrive at the conclusion that $\mathbf{P}_+^{-1}(\zeta)$ belongs to $\mathbf{H}_2$ and hence, by virtue of its boundedness, to $\mathbf{H}_\infty$. It can be shown in precisely the same way that $(\mathbf{P}_-^*)^{-1}(\zeta)$ belongs to $\mathbf{H}_\infty$.

4. Since $\det \mathbf{P}_+(\zeta) \in \mathbf{H}_{2/n_1}$ and $[\det \mathbf{P}_+(\zeta)]^{-1} \in \mathbf{H}_\infty$, we can draw the following conclusions:

1°. $\det \mathbf{P}_+(\zeta)$ is an outer function, and hence $\mathbf{P}_+(\zeta)$ is an outer matrix-valued function, i.e.

$$\det \mathbf{P}_+(z) = \exp\left\{\frac{1}{2\pi}\int_C \frac{\zeta + z}{\zeta - z} \ln|\det \mathbf{P}_+(\zeta)|\,|d\zeta|\right\} \tag{5.7}$$

(here we have made use of the fact that $\det \mathbf{P}_+(0) > 0$). In precisely the same way

$$\det(\mathbf{P}_-^*)(z) = \exp\left\{\frac{1}{2\pi}\int_C \frac{\zeta + z}{\zeta - z} \ln|\det \mathbf{P}_-^*(\zeta)|\,|d\zeta|\right\}. \tag{5.8}$$

By virtue of the second identity in (4.2) and the first identity in (5.1) we have

$$|\det \mathbf{P}_+(\zeta)|^2 = \det(E_1 - \mathbf{Q}_+(\zeta)\mathbf{Q}_+^*(\zeta)) = \det(E_2 - \mathbf{Q}_+^*(\zeta)\mathbf{Q}_+(\zeta)) = |\det \mathbf{P}_-(\zeta)|^2.$$

Equalities (5.7) and (5.8) therefore imply

2°. $$\det \mathbf{P}_+(z) = \det(\mathbf{P}_-^*)(z) \quad (|z| < 1). \tag{5.9}$$

As is well known (see, for example, [17]), condition (5.3) is sufficient for the solvability of the left factorization problem in $\mathbf{H}_\infty$ for the matrix-valued function $E_1 - \chi(\zeta)\chi^*(\zeta)$. The solution of this problem is unique in the class of outer matrix-valued functions taking positive values at $z = 0$. Since

$$E_1 - \chi(\zeta)\chi^*(\zeta) = \mathbf{P}_+^{-1}(\zeta)\mathbf{P}_+^{-1*}(\zeta), \tag{5.10}$$

it follows that $\mathbf{P}_+^{-1}(\zeta)$ and hence $\mathbf{P}_+(\zeta)$ and $\mathbf{Q}_+(\zeta)$ $(= \mathbf{P}_+(\zeta)\chi(\zeta))$ are uniquely determined by $\chi(\zeta)$. Analogously, $(\mathbf{P}_-^*)^{-1}(\zeta)$ is the unique solution of the right factorization problem

$$E_2 - \chi^*(\zeta)\chi(\zeta) = \mathbf{P}_-^{-1}(\zeta)(\mathbf{P}_-^*)^{-1}(\zeta) \tag{5.11}$$

in the class of outer matrix-valued functions taking positive values at $z = 0$, and hence $\mathbf{P}_-(\zeta)$ and $\mathbf{Q}_-(\zeta)$ $(= \mathbf{P}_-(\zeta)\chi^*(\zeta))$ are uniquely determined by $\chi(\zeta)$.

Thus we have

THEOREM 5.1. *Suppose $\Gamma \in H^+(N_1, N_2)$, $\rho > \|\Gamma\|$ and n_k $(= \dim N_k,\ k = 1, 2) < \infty$. Then the matrix-valued function $\zeta(z)$ defined by (5.2) is holomorphic in the disk $|z| < 1$ and has the properties*

$$1)\ \chi(0) = 0,\quad 2)\ \|\chi(z)\| < 1\ (|z| < 1),\quad 3)\ (1 - \|\chi(\zeta)\|)^{-1} \in \mathbf{L}_1. \tag{5.12}$$

The functions $\mathbf{P}_\pm(\zeta)$ and $\mathbf{Q}_\pm(\zeta)$ are uniquely recovered from $\chi(\zeta)$.

5. Inasmuch as the left and right half-radii $\mathbf{R}_{\mathfrak{l}}(\Gamma;\rho)$ and $\mathbf{R}_{\mathfrak{r}}(\Gamma;\rho)$ of the operator quasi-ball (1.2) are expressed in terms of $\mathbf{P}_+(0)$ and $(\mathbf{P}^*_-)(0)$ respectively:

$$\mathbf{R}_{\mathfrak{l}}(\Gamma;\rho)=\rho^{1/2}\mathbf{P}_+(0),\qquad \mathbf{R}_{\mathfrak{r}}(\Gamma;\rho)=\rho^{1/2}(\mathbf{P}^*_-)(0),$$

equality (5.9) implies

3°. $$\det \mathbf{R}_{\mathfrak{l}}(\Gamma;\rho)=\det \mathbf{R}_{\mathfrak{r}}(\Gamma;\rho). \tag{5.13}$$

Letting $\rho\downarrow\|\Gamma\|$, we get

THEOREM 5.2. *Suppose* $\dim N_k<\infty$ $(k=1,2)$. *Then for any* $\Gamma\in H^+(N_1,N_2)$

$$\det \mathbf{R}_{\mathfrak{l}}(\Gamma)=\det \mathbf{R}_{\mathfrak{r}}(\Gamma)$$

and hence, if one of the half-radii $\mathbf{R}_{\mathfrak{l}}(\Gamma)$ *or* $\mathbf{R}_{\mathfrak{r}}(\Gamma)$ *is nondegenerate, so is the other.*

§6. Description of the full $(\Gamma;\rho)$-extensions in the finite-dimensional case

Let $\widetilde{\Gamma}=(\widetilde{\gamma}_{j+k-1})_{-\infty}^{\infty}$ be a full $(\Gamma;\rho)$-extension of a Hankel block matrix $\Gamma=(\gamma_{j+k-1})_1^\infty$, and let $F(\zeta;\widetilde{\Gamma})$ be the $n_2\times n_1$ matrix-valued function with Fourier series

$$F(\zeta;\widetilde{\Gamma})=\sum_{-\infty}^{\infty}\widetilde{\gamma}_j\zeta^{-j}. \tag{6.1}$$

It has already been noted that for any full $(\Gamma;\rho)$-extension $\widetilde{\Gamma}$ the function $F(\zeta;\widetilde{\Gamma})$ is measurable and bounded in the sense that

$$\operatorname*{ess\,sup}_{\zeta\in C}\|F(\zeta;\widetilde{\Gamma})\|\leqslant\rho\qquad(\rho>\|\Gamma\|). \tag{6.2}$$

The matrix-valued functions $F(\zeta;\widetilde{\Gamma})$ connected with the $(\Gamma;\rho)$-extensions $\widetilde{\Gamma}$ by (6.1) will be called $(\Gamma;\rho)$-*functions.* The $(\Gamma;\rho)$-functions corresponding to the canonical $(\Gamma;\rho)$-extensions are naturally called *canonical.* We emphasize that, according to the function-theoretic interpretation made in §3, any $n_2\times n_1$ matrix-valued function $F(\zeta)$ satisfying (6.2) and such that

$$\frac{1}{2\pi}\int_C F(\zeta)\zeta^j|d\zeta|=\gamma_j\qquad(j=1,2,\dots), \tag{6.3}$$

is a $(\Gamma;\rho)$-function. Therefore the problem of describing all of the $(\Gamma;\rho)$-extensions $\widetilde{\Gamma}$ reduces to the problem of describing all of the $(\Gamma;\rho)$-functions, i.e. all of the $n_2\times n_1$ matrix-valued functions satisfying (6.3) and (6.2).

Let $B(n_2\times n_1)$ denote the set of $n_2\times n_1$ matrix-valued functions $\mathbf{E}(\zeta)$ of class $\mathbf{H}_\infty$ such that

$$\operatorname*{ess\,sup}_{\zeta\in C}\|\mathbf{E}(\zeta)\|\leqslant 1. \tag{6.4}$$

THEOREM 6.1. *The formula*

$$F_{\mathbf{E}}(\zeta)=\rho[\mathbf{Q}_-(\zeta)+\mathbf{P}_-(\zeta)\mathbf{E}(\zeta)][\mathbf{P}_+(\zeta)+\mathbf{Q}_+(\zeta)\mathbf{E}(\zeta)]^{-1} \tag{6.5}$$

establishes a one-to-one correspondence between the set of all matrix-valued functions $\mathbf{E}(\zeta)\in B(n_2\times n_1)$ *and the set of all* $(\Gamma;\rho)$-*functions.*

PROOF. Since the matrix-valued function

$$\mathbf{U}(\zeta) = \begin{pmatrix} \mathbf{P}_+(\zeta) & \mathbf{Q}_+(\zeta) \\ \mathbf{Q}_-(\zeta) & \mathbf{P}_-(\zeta) \end{pmatrix}$$

is **J**-unitary for almost all ζ, the transformation (6.5) maps the set of all $n_2 \times n_1$ matrix-valued functions satisfying (6.4) onto the set of all $n_2 \times n_1$ matrix-valued functions satisfying (6.2).(4)

It remains for us to determine the additional properties that the functions $\mathbf{E}(\zeta)$ must have in order for (6.3) to hold for the corresponding matrix-valued functions $F_{\mathbf{E}}(\zeta)$.

Let $\mathbf{E}_1(\zeta)$ and $\mathbf{E}_2(\zeta)$ be arbitrary matrix-valued functions in $B(n_2 \times n_1)$. For the difference of $F_{\mathbf{E}_2}(\zeta)$ and $F_{\mathbf{E}_1}(\zeta)$, in the same way as in §5, we get

$$\begin{aligned}(G_{\mathbf{E}_1,\mathbf{E}_2}(\zeta) &\overset{\text{def}}{=\!=})F_{\mathbf{E}_2}(\zeta) - F_{\mathbf{E}_1}(\zeta) \\ &= \rho[\mathbf{P}^*_-(\zeta) + \mathbf{E}_2(\zeta)\mathbf{Q}^*_-(\zeta)]^{-1}[\mathbf{E}_2(\zeta) - \mathbf{E}_1(\zeta)][\mathbf{P}_+(\zeta) + \mathbf{Q}_+(\zeta)\mathbf{E}_1(\zeta)]^{-1}.\end{aligned} \tag{6.6}$$

Suppose first that ess sup $\|\mathbf{E}_k(\zeta)\| < 1$ $(k = 1, 2)$. Then Theorem 5.1 and statement 1° of §5 will imply that

$$[\mathbf{P}_+(\zeta) + \mathbf{Q}_+(\zeta)\mathbf{E}_1(\zeta)]^{-1} = [E_1 + \chi(\zeta)\mathbf{E}_1(\zeta)]^{-1}\mathbf{P}_+^{-1}(\zeta) \in \mathbf{H}_\infty.$$

Analogously,

$$[(\mathbf{P}^*_-)(\zeta) + \mathbf{E}_2(\zeta)(\mathbf{Q}^*_-)(\zeta)]^{-1} \in \mathbf{H}_\infty.$$

Thus, under the above assumptions concerning $\mathbf{E}_k(\zeta)$ $(k = 1, 2)$ the matrix-valued function $G_{\mathbf{E}_1,\mathbf{E}_2}(\zeta)$ will belong to the Hardy class $\mathbf{H}_\infty$, i.e. the functions $F_{\mathbf{E}_1}(\zeta)$ and $F_{\mathbf{E}_2}(\zeta)$ will have the same "negative" coefficients.

We next put $\mathbf{E}_k(\zeta) = r\widetilde{\mathbf{E}}_k(\zeta)$ $(k = 1, 2)$, where $0 \leqslant r < 1$ and the $\widetilde{\mathbf{E}}_k(\zeta)$ are now arbitrary matrix-valued functions in $B(n_2 \times n_1)$. Since

$$\operatorname*{ess\,sup}_{\zeta\in C}\|G_{r\widetilde{\mathbf{E}}_1, r\widetilde{\mathbf{E}}_2}(\zeta)\| \leqslant 2\rho$$

and for almost all ζ

$$\lim_{r\uparrow 1} G_{r\widetilde{\mathbf{E}}_1, r\widetilde{\mathbf{E}}_2}(\zeta) = G_{\widetilde{\mathbf{E}}_1, \widetilde{\mathbf{E}}_2}(\zeta),$$

it follows that

$$\lim_{r\uparrow 1} \frac{1}{2\pi}\int_C \zeta^j G_{r\widetilde{\mathbf{E}}_1, r\widetilde{\mathbf{E}}_2}(\zeta)|d\zeta| = \frac{1}{2\pi}\int_C \zeta^j G_{\widetilde{\mathbf{E}}_1, \widetilde{\mathbf{E}}_2}(\zeta)|d\zeta| = 0 \qquad (j = 1, 2, \ldots).$$

Therefore the functions $F_{\mathbf{E}_1}(\zeta)$ and $F_{\mathbf{E}_2}(\zeta)$ have the same "negative" Fourier coefficients for any $\mathbf{E}_1(\zeta)$ and $\mathbf{E}_2(\zeta)$ in $B(n_2 \times n_1)$.

(4) We have here made use of the fact that if a matrix $A = (A_{jk})_1^2$, where the blocks A_{jk} are of dimension $n_j \times n_k$, is **J**-unitary, the transformation

$$\mathbf{E} - (A_{21} + A_{22}\mathbf{E})(A_{11} + A_{12}\mathbf{E})^{-1}$$

is a one-to-one mapping of the ball $\mathbf{E}^*\mathbf{E} \leqslant I$ of all $n_2 \times n_1$ matrix-valued contractions onto itself.

Suppose now that for some $n_2 \times n_1$ matrix-valued function $\mathbf{E}_0(\zeta)$ satisfying (6.4) the function $F_{\mathbf{E}_0}(\zeta)$ has the same negative Fourier coefficients as all of the functions $F_{\mathbf{E}}(\zeta)$ for $\mathbf{E}(\zeta) \in B(n_2 \times n_1)$.

Consider relation (6.6) when $\mathbf{E}_1(\zeta) = \mathbf{E}_0(\zeta)$ and $\mathbf{E}_2(\zeta) \equiv 0$. Since the matrix-valued function $G_{\mathbf{E}_0,0}(\zeta)$ belongs by assumption to the class $\mathbf{H}_\infty$, the matrix-valued function

$$(\mathbf{P}_-^*)(\zeta) G_{\mathbf{E}_0,0}(\zeta)\mathbf{P}_+(\zeta) \; (= -\mathbf{E}_0(\zeta)[E_1 + \chi(\zeta)\mathbf{E}_0(\zeta)]^{-1}) \tag{6.7}$$

belongs to the class $\mathbf{H}_1$. Therefore $\mathbf{H}_1$ will also contain the matrix-valued function

$$\begin{aligned}\Phi_{\mathbf{E}_0}(\zeta) &= [E_1 - \chi(\zeta)\mathbf{E}_0(\zeta)]\,[E_1 + \chi(\zeta)\mathbf{E}_0(\zeta)]^{-1} \\ &= E_1 + 2\chi(\zeta)(\mathbf{P}_-^*)(\zeta) G_{\mathbf{E}_0,0}(\zeta)\mathbf{P}_+(\zeta).\end{aligned}$$

Using precisely the same arguments as in subsection 3 of §5, we deduce from this result that $\mathbf{H}_\infty$ contains the product $\chi(\zeta)\mathbf{E}_0(\zeta)$, and hence, in view of (6.7), the matrix-valued function $\mathbf{E}_0(\zeta)$.

For the completion of the proof of the theorem it remains to recall that it was shown back in §3 that in the case of constant isometric $\mathbf{E}$ formula (6.5) gives a description of all of the canonical $(\Gamma; \rho)$-functions. The theorem is proved.

REMARK 6.1. Formula (6.5) establishes a one-to-one correspondence between the set of all matrix-valued functions $\mathbf{E}(\zeta) \in B(n_2 \times n_1)$ satisfying

$$\mathbf{E}^*(\zeta)\mathbf{E}(\zeta) = E_1$$

and the set of all $(\Gamma; \rho)$-functions $F_{\mathbf{E}}(\zeta)$ satisfying

$$F_{\mathbf{E}}^*(\zeta)F_{\mathbf{E}}(\zeta) = \rho^2 E_1.$$

2. The proof of Theorem 6.1 implies the following proposition.

Let $\chi(z)$ be an arbitrary $n_1 \times n_2$ matrix-valued function that is holomorphic in the disk $|z| < 1$ and satisfies (5.12). We define a set of four matrix-valued functions $\mathbf{P}_\pm(\zeta)$ and $\mathbf{Q}_\pm(\zeta)$ in terms of it as follows: $\mathbf{P}_+^{-1}(\zeta)$ $((\mathbf{P}_-^*)^{-1}(\zeta))$ is the outer matrix-valued function that takes a positive value at $z = 0$ and is a solution of the left (right) factorization problem (5.10) ((5.11)), while $\mathbf{Q}_+(\zeta) = \mathbf{P}_+(\zeta)\chi(\zeta)$ and $\mathbf{Q}_-(\zeta) = \mathbf{P}_-(\zeta)\chi^*(\zeta)$. If a family of matrix-valued functions $F_{\mathbf{E}}$ ($\mathbf{E} \in B(n_2 \times n_1)$) is formed from these four functions by means of (6.5) with an arbitrary $\rho > 0$, all of them will have the same "negative" Fourier coefficients $\gamma_k = c_{-k}(F_{\mathbf{E}})$ $(k = 1, 2, \ldots)$. Moreover, (6.5) will describe all of the $(\Gamma; \rho)$-functions for the Hankel block matrix $\Gamma = \Gamma(\chi; \rho) = (\gamma_{j+k-1})_1^\infty$. In this connection, however, it can turn out that $\|\Gamma\| = \rho$.

Formula (6.5) can be rewritten in the form

$$F_{\mathbf{E}}(\zeta) = \rho\Psi_-^{-1}(\zeta)[\chi^*(\zeta) + \mathbf{E}(\zeta)]\,[E_1 + \chi(\zeta)\mathbf{E}(\zeta)]^{-1}\Psi_+(\zeta), \tag{6.8}$$

where $\Psi_\pm(\zeta) = \mathbf{P}_\pm^{-1}(\zeta)$, so that $\Psi_+ \in B(n_1 \times n_1)$ and $\Psi_-^* \in B(n_2 \times n_2)$.

For the scalar case $(n_1 = n_2 = 1)$ the authors originally obtained this formula under more general assumptions on the function χ in [1b], where they also indicated when $\|\Gamma\| < \rho$.

BIBLIOGRAPHY

1. V. M. Adamjan, D. Z. Arov and M. G. Kreĭn, a) *Infinite Hankel matrices and generalized problems of Carathéodory-Fejér and F. Riesz,* Funkcional. Anal. i Priložen. **2** (1968), no. 1, 1–19 = Functional Anal. Appl. **2** (1968), 1–18. MR **38** #2591.

b) *Infinite Hankel matrices and generalized problems of Carathéodory-Fejér and I. Schur,* Funkcional. Anal. i Priložen. **2** (1968), no. 4, 1–17 = Functional Anal. Appl. **2** (1968), 269–281.

c) *Bounded operators which commute with a* C_{00} *class contraction whose rank of nonunitarity is one,* Funkcional. Anal. i Priložen. **3** (1969), no. 3, 86–87 = Functional Anal. Appl. **3** (1969), 242–243. MR **41** #5990.

d) *Analytic properties of Schmidt pairs for a Hankel operator and the generalized Schur-Takagi problem,* Mat. Sb. **86** (128) (1971), 34–75 = Math. USSR Sb. **15** (1971), 31–73. MR **45** #7505.

2. Z. Nehari, *On bounded bilinear forms,* Ann. of Math. (2) **65** (1957), 153–162. MR **18**, 633.

3. I. Schur, *Über potenzreihen, die im Innern des Einheitskreises beschränkt sind,* J. Reine Angew. Math. **147** (1917), 205–232.

4. D. Sarason, *Generalized interpolation in* H^∞, Trans. Amer. Math. Soc. **127** (1967), 179–203. MR **34** #8193.

5. Lavon B. Page, *Bounded and compact vectorial Hankel operators,* Trans. Amer. Math. Soc. **150** (1970), 529–539. MR **42** #8327.

6. R. Nevanlinna, *Über beschränkte analytische Funktionen,* Ann. Acad. Sci. Fenn. Ser. A **32** (1929), no. 7.

7. M. G. Kreĭn, *On a generalized problem of moments,* C. R. (Dokl.) Acad. Sci. URSS **44** (1944), 219–222. MR **6**, 270.

8. B. Sz.-Nagy and A. Korányi, *Relations d'un problème de Nevanlinna et Pick avec la théorie d'opérateurs de l'espace hilbertien,* Acta Math. Acad. Sci. Hungar. **7** (1956), 295–303. MR **19**, 296.

9. M. G Kreĭn, *Sur le problème du prolongement des fonctions hermitiennes positives et continues,* C. R. (Dokl.) Acad. Sci. URSS **26** (1940), 17–22. MR **2**, 361.

10. ———, *Fundamental aspects of the representation theory of Hermitian operators with deficiency index* (*m, m*), Ukrain. Mat. Ž. **1** (1949), no. 2, 3–66; English transl., Amer. Math. Soc. Transl. (2) **97** (1971), 75–143. MR **14**, 56.

11. Ju. M. Berezanskiĭ, *Expansion in eigenfunctions of selfadjoint operators,* "Naukova Dumka", Kiev, 1965; English transl., Amer. Math. Soc., Providence, R. I., 1968. MR **36** #5768, 5769.

12. M. G. Kreĭn and F. È. Melik-Adamjan, *Certain applications of the theorem on the factorization of a unitary matrix,* Funkcional. Anal. i Priložen. **4** (1970), no. 4, 73–75 = Functional Anal. Appl. **4** (1970), 327–329. MR **48** #4396.

13. V. P. Potapov, *General theorems in the theory of analytic J-nonexpansive matrix-valued functions,* All-Union Conf. Theory of Functions of a Complex Variable, Abstracts of Reports, Kharkov, 1971. (Russian)

14. I. V. Kovališina, *Analytic and J-nonexpansive matrix-valued functions, and classical problems of analysis,* All-Union Conf. Theory of Functions of a Complex Variable, Abstracts of Reports, Kharkov, 1971. (Russian)

15. Ju. L. Šmul'jan, *Operator balls,* Teor. Funkciĭ, Funkcional. Anal. i. Priložen. Vyp. 6 (1968), 68–81. (Russian) MR **42** #8256.

16. B. Sz.-Nagy and C. Foiaş, *Analyse harmonique des opérateurs de l'espace de Hilbert,* Masson, Paris; Akad. Kiadó, Budapest, 1967; rev. English transl., North-Holland, Amsterdam; Amer. Elsevier, New York; Akad. Kiadó, Budapest, 1970. MR **37** #778; **43** #947.

17. Ju. A. Rozanov, *Stationary random processes,* Fizmatgiz, Moscow, 1963; English transl., Holden-Day, San Francisco, Cal., 1967. MR **28** #2580; **35** #4985.

Translated by S. SMITH

Amer. Math. Soc. Transl.
(2) Vol. 111, 1978

STUDY OF DIFFERENCE SCHEMES FOR NONLINEAR EQUATIONS BY A VARIATIONAL METHOD*

A. D. LJAŠKO AND M. M. KARČEVSKIĬ

In this paper, leaning heavily on the results of Langenbach [1]–[3], we present a sufficiently general method of constructing solvable and convergent difference schemes for nonlinear elliptic equations. As a preliminary, we give the generalization we need of the results of Langenbach. This generalization widens the class of nonlinear problems which are solvable by the variational method. The proofs of our theorems are like the proofs of the theorems of Langenbach, so the proof of the equivalence theorem only is given. (For similar constructions in the linear case see [4]–[6].) Of the papers devoted to the method of nets for nonlinear elliptic equations, we note [7], in which an investigation of a quasi-linear second order elliptic equation is carried out by the method of successive approximations proposed by A. I. Koševelev [8]. We note also the paper of Lin Qun [9], which contains theorems on the convergence and solvability of nonlinear difference schemes.

1. Let us consider a generalization of the results of Langenbach. Let the following equation with a nonlinear operator be given in the Hilbert space H:

$$P(x) = y. \tag{1}$$

Further, let the following conditions be satisfied.

A. The domain of definition $D(P) = M$ is a linear manifold dense in H; for any $x \in M$ the Gâteaux differential $P'(x)y$ [10], linear in H, exists and is continuous for fixed y in any plane of two variables containing the point x; and $P(0) = 0$.

B. A linear operator Q exists for which

$$(P'(x)u,\ Qv) = (P'(x)v,\ Qu),\quad x,\ u,\ v \in M.$$

C. $(P'(x)u, Qu) \geqslant \alpha^2 (P'(0)u, u) \geqslant c^2 \|Qu\|^2 \geqslant \gamma^2 \|u\|^2$.

D. $D(Q) \supset M$; the range of Q is dense in H.

We note that the conditions of Langenbach are obtained from ours for $Q = E$, the identity operator.

Equation (1) cannot have more than one solution. This follows from the equality

$$(P(x_1) - P(x_2),\ Qx_1 - Qx_2) = \int_0^1 (P'(x_2 + t(x_1 - x_2))(x_1 - x_2)\, Q(x_1 - x_2))\, dt$$

and condition C.

AMS (MOS) subject classifications (1970). Primary 65N05, 65N10, 65N30; Secondary 65L10.

*Translation of Izv. Vysš. Učebn. Zaved. Matematika 1967, no. 3 (58), 59–65. MR 35 #5149.

THEOREM 1. *The problem of finding the solution of* (1) *is equivalent to the problem of minimizing the functional*

$$\Phi(x) = \int_0^1 (Ptx,\ Qx)\,dt - (y,\ Qx). \tag{2}$$

PROOF (see [1]). Let x_0 be the solution of (1). We shall prove that x_0 yields the minimum of the functional (2). We have

$$\Phi(x+h) - \Phi(x) = \int_0^1 (Pt(x+h),\ Qh)\,dt$$
$$+ \int_0^1 (Pt(x+h) - Ptx,\ Qx)\,dt - (y,\ Qh).$$

We transform the second integral, using property B:

$$\int_0^1 (Pt(x+h) - Ptx,\ Qx)\,dt = \int_0^1 dt \int_0^t \frac{d}{ds}(P(tx+sh),\ Qx)\,ds$$
$$= \int_0^1 ds \int_s^1 (P'(tx+sh)\,h,\ Qx)\,dt = \int_0^1 ds \int_s^1 (P'(tx+sh)\,x,\ Qh)\,dt$$
$$= \int_0^1 (P(x+th) - P(tx+th),\ Qh)\,dt.$$

Consequently,

$$\Phi(x+h) - \Phi(x) = \int_0^1 (P(x+th),\ Qh)\,dt - (y,\ Qh).$$

Since $Px_0 = y$, it follows that

$$\Phi(x_0+h) = \Phi(x_0) + \int_0^1 (P(x_0+th) - P(x_0),\ Qh)\,dt$$
$$= \Phi(x_0) + \int_0^1 \frac{dt}{t} \int_0^1 (P'(x_0+sth)\,th,\ Qth)\,ds.$$

By property C the integral is nonnnegative, and hence $\Phi(x_0 + h)$ attains a minimum for $h = 0$, i.e. x_0 yields the minimum of the functional (2).

Now let x_0 yield the minimum of (2); then $d\Phi(x_0 + th)/dt|_{t=0} = 0$, $h \in M$. Transforming this equality by the usual calculations in the theory of potential operators [10], we obtain $(Px_0 - y, Qh) = 0$, whence by property D we find that $Px_0 = y$. The theorem is proved.

We introduce in M the metric $[u, v] = (P'(0)u, Qv)$, $[u]^2 = [u, u]$. The completion of M in this metric is a Hilbert space which we shall denote by H_Q. The inclusion $H_Q \subset H$ follows from conditions A–D (see [4]–[6]).

THEOREM 2. *The functional* (2) *is bounded below. Any minimizing sequence converges in* H_Q *and in* H.

The proof is carried out similarly to the corresponding arguments of Langenbach, and we shall not give it. We note that any minimizing sequence converges to the element x_0, independent of the choice of this sequence. We shall call the element x_0 the generalized solution.

2. Let us pass on to the investigation of difference schemes. We shall write the boundary value problem for a nonlinear equation with linear boundary conditions in the form

$$Lu \equiv F(x,\ u,\ D^i u) = f(x),\ l_j u = \varphi_j \quad (j = 1, \dots, s), \tag{3}$$

where $u(x)$ is a function of n variables defined in the bounded domain $\Omega \subset R_n$, and $D^i u$ is the ith derivative of u. We construct a net Ω_h on the domain Ω; we replace the derivatives which appear in (3) by difference quotients. In this way, we obtain for (3) the difference boundary value problem

$$L_h u \equiv F(x,\ u.\ D_h^i u) = f(x), \quad l_j^h u = \varphi_j \quad (j = 1, \dots, s). \tag{4}$$

Here we retain the previous notation for the functions defined on the net Ω_h. We introduce a Hilbert metric on the set of net functions. Then the operator Δ_h, defined by problem (4), will be considered in Hilbert space.

THEOREM 3. *If the operator* Δ_h *satisfies conditions* A–D, *then problem* (4) *has a unique solution for arbitrary* f *and* φ_i.

The proof of the theorem follows directly from Theorems 1 and 2 if one notes that the limit of a minimizing sequence automatically belongs to the domain of definition of Δ_h.

Let us investigate the convergence of the difference scheme. Let u and $\bar{u}$ be solutions of problems (3) and (4), respectively. We shall assume that u belongs to the domain of definition of Δ_h. Further, let the approximation condition be satisfied, i.e., for unlimited refinement of the net, $\|\psi\| = \|\Delta_h u - \Delta_h \bar{u}\| \longrightarrow 0$.

THEOREM 4. *If conditions* A–D *are satisfied for the operator* Δ_h *with some difference operator* Q_h, *then the following bounds hold*:

$$\|u - \bar{u}\| \leqslant \frac{\|\psi\|}{c\gamma}, \quad \|Q_h(u - \bar{u})\| \leqslant \frac{\|\psi\|}{c^2}.$$

PROOF. We shall use the following result of [10]. Let the nonlinear operator P have a Gâteaux derivative on some convex set. Then for any pair of points in this set and any linear functional g the following equality is valid:

$$g(Pu_1 - Pu_2) = g(P'(\xi_g)(u_1 - u_2)),$$

where ξ_g is some point on the segment (u_1, u_2), determined by the choice of the functional g. As the element defining a linear functional in the Hilbert space of net functions, we choose the element $Q_h(u - \bar{u})$. Then we have

$$(\Delta_h u - \Delta_h \bar{u},\ Q_h(u - \bar{u})) = (\Delta_h'(\xi_{u-\bar{u}})(u - \bar{u}),\ Q_h(u - \bar{u})).$$

Using C and the Cauchy-Schwarz-Bunjakowskiĭ inequality, we get

$$(\Delta_h(\xi_{u-\bar{u}})(u-u),\ Q_h(u-\bar{u}))\geqslant c^2\|Q_h(u-\bar{u})\|^2\geqslant\gamma^2\|u-\bar{u}\|^2,$$
$$(\Delta_h u-\Delta_h\bar{u},\ Q_h(u-\bar{u}))\leqslant\|\Delta_h u-\Delta_h\bar{u}\|\,\|Q_h(u-\bar{u})\|,$$

whence $\|Q_h(u-\bar{u})\|\leqslant\|\psi\|/c^2$ and $\|u-\bar{u}\|\leqslant\|\psi\|/c\gamma$. The theorem is proved.

REMARK 1. In proving Theorem 4 we did not actually use the fact that Δ_h is a difference operator. Therefore one may use the theorem to investigate other approximation methods, for example, the method of lines and Galerkin-type methods.

REMARK 2. If conditions A–D are not satisfied for the operator Δ_h, then one may investigate the convergence in the following way. The difference operator Δ_h is obviously defined by some system of functions $\psi_i(u_1,\ldots,u_N)$ $(i=1,\ldots,N)$; here N is the number of points in the net. Further, one may write

$$\Delta_h u-\Delta_h\bar{u}=A(\xi)(u-\bar{u})=\psi, \tag{$*$}$$

where $A(\xi)$ denotes the matrix of elements $a_{ij}=\partial\psi_i(\xi_i)/\partial u_j$, and ξ_i is a point on the segment $(u,\bar{u})$. Now it is clear that requiring correctness of the linear system $(*)$ uniformly in ξ ensures the convergence of the nonlinear difference scheme.

3. In this section a method of constructing solvable and convergent difference schemes for a certain class of problems written in the form (3) is proposed.

Let the operator L satisfy the conditions of Langenbach, namely, conditions A–D with $Q=E$. The Gâteaux differential of L has the form

$$L'(u)v=\sum_{i=0}^{m}\frac{\partial F}{\partial D^iu}D^iv,\quad D^0v=v.$$

For the positive definite linear operator $L'(u)$ let the following positive definite difference approximation be constructed:

$$L'_h(u)=\sum_{i=0}^{m}\frac{\partial F}{\partial D^iu}D^i_h v.$$

Putting the expressions obtained for D^i_h in (3), we find a difference approximation of the operator L: $L_hu=F(x,u,D^i_hu)$. The operator L_h, by its very construction, satisfies conditions A–D with $Q_h=E$, and consequently the difference scheme obtained is convergent and uniquely solvable, by Theorems 3 and 4.

To construct yet another positive definite approximation of $L'(u)$ we use the following sufficiently well-known method which, as a rule, leads to the goal. Let the equation $Au=f$ be given, where A is a positive definite linear operator in some Hilbert space. We construct the functional $\Phi(u)=(Au,u)-2(u,f)$; then we replace this functional by the finite difference one $\Phi_h(u)$, using a quadrature formula and an approximation of the derivatives. The finite-difference approximation for the equation $Au=f$ is now obtained from the conditions for a minimum of the functional $\Phi_h(u)$.

4. Let us consider some examples.

a). *The Dirichlet problem for a quasi-linear elliptic equation*:

$$L(u) = -\sum_{i=1}^{m} \frac{\partial}{\partial x_i} a_i(x, u, p_j) + a_0(x, u, p_j) = f(x), \quad u|_\Gamma = 0;$$
$$p_j = \frac{\partial u}{\partial x_j}, \quad p_0 = u; \quad L(0) = 0. \tag{5}$$

Let the following conditions be satisfied:

$$\sum_{i,j=0}^{m} \frac{\partial a_i}{\partial p_j} \xi_i \xi_j \geqslant \varkappa^2 \sum_{i=0}^{m} \xi_i^2; \quad \frac{\partial a_i}{\partial p_j} = \frac{\partial a_j}{\partial p_i}, \quad i, j = 0, 1, \dots, m. \tag{6}$$

Taking $L_2(\Omega)$ as the space H, let us verify that the conditions of Langenbach are satisfied for problem (5). We calculate the Gâteaux derivative of L:

$$L'(u)v = -\sum_{i=1}^{m} \frac{\partial}{\partial x_i} \left[\sum_{k=0}^{m} \frac{\partial a_i}{\partial p_k} \frac{\partial v}{\partial x_k} \right] + \sum_{k=0}^{m} \frac{\partial a_0}{\partial p_k} \frac{\partial v}{\partial x_k}; \quad \frac{\partial v}{\partial x_0} = v.$$

Applying the Green's formula and using (6) and the Friedrichs inequality, we have

$$\Phi(v) = (L'(u)v, v) = \int_\Omega L'(u) v v \, d\Omega = \int_\Omega \sum_{i,j=0}^{m} \frac{\partial a_i}{\partial p_j} \frac{\partial v}{\partial x_i} \frac{\partial v}{\partial x_j} d\Omega,$$

$$\Phi(v) \geqslant \varkappa^2 \int_\Omega \sum_{i=0}^{m} \left(\frac{\partial v}{\partial x_i} \right)^2 \geqslant \frac{\varkappa^2}{\gamma^2} \|v\|^2.$$

Thus the conditions of Langenbach are satisfied. Let us construct on the domain Ω a square net with step h; also, set $\overline{\Omega}_h = \Omega_h + \Gamma_h$ (Γ_h is the set of boundary nodes). In accordance with §3 we take as a basis the quadrature formula

$$\int_\Omega f(x)\, dx = \sum_{\Omega_h} f(x) h^m$$

and the approximation of the derivative

$$u_{x_i} = \frac{1}{h} [u(x_i + h) - u(x_i)].$$

Now we form the finite-difference functional

$$\Phi_h(v) = h^m \sum_{\Omega_h} \left(\sum_{i,j=0}^{m} \frac{\partial a_i}{\partial p_j} v_{x_i} v_{x_j} \right).$$

Applying (6) and the Friedrichs inequality in difference form, we get

$$[\Phi_h(v) \geqslant \frac{\varkappa^2}{a^2} \|v\|^2_{L_h^2(\Omega_h)}, \tag{7}$$

where the constants κ and a are independent of v and of the net step, and the norm is defined in accordance with the quadrature formula chosen: $\|v\|^2_{L_h^2(\Omega_h)} = \Sigma_{\Omega_h} v^2(x) h^m$. Differentiating $\Phi(v)$ with respect to $v(x)$, we obtain the approximation to the Gâteaux derivative:

$$L'_h(u)v = -\sum_{i=1}^{m}\Big(\sum_{k=0}^{m}\frac{\partial a_i}{\partial p_k}v_{x_k}\Big)_{\bar{x}_i} + \sum_{k=0}^{m}\frac{\partial a_0}{\partial p_k}v_{x_k}, \tag{8}$$

$$z_{\bar{x}_i} = \frac{1}{h}(z(x_i) - z(x_i - h)).$$

The equality $(L'_h(u)v,\ v) = \Phi_h(v)$, which is easily obtained by the Green's formula in difference form, and inequality (7) show that the operator $L'_h(u)$ is positive definite uniformly with respect to u and h. Recovering the approximation for the operator L, we obtain for L_h the expression

$$L_h(u) = -\sum_{i=1}^{m}[a_i(x,\ u,\ u_{x_i})]_{\bar{x}_i} + a_0(x,\ u,\ u_{x_j}).$$

The unique solvability of the problem

$$L_h u = f(x), \quad u|_{\Gamma_h} = 0 \tag{9}$$

follows from Theorem 3. The convergence of the solution of (9) to the solution of (5) in the metric of $L_h^2(\Omega_h)$ with the approximation condition (see §2) follows from Theorem 4.

REMARK 3. Modifying the argument in Theorem 4 somewhat, it is easy to prove the convergence of the difference scheme (9) in the metric

$$\|u\|^2 = \sum_{\Omega_h}\Big(u^2 + \sum_{i=1}^{m}u_{x_i}^2\Big)h^m,$$

also, which corresponds to the metric of the space $W_2^{(1)}$.

REMARK 4. Questions of the solvability of difference schemes for higher order quasi-linear elliptic equations are examined in a similar way. Inequalities of type (7) are obtained in this case by imbedding theorems in difference form (see, for example, [11]).

b). Let us consider the third-order ordinary differential equation

$$P(x) = -\frac{d}{dt}[f(x'')x''] + a(t)x' = \varphi(t),\ t \in (0,\ 1),$$

$$x(0) = x'(0) = x'(1) = 0; \quad a(t) \geqslant 0; \quad (f(\xi)\xi)'_\xi \geqslant \gamma^2 > 0.$$

Let us ascertain that the operator P satisfies conditions A–D with the operator $Q(x) = x'(t)$:

$$P'(x)h = -\frac{d}{dt}\{[f'(x'')x'' + f(x'')]h''\} + a(t)h';$$

$$\int_0^1 P'(x)hh'dt = \int_0^1[f'(x'')x'' + f(x'')]h''^2dt + \int_0^1 a(t)h'^2dt$$

$$\geqslant \gamma^2\int_0^1 h''^2dt \geqslant \gamma^2\int_0^1 h'^2dt \geqslant \gamma^2\int_0^1 h^2dt.$$

As a difference approximation for our problem we take

$$P_\tau(x) = -\{[f(x_{\bar{t}t})]\, x_{\bar{t}t}\}_{\bar{t}} + a(t)\, x_{\bar{t}} = \varphi(t);$$
$$x(0) = x(\tau) = 0, \quad x(1-\tau) = x(1); \quad x_t(t_k) = x_{\bar{t}}(t_{k+1})$$
$$= \frac{x(t_{k+1}) - x(t_k)}{\tau}, \quad N\tau = 1.$$

Let us verify that conditions A–D are satisfied for the operator $P_\tau(x)$, putting $Q_\tau(x) = x_{\bar{t}}$ and defining the scalar product by

$$(u,\ v) = \sum_{k=1}^{N} u(k\tau)\, v(k\tau).$$

Applying the formula for summation by parts, we have

$$(P'_\tau(x)\, h,\quad \dot{Q}_\tau h) = ([f'(x_{\bar{t}t})\, x_{\bar{t}t} + f(x_{\bar{t}t})]^2_{\bar{t}t},\ 1) + (a(t)\, h_{\bar{t}},\ h_{\bar{t}})$$
$$- [f'(x_{\bar{t}t})\, x_{\bar{t}t} + f(x_{\bar{t}t})]\, h_{\bar{t}t}\, h_{\bar{t}}\Big|_0^1 \geqslant \gamma^2 \|h_{\bar{t}t}\|^2 \geqslant \gamma^2 \|h_{\bar{t}}\|^2 \geqslant \gamma^2 \|h\|^2.$$

In these calculations we assume the functions x and h are constant outside the interval (0, 1). Thus conditions A–D are satisfied.

The authors express their gratitude to N. N. Janenko for his attention to and discussion of the paper.

BIBLIOGRAPHY

1. Arno Langenbach, *Variationsmethoden in der nichtlinearen Elaztizitäts- und Plastizitätstheorie,* Wiss. Z. Humboldt-Univ. Berlin Math.-Nat. Reihe 9 (1959/60), 145–164. MR **25** #4670.

2. ———, *On the application of a variational principle to some nonlinear differential equations,* Dokl. Akad. Nauk SSSR **121** (1958), 214–217. (Russian) MR **20** #5361.

3. ———, *Die Regularisierung nichtlinearer Gleichungen,* Math. Nachr. 24 (1962), 33–51. MR **25** #4359.

4. A. E. Martynjuk, *On a generalization of a variational method,* Dokl. Akad. Nauk SSSR **117** (1957), 374–377. (Russian) MR **21** #7439.

5. A. D. Ljaško, *The convergence of Galerkin-type methods,* Dokl. Akad. Nauk SSSR **120** (1958), 242–244. (Russian) MR **20** #4346.

6. ———, *Some versions of the Galerkin-Krylov method,* Dokl. Akad. Nauk SSSR **128** (1959), 468–470. (Russian) MR **22** #1063.

7. M. P. Sapagovas, *The method of finite differences for the solution of quasilinear elliptic equations with discontinuous coefficients,* Ž. Vyčisl. Mat. i Mat. Fiz. **5** (1965), 638–647 = USSR Comput. Math. and Math. Phys. **5** (1965), no. 4, 72–85. MR **34** #5321.

8. A. I. Košelev, *Convergence of the method of successive approximations for quasilinear elliptic equations,* Dokl. Akad. Nauk SSSR 142 (1962), 1007–1010 = Soviet Math. Dokl. **3** (1962), 219–222. MR **24** #A3411.

9. Lin Chün [Lin Qun], *A discussion on the difference method for the solution of nonlinear differential equations,* Sci. Sinica **10** (1961), 414–419. MR **26** #7152.

10. M. M. Vaĭnberg, *Variational methods for the study of nonlinear operators,* GITTL, Moscow, 1956; English transl., Holden-Day, San Francisco, Cal., 1964. MR **19**, 367; **31** #638.

11. V. K. Saul'ev, *On a class of elliptic equations solvable by the method of finite differences,* Vyčisl. Mat. **1** (1957), 81–86. (Russian) MR **20** #2541.

Translated by H. N. BROWNE, JR.

Amer. Math. Soc. Transl.
(2) Vol. 111 (1978)

FOUNDATIONS OF MULTISTAGE SYSTEM OPTIMIZATION*

PETR MANDL

CONTENTS

Introduction 165
Chapter I. The Bellman Principle 166
§1. Basic topics 166
§2. The Bellman principle of optimality 167
§3. Problems with an infinite planning horizon 170
§4. Combination of dynamic programming with the Lagrange multipliers method 172
§5. The Hamilton-Jacobi equation 173
Chapter II. The Discrete Optimum Principle 176
§1. Lagrange multipliers in multistage optimization 176
§2. The discrete optimum principle 178
§3. Examples 182
Chapter III. The Pontrjagin Principle 186
§1. Introduction 186
§2. Minimization of ψ when T and $x(0)$ are given in advance 188
§3. Linear systems with quadratic objective functional 193
§4. Minimization of ψ under the conditions $\mathbf{x}(0) \in I$, $\mathbf{x}(T) \in F$ 195
§5. The minimum-time problem 201
§6. The Pontrjagin principle 204
§7. Examples 205
Bibliography 211

Introduction

This article treats multistage systems with discrete evolution (e.g. time) parameter and systems with continuous parameter, described by differential equations. The latter may be considered as the limit of systems with discrete parameter. We concentrate upon general principles, expressed as necessary and sufficient conditions for an optimum. Complete solution of mathematically formulated optimization problems then employs methods of mathematical programming, numerical mathematics and computer programming. The material is divided into three chapters, dealing with the Bellman principle, the discrete optimum principle and the Pontrjagin principle. Basic concepts are given at the beginning of Chapter I. A list of references is placed at the end of the paper. Examples are part of the presentation. The author of an example

AMS (MOS) subject classifications (1970). Primary 49-01, 49B10, 49B15; Secondary 49D05, 49C05.

*Translation of Kybernetika (Prague) 8 (1972), Suppl., 3–52.

is mentioned in parentheses. Formulas, sections and theorems within a chapter are referred to by their numbers; outside of a chapter also by the number of the chapter. The symbol □ marks the end of a proof. Vectors and matrices are set in boldface type. A state vector $\boldsymbol{x}$ is distinguished from an extended state vector x; there is a similar distinction between the corresponding maps $\boldsymbol{F}$ and F. Material in this article together with an exposition about controlled Markov chains made up a course entitled "Foundations of Optimization" in theoretical cybernetics given under the Faculty of Nuclear Physics at the Technical University of Prague.

CHAPTER I
THE BELLMAN PRINCIPLE

§1. Basic topics

EXAMPLE 1 (L. C. MITTEN). In a certain (highly simplified) economy, commodity V can be produced by either of two industries A and B. Each dollar invested in industry A at the beginning of the year yields, at the end of that year and each succeeding year, 2 kilograms of commodity V and \$0.50 profit. Corresponding values for industry B are 1 kilogram of commodity V and \$0.90. The problem is: given an initial capital \$$y$, how should the available capital at the beginning of each year be allocated between the two industries so as to maximize the total amount of commodity V produced during the next five years?

This is a multistage process, and we can describe it using the following notation (coordinates are numbered by upper indices):

x_n^1 – total capital invested in industry A at the end of the nth year,

x_n^2 – total capital in B at the end of the nth year,

x_n^3 – capital available at the end of the nth year,

u_n – proportional part of the available capital reinvested in industry A, $0 \leqslant u_n \leqslant 1$.

Hence

$$x_n^1 = x_{n-1}^1 + u_{n-1}x_{n-1}^3,$$
$$x_n^2 = x_{n-1}^2 + (1 - u_{n-1})\, x_{n-1}^3,$$
$$x_n^3 = 0{,}5(x_{n-1}^1 + u_{n-1}x_{n-1}^3) + 0{,}9(x_{n-1}^2 + (1 - u_{n-1})\, x_{n-1}^3),$$
$$n = 1, 2, \ldots .$$

The amount of commodity V produced during the nth year is $(2x_n^1 + x_n^2)$kg. Therefore we have to maximize

$$\sum_{n=1}^{5} (2x_n^1 + x_n^2) = \sum_{k=0}^{4} [2(x_k^1 + u_k x_k^3) + x_k^2 + (1 - u_k)\, x_k^3].$$

Let us proceed to general definitions. A *multistage process* describes the evolution of a system whose state is given by a state vector $\boldsymbol{x} = (x^1, \ldots, x^r)$. The evolution is described by a sequence of transformations $\boldsymbol{F}_n(\boldsymbol{x}) = (F_n^1(\boldsymbol{x}), \ldots, F_n^r(\boldsymbol{x}))$ of state vectors. If $\boldsymbol{x}_n$ is the state of the system at the nth stage, then $\boldsymbol{x}_n = \boldsymbol{F}_{n-1}(\boldsymbol{x}_{n-1})$; here n is the evolution parameter. The word "evolution" is only symbolic, for the evolution parameter has often no connection with time.

Multistage decision processes are those where the transformations of state vectors depend on another parameter $u = (u^1, \ldots, u^r) \in U$, the so-called *control parameter* (decision variable, etc.). U is the control parameter region (decision space). Thus the process is characterized by maps

$$F_n(x, u) = (F_n^1(x, u), \ldots, F_n^r(x, u)), \quad n = 0, 1, \ldots.$$

For a given initial state x_0, the system evolution $\{x_0, \ldots, x_N\}$ up to the Nth stage is determined by the choice of a sequence of decision variables $\{u_0, \ldots, u_{N-1}\}$. We have $x_n = F_{n-1}(x_{n-1}, u_{n-1})$, $n = 1, \ldots, N$. The sequence $\{u_0, \ldots, u_{N-1}\}$ is called a *control.*

Different controls are compared by means of an *objective* (utility, criterion etc.) *function* $\psi(x_0; u_0, \ldots, u_{N-1}; N)$. The function will be mostly assumed to be in the additive form

$$\psi(x_0; u_0, \ldots, u_{N-1}; N) = \sum_{k=0}^{N-1} G_k(x_k, u_k). \tag{1}$$

The problem is to optimize (that is, either maximize or minimize) the objective function. Solution of the optimization problem is given by an optimal control $\{u_0, \ldots, u_{N-1}\}$ and the corresponding trajectory $\{x_0, \ldots, x_N\}$. When F and G do not depend on the evolution parameter, the process is called *homogeneous* (*autonomous*).

§2. The Bellman principle of optimality

Suppose that U is a closed bounded set and the maps $F_n(x, u)$ and $G_n(x, u)$ are continuous. Then it can be proved that there exists an optimal control, and we have the Bellman principle of optimality:

The optimal control has the property that with regard to the state resulting from the first decision, the remaining decisions form an optimal control.

The principle is intuitively self-evident, and in this simple situation it does not require a formal proof. Denote the optimal value of the objective function within the last $N - m$ stages by

$$\hat{\psi}(x; m, N) = \operatorname*{opt}_{\{x_m = x, u_m, \ldots, u_{N-1}\}} \sum_{k=m}^{N-1} G_k(x_k, u_k).$$

The symbol opt denotes either maximum or minimum, depending on the context. If a control begins with u as the first decision and then uses an optimal control with respect to the state $F_0(x, u)$, then the value of the objective function at $x_0 = x$ is equal to

$$G_0(x, u) + \hat{\psi}(F_0(x, u); 1, N).$$

According to the principle of optimality, the optimal control satisfies this condition. Therefore

$$\hat{\psi}(x; 0, N) = \operatorname*{opt}_{u \in U} \{G_0(x, u) + \hat{\psi}(F_0(x, u); 1, N)\}.$$

The same reasoning applied to the mth stage gives

$$\hat{\psi}(x; m, N) = \operatorname*{opt}_{u \in U} \{G_0(x, u) + \hat{\psi}(F_m(x, u); m + 1, N)\}, \tag{2}$$
$$m = N - 1, N - 2, \ldots, 0, \quad \hat{\psi}(x; N, N) \equiv 0.$$

Equations (2) are recurrence relations for $\hat{\psi}(x; m, N)$. Several remarks are in order. Suppose that the initial state x_0 is given and one has to find an optimal control $\{\hat{u}_0, \ldots, \hat{u}_{N-1}\}$ and the corresponding trajectory $\{\hat{x}_0 = x_0, \ldots, \hat{x}_N\}$. Begin by solving (2) for $m = N - 1, \ldots, 1$. Then find a $\hat{u}_0$ optimizing the expression within braces in (2) for $m = 0$, use $\hat{u}_0$ to determine $\hat{x}_1 = F_0(x_0, \hat{u}_0)$, and find $\hat{u}_1$ optimizing the expression within braces for $m = 1$. Put $\hat{x}_2 = F_1(\hat{x}_1, \hat{u}_1)$, etc. To compute $2N - 1$ vectors, we must tabulate $N - 1$ functions $\hat{\psi}(x; m, N)$, $m = 1, \ldots, N - 1$. The amount of necessary computations is therefore quite large. However, when solving (2), we also obtain the sequence of functions $\{\hat{u}_m(x), m = 0, \ldots, N - 1\}$. These show how the values of the control parameter, optimizing the expression within braces in (2), depend on x.

Thus we come to a rather different view at the multistage process control. A *policy** is a sequence $\{u_0(x), \ldots, u_{N-1}(x)\}$ of maps from the state space R^r into the decision space U. Then $u_m(x)$ gives the value of the control parameter chosen at the mth stage if the system is in a state x. Hence the trajectory corresponding to the policy $\{u_0(x), \ldots, u_{N-1}(x)\}$ is given by

$$x_n = F_{n-1}(x_{n-1}, u_{n-1}(x_{n-1})), \quad n = 1, \ldots, N.$$

It is easy to check that the policy $\{\hat{u}_0(x), \ldots, \hat{u}_{N-1}(x)\}$ is optimal. (Compare Chapter III, §3, about the feedback control.)

EXAMPLE 2 (R. BELLMAN AND S. DREYFUS). Let the reliability of a certain system (e.g. a part of a computer), consisting of N aggregates $A_1, \ldots, A_N$, be defined as the probability that it will operate successfully. This probability is assumed to be $P = P_1 \cdots P_N$, where P_i is the probability of successful operation of A_i. Aggregate A_i contains k_i identical components, and it operates successfully if at least one of the components does not fail. If p_i is the probability of failure for a single component in A_i, then, assuming independence, we have $P_i = 1 - p_i^{k_i}$. Let c_i be the cost of a single component in A_i. The problem is to maximize the reliability of the system subject to the condition that the total cost of components used does not exceed c.

For the solution, denote by x_i the resources remaining after $A_1, \ldots, A_{i-1}$ are constructed. Thus

$$x_1 = c, \quad x_i = x_{i-1} - k_{i-1}c_{i-1}, \quad i = 2, \ldots, N.$$

The objective function is $(1 - p_1^{k_1})(1 - p_2^{k_2}) \cdots (1 - p_N^{k_N})$. The difference between multiplicative and additive form is not important—we may pass to the additive form, using logarithms. The maximal reliability of the subsystem consisting of aggregates $A_m, \ldots, A_N$, with cost not exceeding x, will be

**Translator's note.* The author's term 'strategy' is translated as 'policy'.

$$\hat{\psi}(x; m, N)$$
$$= \max\{(1 - p_m^{k_m})(1 - p_{m+1}^{k_{m+1}})\dots(1 - p_N^{k_N}) : k_m c_m + k_{m+1}c_{m+1} + \dots + k_N c_N \leqq x\}.$$

From (2) we obtain

$$\hat{\psi}(x; m, N) = \max_{k=0,1,\dots} (1 - p_m^k)\,\hat{\psi}(x - k_m c_m; m + 1, N),$$
$$m = N - 1, N - 2, \dots, 1.$$

Obviously

$$\hat{\psi}(x; N, N) = 1 - p_N^{[x/c_N]}.$$

EXAMPLE 3 (S. M. ROBERTS)—*optimal catalyst replacement.* Consider the endothermic catalytic reaction illustrated in Figure 1. A substance A in gas phase passes through a reactor, with flow rate q. A fraction c of the substance A is converted to products X and Y which are then taken off in a separating unit. The remaining A is returned to the reactor. Hence the quantity of A supplied

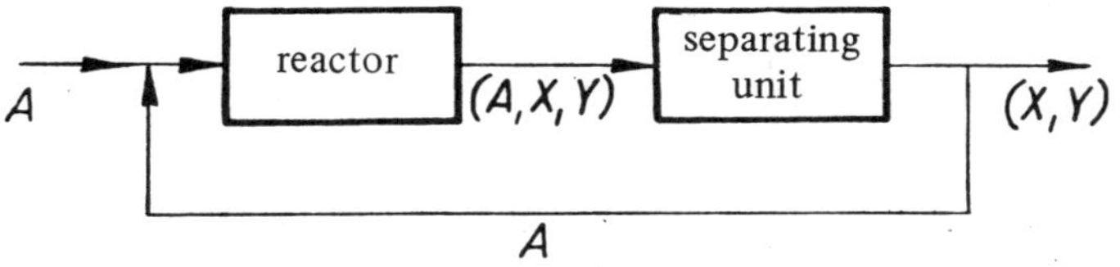

FIGURE 1.

during a time unit is qc. The conversion c is assumed to be a linear function of the operating temperature T, the flow rate q and the total quantity s of the substance A that has passed through the reactor since the last replacement of the catalyst; namely

$$c = c_0 + c_1 T - c_2 q - c_3 s. \tag{3}$$

Let the time unit be one day. The state of the system will be the value s_n of the parameter s at the end of the nth day. The reaction requires the daily supply of heat

$$Q = qc_p(T - T_0) + (\Delta H)\,qc, \tag{4}$$

where T_0 is the temperature of A at input, c_p is the specific heat of the reactants and ΔH is the heat consumed during the reaction. The daily profit is assumed to be

$$G = v_1 qc - v_2 Q - v_3(1 - c)\,q - v_4. \tag{5}$$

The first term on the right-hand side represents the net increase of the value of the products X and Y over the material A. The other terms are the cost of heating, the cost of separation and return of the unreacted A, and the fixed charges. Equations (3)–(5) describe the daily profit as a function $G(s, T, q)$ of the state s, the temperature T and the flow rate q. Let R denote the cost of a catalyst replacement which takes one day. This description yields a multistage time-homogeneous process. The problem is to maximize the total profit over a period of N days, by means of determining the daily operating values of T_n and q_n and deciding about the catalyst replacement. Denote by $\hat{\psi}(s; N)$ this maximum profit in N days, starting in a state s. By virtue of (2) we have

$$\hat{\psi}(s; N) = \max_{T,q} \{\max [-R + \hat{\psi}(0; N-1), G(s, T, q) + \hat{\psi}(s + q; N-1)]\}.$$

§3. Problems with an infinite planning horizon

The number N in (1), indicating how many stages are considered for the optimization (duration of the process), is called a *planning horizon.* It is often advantageous to consider an infinite horizon. We shall investigate homogeneous processes. When defining the objective function

$$\psi(\boldsymbol{x}_0; \boldsymbol{u}_0, \boldsymbol{u}_1, \ldots) = \sum_{k=0}^{\infty} G(\boldsymbol{x}_k, \boldsymbol{u}_k), \tag{6}$$

we have to consider the problem of convergence of the infinite series on the right-hand side in (6). To avoid it, we may use the objective function

$$\phi(\boldsymbol{x}_0; \boldsymbol{u}_0, \boldsymbol{u}_1, \ldots) = \sum_{k=0}^{\infty} \beta^k G(\boldsymbol{x}_k, \boldsymbol{u}_k). \tag{7}$$

In this case we speak about *discounted* cost (or profit), since β may be interpreted as a discount ratio reducing the future payments to the value at the end of the first period, for the given interest rate.

For definiteness, assume opt = min. We restrict our attention to two cases:

(i) $G(\boldsymbol{x}, \boldsymbol{u}) \geqslant 0$, the objective function is defined by (6), and

$$\hat{\psi}(\boldsymbol{x}) = \min_{\{\boldsymbol{u}_0, \boldsymbol{u}_1, \ldots\}} \psi(\boldsymbol{x}; \boldsymbol{u}_0, \boldsymbol{u}_1, \ldots) < \infty$$

for all $\boldsymbol{x}$.

(ii) $|G(\boldsymbol{x}, \boldsymbol{u})| \leqslant K < \infty$, the objective function is defined by (7), and $0 < \beta < 1$. Put

$$\hat{\phi}(\boldsymbol{x}) = \min_{\{\boldsymbol{u}_0, \boldsymbol{u}_1, \ldots\}} \phi(\boldsymbol{x}; \boldsymbol{u}_0, \boldsymbol{u}_1, \ldots).$$

Applying the Bellman principle, we obtain

$$\hat{\psi}(\boldsymbol{x}) = \min_{\boldsymbol{u} \in U} \{G(\boldsymbol{x}, \boldsymbol{u}) + \hat{\psi}(\boldsymbol{F}(\boldsymbol{x}, \boldsymbol{u}))\}, \tag{8}$$

$$\hat{\phi}(\boldsymbol{x}) = \min_{\boldsymbol{u} \in U} \{G(\boldsymbol{x}, \boldsymbol{u}) + \beta\, \hat{\phi}(\boldsymbol{F}(\boldsymbol{x}, \boldsymbol{u}))\}. \tag{9}$$

The uniqueness of the bounded solution of (9) may be easily checked. The solution of (8) is not unique. Therefore further conditions have to be added to (8). $\hat{\psi}(\boldsymbol{x})$ represents the smallest nonnegative solution.

Let $\hat{\boldsymbol{u}}(\boldsymbol{x})$ be the value of the parameter $\boldsymbol{u}$ yielding the minimal value of the expression in braces in (8) or (9), respectively. It is easy to prove by successive substitutions that $\{\hat{\boldsymbol{u}}_n(\boldsymbol{x}) = \hat{\boldsymbol{u}}(\boldsymbol{x}), n = 0, 1, \ldots\}$ is an optimal policy. We identify the policy $\{\boldsymbol{u}_n(\boldsymbol{x}), n = 0, 1, \ldots\}$ such that $\boldsymbol{u}_n(\boldsymbol{x}) = \boldsymbol{u}(\boldsymbol{x}), n = 0, 1, \ldots,$ with the function $\boldsymbol{u}(\boldsymbol{x})$. Under assumptions (i) we can construct $\hat{\psi}(\boldsymbol{x})$, putting $\hat{\psi}(\boldsymbol{x}; 0) \equiv 0$ and successively

$$\hat{\psi}(x; N) = \min_{u \in U} \{G(x, u) + \hat{\psi}(F(x, u); N - 1)\}, \quad N = 1, 2, \ldots .$$

Comparing with (2), we get

$$\hat{\psi}(x; N) = \min_{\{x_0 = x, u_0, \ldots, u_{N-1}\}} \sum_{k=0}^{N-1} G(x_k, u_k).$$

Since $G(x, u)$ is nonnegative,

$$\hat{\psi}(x; N) \leqq \hat{\psi}(x; N + 1) \leqq \hat{\psi}(x).$$

If $\lim_{N \to \infty} \hat{\psi}(x; N)$ is continuous, then it gives a solution of (8); hence $\hat{\psi}(x) = \lim_{N \to \infty} \hat{\psi}(x; N)$. This exemplifies *approximation in the space of objective functions.*

We illustrate *approximation in the policy space* under conditions (ii). Here we begin by choosing an arbitrary policy ${}^0u(x)$. Then we find the corresponding objective function $\phi_0(x)$ from the equation

$$\phi_0(x) = G(x, {}^0u(x)) + \beta\, \phi_0(F(x, {}^0u(x))).$$

We determine ${}^1u(x)$ such that

$$G(x, {}^1u(x)) + \beta\, \phi_0(F(x, {}^1u(x))) = \min_{u \in U} \{G(x, u) + \beta\, \phi_0(F(x, u))\}.$$

In general,

$$\phi_n(x) = G(x, {}^nu(x)) + \beta\, \phi_n(F(x, {}^nu(x))),$$

$$G(x, {}^{n+1}u(x)) + \beta\, \phi_n(F(x, {}^{n+1}u(x))) = \min_{u \in U} \{G(x, u) + \beta\, \phi_n(F(x, u))\},$$

$$n = 0, 1, \ldots .$$

We have

$$\phi_n(x) \geqq \phi_{n+1}(x), \quad \lim_{n \to \infty} \phi_n(x) = \hat{\phi}(x).$$

EXAMPLE 4. Consider r places, denoted by $1, \ldots, r$, and positive numbers g_{ij} ($i \neq j; i = 1, \ldots, r - 1; j = 1, \ldots, r$). The number g_{ij} represents the cost of moving from i to j. The aim is to arrive at r with the minimal cost, starting from an arbitrary place $1, \ldots, r - 1$. A policy is a function $u(i)$ indicating the place which one goes to from place i. The minimal cost for arriving at r from i, denoting by $\hat{\psi}(i)$, satisfies

$$\hat{\psi}(i) = \min_{j \neq i} \{g_{ij} + \hat{\psi}(j)\}, \quad i = 1, \ldots, r - 1, \quad \hat{\psi}(r) = 0.$$

The quantities g_{ij} are given in Table 1, where $r = 7$ and a dash denotes a number so large that the corresponding move need not be considered. Tables 2 and 3 show successive approximations in the space of objective functions and in the policy space ($\hat{\psi}(j) = \hat{\psi}(j; 5) = \psi_2(j)$).

TABLE 1.

$i \backslash j$	1	2	3	4	5	6	7
1	—	4	—	9	—	—	—
2	—	—	4	6	—	—	—
3	—	—	—	—	—	—	12
4	—	—	—	—	4	—	16
5	—	—	—	—	—	2	8
6	—	—	—	—	—	—	2

TABLE 2.

j	1	2	3	4	5	6	7
$\hat{\psi}(j; 0)$	0	0	0	0	0	0	0
$\hat{\psi}(j; 1)$	4	4	12	4	2	2	0
$\hat{\psi}(j; 2)$	8	10	12	6	4	2	0
$\hat{\psi}(j; 3)$	14	12	12	8	4	2	0
$\hat{\psi}(j; 4)$	16	14	12	8	4	2	0
$\hat{\psi}(j; 5)$	17	14	12	8	4	2	0

TABLE 3.

j	1	2	3	4	5	6	7
${}^0u(j)$	7	7	7	7	7	7	7
${}^1u(j)$	4	3	7	5	6	7	7
${}^2u(j)$	4	4	7	5	6	7	7

j	1	2	3	4	5	6	7
$\psi_0(j)$	—	—	12	16	8	2	0
$\psi_1(j)$	17	16	12	8	4	2	0
$\psi_2(j)$	17	14	12	8	4	2	0

§4. Combination of dynamic programming with the Lagrange multipliers method

Let us return to the problem in §1. Suppose that we are given, in addition, a sequence of functions $H_k(x, u)$, $k = 0, \dots, n - 1$, and a number Δ. We seek to minimize the objective function (1) under the condition

$$\sum_{k=0}^{N-1} H_k(\boldsymbol{x}_k, \boldsymbol{u}_k) = \Delta . \tag{10}$$

When using the dynamic programming method, we denote

$$\hat{\phi}(\boldsymbol{x}, y; m, N) = \min \{ \sum_{k=m}^{N-1} G_k(\boldsymbol{x}_k, \boldsymbol{u}_k) : \sum_{k=m}^{N-1} H_k(\boldsymbol{x}_k, \boldsymbol{u}_k) = y \} .$$

The Bellman principle then yields

$$\hat{\phi}(\boldsymbol{x}, y; m, N) = \min_{\boldsymbol{u} \in U} \{ G_m(\boldsymbol{x}, \boldsymbol{u}) + \hat{\phi}(F_m(\boldsymbol{x}, \boldsymbol{u}), y - H_m(\boldsymbol{x}, \boldsymbol{u}); m + 1, N) \} ,$$

and for $m = N - 1, N - 2, \dots, 0$ this gives a recurrence relation for finding the minimum $\hat{\phi}(\boldsymbol{x}_0, \Delta; 0, N)$. The computation requires us to tabulate $r + 1$ variables in the functions $\hat{\phi}(\boldsymbol{x}, y; m, N)$. We may avoid this increase in the number of variables by introducing an *undetermined multiplier* and looking for the (unconditional) minimum of the expression

$$\Xi(\boldsymbol{x}_0; \boldsymbol{u}_0, \dots, \boldsymbol{u}_{N-1}; N) = \sum_{k=0}^{N-1} (G(\boldsymbol{x}_k, \boldsymbol{u}_k) + \lambda H_k(\boldsymbol{x}_k, \boldsymbol{u}_k)) .$$

If for $\lambda = \hat{\lambda}$ we have

$$\Xi(x_0; \hat{u}_0, \ldots, \hat{u}_{N-1}; N) = \min_{\{u_0, \ldots, u_{N-1}\}} \Xi(x_0; u_0, \ldots, u_{N-1}; N)$$

and also

$$\sum_{k=1}^{N-1} H_k(\hat{x}_k, \hat{u}_k) = \Delta ,$$

where $\{\hat{x}_0 = x_0, \ldots, \hat{x}_N\}$ is the trajectory corresponding to the control $\{\hat{u}_0, \ldots, \hat{u}_{N-1}\}$, then $\{\hat{u}_0, \ldots, \hat{u}_{N-1}\}$, $\{\hat{x}_0, \ldots, \hat{x}_N\}$ is a solution of our problem. For if a control $\{u_0, \ldots, u_{N-1}\}$ satisfies (10), then

$$\sum_{k=0}^{N-1} G_k(\hat{x}_k, \hat{u}_k) + \hat{\lambda}\Delta = \sum_{k=0}^{N-1} (G_k(\hat{x}_k, \hat{u}_k) + \hat{\lambda} H_k(\hat{x}_k, \hat{u}_k))$$

$$\leqq \sum_{k=0}^{N-1} (G_k(x_k, u_k) + \hat{\lambda} H_k(x_k, u_k)) = \sum_{k=0}^{N-1} G_k(x_k, u_k) + \hat{\lambda}\Delta .$$

Hence

$$(11) \qquad \sum_{k=0}^{N-1} G(\hat{x}_k, \hat{u}_k) \leqq \sum_{k=0}^{N-1} G_k(x_k, u_k) .$$

Note that here the Lagrange multiplier appears in a sufficient condition for an extremum. If $\hat{\lambda} \geqslant 0$, then (11) holds for all controls satisfying

$$\sum_{k=0}^{N-1} H(x_k, u_k) \leqq \Delta .$$

EXAMPLE 5. In addition to the assumptions in Example 2, we require that the total weight does not exceed v. Let v_i be the weight of a single component in A_i. The method of Lagrange multipliers leads to the quantity

$$\hat{\Xi}(x; m, N) = \max \{(1 - p_m^{k_m}) \ldots (1 - p_N^{k_N}) e^{-\lambda \sum_{i=m}^{N} k_i v_i} : k_m c_m + \ldots + k_N c_N \leqq x\} ,$$

satisfying

$$\hat{\Xi}(x; m, N) = \max_{k=0,1,\ldots} (1 - p_m^k) e^{-\lambda k v_m} \hat{\Xi}(x - k c_m; m + 1, N) ,$$

$$m = N - 1, \ldots, 1 .$$

We first find the extent of aggregates $\{\hat{k}_1, \ldots, \hat{k}_N\}$ such that

$$\hat{\Xi}(c; 1, N) = (1 - p_1^{\hat{k}_1}) \ldots (1 - p_N^{\hat{k}_N}) e^{-\lambda \sum_{i=1}^{N} \hat{k}_i v_i} ,$$

and then compare the total weight $\Sigma_1^N \hat{k}_i v_i$ with v. If $\Sigma_1^N \hat{k}_i v_i > v$, increase λ. If $\Sigma_1^N \hat{k}_i v_i < v$ and the solution is not yet reached because the control parameter is assumed to be integer-valued, decrease λ. The successive approximations lead to the solution.

§5. The Hamilton-Jacobi equation

Continuous decision processes are discussed in Chapter III. Here we only informally consider passing to the limit in equation (2), without attempting to preserve mathematical rigor. Let Δt be the duration of a single stage ($\Delta t \longrightarrow 0$), $N\Delta t = T$. In order to obtain a continuous process, we assume that the change is proportionate to Δt. That is,

$$\boldsymbol{F}_n(\boldsymbol{x}, \boldsymbol{u}) = \boldsymbol{x} + \boldsymbol{f}(n\,\Delta t, \boldsymbol{x}, \boldsymbol{u})\,\Delta t$$

and similarly

$$G_n(\boldsymbol{x}, \boldsymbol{u}) = g(n\,\Delta t, \boldsymbol{x}, \boldsymbol{u})\,\Delta t\,.$$

From (2) it follows that

$$\hat{\psi}(\boldsymbol{x};\, m\,\Delta t, T) = \operatorname*{opt}_{\boldsymbol{u}\in U} \{g(m\,\Delta t, \boldsymbol{x}, \boldsymbol{u})\,\Delta t + \hat{\psi}(\boldsymbol{x} + \boldsymbol{f}(m\,\Delta t, \boldsymbol{x}, \boldsymbol{u})\,\Delta t;\, \overline{m+1}\,\Delta t, T)\}\,,$$

or, equivalently,

$$\operatorname*{opt}_{\boldsymbol{u}\in U} \{g(m\,\Delta t, \boldsymbol{x}, \boldsymbol{u})\,\Delta t + \frac{\partial}{\partial t}\hat{\psi}\,\Delta t + \overline{\nabla}\hat{\psi}\,.\,\boldsymbol{f}(m\,\Delta t, \boldsymbol{x}, \boldsymbol{u})\,\Delta t + o(\Delta t)\} = 0\,. \tag{12}$$

Here $\overline{\nabla}$ is the gradient with respect to the state variables. The gradient with respect to the decision variables is denoted by ∇. The dot denotes the scalar product. As $\Delta t \longrightarrow 0$ and $m\Delta t \longrightarrow t$, (12) gives

$$\frac{\partial}{\partial t}\hat{\psi}(\boldsymbol{x};\, t, T) + \operatorname*{opt}_{\boldsymbol{u}\in U} \{g(t, \boldsymbol{x}, \boldsymbol{u}) + \overline{\nabla}\hat{\psi}(\boldsymbol{x};\, t, T)\,.\,\boldsymbol{f}(t, \boldsymbol{x}, \boldsymbol{u})\} = 0\,, \tag{13}$$

$$\hat{\psi}(\boldsymbol{x};\, T, T) = 0\,. \tag{14}$$

In the continuous case, a control is described by a function $\{\boldsymbol{u}(t), 0 \leqslant t \leqslant T\}$. A typical class of admissible controls is the set of piecewise continuous functions. Passing to the limit, we obtain a system of differential equations for the trajectory of the process $\{\boldsymbol{x}(t), 0 \leqslant t \leqslant T\}$:

$$\frac{\mathrm{d}\boldsymbol{x}}{\mathrm{d}t} = \boldsymbol{f}(t, \boldsymbol{x}, \boldsymbol{u}(t))\,.$$

The objective function is $\int_0^T g(t, \boldsymbol{x}(t), \boldsymbol{u}(t))\,dt$. An admissible policy is a function $\{\boldsymbol{u}(t, \boldsymbol{x}), 0 \leqslant t \leqslant T, \boldsymbol{x} \in R^r\}$ such that for every $\boldsymbol{x}_0 \in R^r$ there exists a solution of the initial-value problem

$$\frac{\mathrm{d}\boldsymbol{x}}{\mathrm{d}t} = \boldsymbol{f}(t, \boldsymbol{x}, \boldsymbol{u}(t, \boldsymbol{x}))\,, \quad t \in [0, T]\,, \quad \boldsymbol{x}(0) = \boldsymbol{x}_0\,,$$

and $\boldsymbol{u}(t, \boldsymbol{x}(t))$ is an admissible control. The functions $\boldsymbol{f}(t, \boldsymbol{x}, \boldsymbol{u})$ and $g(t, \boldsymbol{x}, \boldsymbol{u})$ are assumed to be continuous and continuously differentiable with respect to x^i, $i = 1, \ldots, r$.

Equation (13) is called the *Hamilton-Jacobi equation* of the problem. The assumption about the differentiability of $\hat{\psi}(\boldsymbol{x};\, t, T)$ often does not hold, and the applicability of the equation is therefore limited. However, its solvability gives a sufficient condition for an optimum.

THEOREM 1. *Let $\hat{\psi}(\boldsymbol{x}; t, T)$ be a solution of problem* (13), (14) *which is continuously differentiable on $R^r \times [0, T]$. Suppose there exists an admissible policy $\hat{\boldsymbol{u}}(t, \boldsymbol{x})$ such that*

$$g(t, \boldsymbol{x}, \hat{\boldsymbol{u}}(t, \boldsymbol{x})) + \nabla\hat{\psi}(\boldsymbol{x}; t, T) \,.\, \boldsymbol{f}(t, \boldsymbol{x}, \hat{\boldsymbol{u}}(t, \boldsymbol{x}))$$
$$= \operatorname*{opt}_{\boldsymbol{u}\in U} \{g(t, \boldsymbol{x}, \boldsymbol{u}) + \nabla\hat{\psi}(\boldsymbol{x}; t, T) \,.\, \boldsymbol{f}(t, \boldsymbol{x}, \boldsymbol{u})\} \,.$$

Then

$$\hat{\psi}(\boldsymbol{x}; t, T) = \operatorname{opt}\left\{\int_t^T g(s, \boldsymbol{x}(s), \boldsymbol{u}(s))\, \mathrm{d}s : \boldsymbol{x}(t) = \boldsymbol{x}\right\},$$

and $\hat{\boldsymbol{u}}(t, \boldsymbol{x})$ is an optimal policy.

PROOF. For definiteness, let opt = max. Choose any t and $\boldsymbol{x}$. For an arbitrary control $\{\boldsymbol{u}(s), t \leqslant s \leqslant T\}$, let $\{\boldsymbol{x}(s), t \leqslant s \leqslant T\}$, $\boldsymbol{x}(t) = \boldsymbol{x}$, be the corresponding trajectory. By (13),

$$\frac{\partial}{\partial s}\hat{\psi}(\boldsymbol{x}(s); s, T) + g(s, \boldsymbol{x}(s), \boldsymbol{u}(s)) + \nabla\hat{\psi}(\boldsymbol{x}(s); s, T) \,.\, f(s, \boldsymbol{x}(s), \boldsymbol{u}(s)) \leqq 0 \,,$$

or, equivalently,

$$\frac{\partial}{\partial s}\hat{\psi} + \sum_{i=1}^{r} \frac{\partial\hat{\psi}}{\partial x^i}\frac{\mathrm{d}x^i}{\mathrm{d}s} + g \leqq 0 \,.$$

Integrating over $s \in [t, T]$, we get

$$\hat{\psi}(\boldsymbol{x}(T); T, T) - \hat{\psi}(\boldsymbol{x}; t, T) + \int_t^T g(s, \boldsymbol{x}(s), \boldsymbol{u}(s))\, \mathrm{d}s \leqq 0 \,,$$

and this together with (14) yields

$$\int_t^T g(s, \boldsymbol{x}(s), \boldsymbol{u}(s))\, \mathrm{d}s \leqq \hat{\psi}(\boldsymbol{x}; t, T) \,.$$

Equality holds for $\boldsymbol{u}(s) = \hat{\boldsymbol{u}}(s, \boldsymbol{x}(s))$.

EXAMPLE 6. Let

$$\hat{\psi}(x; t, T) = \min_{\{u(s), t \leqq s \leqq T\}} \frac{1}{2}\int_t^T [q\, x(s)^2 + r\, u(s)^2]\, \mathrm{d}s \,,$$

where

$$\frac{\mathrm{d}}{\mathrm{d}s}x(s) = a\, x(s) - b\, u(s) \,, \quad x(t) = x \,; \quad q \geqq 0 \,, \quad a, b, r > 0 \,.$$

Now (13) becomes

$$0 = \frac{\partial}{\partial t}\hat{\psi} + \min_u \left\{ \tfrac{1}{2}(qx^2 + ru^2) + (ax - bu)\frac{\partial}{\partial x}\hat{\psi} \right\}$$

$$= \frac{\partial}{\partial t}\hat{\psi} + \min_u \left\{ \frac{r}{2}\left(u - \frac{b}{r}\frac{\partial}{\partial x}\hat{\psi}\right)^2 + \frac{1}{2}\left(qx^2 - \frac{b^2}{r}\left(\frac{\partial}{\partial x}\hat{\psi}\right)^2\right) + ax\frac{\partial}{\partial x}\hat{\psi} \right\}$$

$$= \frac{\partial}{\partial t}\hat{\psi} + \frac{1}{2}\left(qx^2 - \frac{b^2}{r}\left(\frac{\partial}{\partial x}\hat{\psi}\right)^2\right) + ax\frac{\partial}{\partial x}\hat{\psi}\,.$$

It is not difficult to check that $\hat{\psi}(x; t, T) = \frac{1}{2}x^2 k(t)$ satisfies this equation when

$$\text{(15)} \qquad \frac{\mathrm{d}}{\mathrm{d}t}k = -q + \frac{b^2}{r}k^2 - 2ak\,, \quad k(T) = 0\,.$$

Equation (15) has a solution

$$k(t) = (A + B)\left(1 - \mathrm{e}^{2B(t-T)}\right)\left[1 + \frac{B + A}{B - A}\mathrm{e}^{2B(t-T)}\right]^{-1},$$

where

$$A = arb^{-2}\,, \quad B = b^{-2}\sqrt{(a^2r^2 + b^2qr)}\,.$$

The optimal policy is

$$\hat{u}(t, x) = \frac{b}{r}\frac{\partial}{\partial x}\hat{\psi}(x; t, T) = x\,k(t)\,b/r$$

(cf. Example 2 in Chapter III).

CHAPTER II. THE DISCRETE OPTIMUM PRINCIPLE

§1. Lagrange multipliers in multistage optimization

Let us first reformulate the original problem, extending the state vector by a new coordinate

$$x_n^0 = \sum_{k=0}^{n-1} G_k(x_k, u_k)\,, \quad n = 1, 2, \ldots, \quad x_0^0 = 0\,.$$

Write $\boldsymbol{x} = (x^0, x)$. Now the problem is to optimize x_N^0, the transformations of state vectors being

$$\boldsymbol{F}_n(\boldsymbol{x}, u) = (F_n^0(\boldsymbol{x}, u), \ldots, F_n^r(\boldsymbol{x}, u)),$$

where

$$F_n^0(\boldsymbol{x}, u) = x^0 + G_n(x, u)\,, \quad F_n^j(\boldsymbol{x}, u) = F_n^j(x, u), \quad j = 1, \ldots, r\,.$$

The value $\boldsymbol{x}_0$ is given, and the relations

$$\boldsymbol{x}_n = \boldsymbol{F}_{n-1}(\boldsymbol{x}_{n-1}, u_{n-1})\,, \quad n = 1, \ldots, N\,,$$

are conditions of the optimization problem.

Assume that the mappings F_n are continuously differentiable for all x and for u

in an open set containing U, and use the Lagrange multipliers method. Put $\lambda_n = (\lambda_n^0, \ldots, \lambda_n^r)$, $n = 1, \ldots, N$, and consider the Lagrange function

$$L = x_n^0 - \sum_{n=1}^{N} \boldsymbol{\lambda}_n \cdot (\mathbf{x}_n - \mathbf{F}_{n-1}(\mathbf{x}_{n-1}, \mathbf{u}_{n-1})) .$$

If a relative extremum appears at an interior point of U, then the Lagrange function is stationary. Stationarity conditions with respect to the state variables are

$$\frac{\partial L}{\partial x_N^0} = 1 - \lambda_N^0 = 0 , \quad \frac{\partial L}{\partial x_N^j} = -\lambda_N^j = 0 ,$$

$$\frac{\partial L}{\partial x_n^j} = -\lambda_n^j + \boldsymbol{\lambda}_{n+1} \cdot \frac{\partial \mathbf{F}_n}{\partial x^j} (\mathbf{x}_n, \mathbf{u}_n) ,$$

$$j = 0, \ldots, r , \quad n = 1, \ldots, N - 1 .$$

Introducing the Jacobi matrices

$$\mathbf{J}_n(\mathbf{x}, \mathbf{u}) = \left\| \frac{\partial F_n^i(\mathbf{x}, \mathbf{u})}{\partial x^j} \right\|_{i,j=0}^{r} ,$$

we can write

$$\boldsymbol{\lambda}_N = (1, 0, \ldots, 0) , \quad \boldsymbol{\lambda}_n = \boldsymbol{\lambda}_{n+1} \mathbf{J}_n(\mathbf{x}_n, \mathbf{u}_n) , \quad n = N - 1, \ldots, 1 . \tag{1}$$

This is a recurrence relation for Lagrange multipliers. Another interpretation of Lagrange multipliers should be mentioned. We show that $\lambda_n^j = \partial x_N^0 / \partial x_n^j$.

Consider the increment of the nth state vector

$$^*\mathbf{x}_n = \mathbf{x}_n + \varepsilon\, \delta \mathbf{x}_n + o(\varepsilon) , \quad \varepsilon \to 0 , \quad \varepsilon > 0 ,$$

for fixed values of the decision variables $\mathbf{u}_n, \mathbf{u}_{n+1}, \ldots, \mathbf{u}_{N-1}$. The linear part $\delta \mathbf{x}_n$ of the increment is called the *variation*. This increment causes a change of $\mathbf{x}_{n+1}$ to

$$^*\mathbf{x}_{n+1} = \mathbf{x}_{n+1} + \varepsilon\, \delta \mathbf{x}_{n+1} + o(\varepsilon) ,$$

where

$$\delta \mathbf{x}_{n+1} = \delta \mathbf{x}_n \mathbf{J}_n(\mathbf{x}_n, \mathbf{u}_n)' .$$

The prime denotes the transposed matrix (or the column vector, in the sequel). Similarly the variation $\delta \mathbf{x}_{n+1}$ is transformed into $\delta \mathbf{x}_{n+2}$, etc. By (1) we have

$$\boldsymbol{\lambda}_n \cdot \delta \mathbf{x}_n = \boldsymbol{\lambda}_{n+1} \mathbf{J}_n(\mathbf{x}_n, \mathbf{u}_n)\, \delta \mathbf{x}_n' = \boldsymbol{\lambda}_{n+1} \cdot \delta \mathbf{x}_{n+1} = \ldots = \boldsymbol{\lambda}_N \cdot \delta \mathbf{x}_N = \delta x_N^0 .$$

Thus λ_n is the gradient of the objective function with respect to $\mathbf{x}_n$.

To determine the gradient with respect to the decision variables, consider the increment

$$^*\mathbf{u}_n = \mathbf{u}_n + \varepsilon\, \delta \mathbf{u}_n + o(\varepsilon)$$

for fixed $\mathbf{x}_n$ and $\mathbf{u}_{n+1}, \ldots, \mathbf{u}_{N-1}$. The corresponding increment in $\mathbf{x}_{n+1}$ is

$$^{*}\boldsymbol{x}_{n+1} = \boldsymbol{F}_n(\boldsymbol{x}_n, \boldsymbol{u}_n) + \varepsilon \sum_{j=1}^{s} \frac{\partial}{\partial u^j} \boldsymbol{F}_n(\boldsymbol{x}_n, \boldsymbol{u}_n)\, \delta u_n^j + o(\varepsilon)\,.$$

Hence the variation of the objective function is

$$\delta x_N^0 = \boldsymbol{\lambda}_{n+1} \,.\, \delta \boldsymbol{x}_{n+1} = \boldsymbol{\lambda}_{n+1} \,.\, \sum_{j=1}^{s} \frac{\partial \boldsymbol{F}_n}{\partial u^j}\, \delta u_n^j\,.$$

Put

$$\mathscr{H}_n(\boldsymbol{\lambda}_{n+1}, \boldsymbol{x}_n, \boldsymbol{u}) = \boldsymbol{\lambda}_{n+1} \,.\, \boldsymbol{F}_n(\boldsymbol{x}_n, \boldsymbol{u})\,, \quad n = 0, 1, \ldots, N-1\,.$$

The function $H_n(\lambda_{n+1}, x_n, \boldsymbol{u})$ is called the *Hamiltonian.* Its gradient with respect to the decision variables coincides with the gradient of the objective function.

Let $\{\hat{\boldsymbol{u}}_0, \ldots, \hat{\boldsymbol{u}}_{N-1}\}$ be a policy maximizing the result, and let $\delta\hat{\boldsymbol{u}}_n$ satisfy

$$\hat{\boldsymbol{u}}_n + \varepsilon\, \delta\hat{\boldsymbol{u}}_n + o(\varepsilon) \in U \tag{2}$$

as $\epsilon \longrightarrow 0$ ($\epsilon > 0$). A variation satisfying (2) will be called *admissible.* We have $\delta\hat{x}_N^0 \leqslant 0$, i.e.

$$\nabla \mathscr{H}_n(\hat{\boldsymbol{\lambda}}_{n+1}, \hat{\boldsymbol{x}}_n, \hat{\boldsymbol{u}}_n) \,.\, \delta\hat{\boldsymbol{u}}_n \leqq 0\,, \quad n = 0, 1, \ldots, N-1\,.$$

If $\hat{\boldsymbol{u}}_n$ is an interior point of U, and hence all the variations are admissible, then

$$\nabla \mathscr{H}_n(\hat{\boldsymbol{\lambda}}_{n+1}, \hat{\boldsymbol{x}}_n, \hat{\boldsymbol{u}}_n) = 0\,.$$

(This is actually the stationarity condition for the Lagrange function.)

Both conditions are *local*, because they take into account only the behavior of the Hamiltonian in a neighborhood of the optimal control. A *global* condition is provided by the discrete optimum principle.

§2. The discrete optimum principle

This principle was formulated in analogy with the Pontrjagin principle. It says that "*the optimal control optimizes the Hamiltonian*". More exactly:

If $\{\boldsymbol{u}_0, \ldots, \boldsymbol{u}_{N-1}\}$ is an optimal control and $\{x_0 = \hat{x}_0, \ldots, \hat{x}_N\}$ is the corresponding trajectory, then

$$\mathscr{H}_n(\hat{\boldsymbol{\lambda}}_{n+1}, \hat{\boldsymbol{x}}_n, \hat{\boldsymbol{u}}) = \operatorname*{opt}_{\boldsymbol{u} \in U} \mathscr{H}_n(\hat{\boldsymbol{\lambda}}_{n+1}, \hat{\boldsymbol{x}}_n, \boldsymbol{u})\,, \quad n = 0, \ldots, N-1\,. \tag{3}$$

We prove (3) without assuming the differentiability of $F_n(x, \boldsymbol{u})$ with respect to $\boldsymbol{u}$. Instead, we assume that for all x and $n = 0, \ldots, N-1$ the range $\{F_n(x, \boldsymbol{u}): \boldsymbol{u} \in U\}$ of F_n is convex.

Let us investigate a more general problem, namely to find a control optimizing $g_0(x_N)$ subject to the condition that x_0 belongs to an initial set $I = \{x: h_i(x) = 0, i = 1, \ldots, l\}$ and x_N to a final set $F = \{x: g_i(x) = 0, i = 1, \ldots, m\}$. Assume $h_i(x)$ and $g_i(x)$ are continuously differentiable in R^{r+1}, the vectors $\overline{\nabla} h_i(x)$, $i = 1, \ldots, l$, are linearly independent for $x \in I$, and the vectors $\overline{\nabla} g_i(x)$, $i = 0, \ldots, m$, are linearly independent for $x \in F$. We make the further restriction $l + m \leqslant r + 1$.

Let $\{\hat{u}_0, \ldots, \hat{u}_{N-1}\}\{x_0, \ldots, x_N\}$ be a solution of this problem, and let $\{u_0, \ldots, u_{N-1}\}$ be an arbitrary control. Consider the following optimization problem: To find $\{a_0, \ldots, a_{N-1}\}$, $0 \leqslant a_n \leqslant 1$, such that $g_0(y_N)$ is optimal (say maximal, for definiteness), when

$$\text{(4)}\qquad \begin{aligned} \mathbf{y}_n = \Phi_{n-1}(\mathbf{y}_{n-1}, a_{n-1}) &= a_{n-1}\, \mathbf{F}_{n-1}(\mathbf{y}_{n-1}, \mathbf{u}_{n-1}) \\ &+ (1 - a_{n-1})\, \mathbf{F}_{n-1}(\mathbf{y}_{n-1}, \hat{\mathbf{u}}_{n-1}), \quad n = 1, \ldots, N, \end{aligned}$$

$$\text{(5)}\qquad \begin{aligned} h_i(\mathbf{y}_0) &= 0, \quad i = 1, \ldots, l, \\ g_i(\mathbf{y}_N) &= 0, \quad i = 1, \ldots, m. \end{aligned}$$

Under conditions (4) and (5) we obviously have $\max g_0(y_N) \geqslant g_0(\hat{x}_N)$, because the control $\{a_n = 0, n = 0, \ldots, N-1\}$ and the initial value $y_0 = \hat{x}_0$ lead to $g_0(\hat{x}_N)$. Further, since the range of F_n is convex, for any control $\{a_0, \ldots, a_{N-1}\}$ there exists a control $\{\bar{u}_0, \ldots, \bar{u}_{N-1}\}$ such that

$$\mathbf{y}_{n+1} = \mathbf{F}_n(\mathbf{y}_n, \bar{\mathbf{u}}_n), \quad n = 0, \ldots, N-1.$$

Therefore every $\{y_0, \ldots, y_{N-1}\}$ satisfying (4) is also a trajectory of the original problem. Hence $g_0(y_N) \leqslant g_0(\hat{x}_N)$ and $\{a_n = 0, n = 0, \ldots, N-1; \hat{y}_0 = \hat{x}_0\}$ is an optimal control.

The problem contains $(N+1)(r+1) + N$ variables $y_0^0, \ldots, y_0^r, \ldots, y_N^0, \ldots, y_N^r, a_0, \ldots, a_{N-1}$. These are subject to $N(r+1) + l + m$ conditions (4), (5). We want to use the implicit function theorem to express the variables $y_1^0, \ldots, y_1^r, \ldots, y_N^0, \ldots, y_N^r$ and $l + m$ of the variables $y_0^0, \ldots, y_0^r$ as functions of $a_0, \ldots, a_{N-1}$ and the remaining $r + 1 - l - m$ of $y_0^0, \ldots, y_0^r$ in a neighborhood of the optimal solution. With this in mind, consider the Jacobi matrix $\hat{J}$ of functions $h_1, \ldots, h_l$, $y_1^0 - \Phi_0^0, y_1^1 - \Phi_0^1, \ldots, y_N^r - \Phi_{N-1}^r, g_1, \ldots, g_m$ with respect to the variables $y_0^0, \ldots, y_N^r$ at the point $y_0 = \hat{x}_0, \ldots, y_N = \hat{x}_N, a_0 = 0, \ldots, a_{N-1} = 0$. It is not difficult to see that $\hat{J}$ may be written as

$$\hat{\mathbf{J}} = \begin{pmatrix} \begin{matrix}\nabla h_1, \\ \vdots \\ \nabla h_l,\end{matrix} & \mathbf{0}, & \mathbf{0}, & \ldots & \mathbf{0}, & \mathbf{0} \\ -\mathbf{J}_0(\hat{\mathbf{x}}_0, \hat{\mathbf{u}}_0), & \mathbf{E}, & \mathbf{0}, & \ldots & \mathbf{0}, & \mathbf{0} \\ \mathbf{0}, & -\mathbf{J}_1(\hat{\mathbf{x}}_1, \hat{\mathbf{u}}_1), & \mathbf{E}, & \ldots & \mathbf{0}, & \mathbf{0} \\ \cdots & \cdots & \cdots & \cdots & \cdots & \cdots \\ \mathbf{0}, & \mathbf{0}, & \mathbf{0}, & \ldots & -\mathbf{J}_{N-1}(\hat{\mathbf{x}}_{N-1}, \hat{\mathbf{u}}_{N-1}), & \mathbf{E} \\ \mathbf{0}, & \mathbf{0}, & \mathbf{0}, & \ldots & \mathbf{0}, & \begin{matrix}\nabla g_1 \\ \vdots \\ \nabla g_m\end{matrix} \end{pmatrix}$$

Here E are unit matrices of order $r + 1$. Rows in the matrix $\hat{J}$ are linearly independent if

$$\text{(6)}\qquad 0 = \lambda\hat{\mathbf{J}}, \quad \lambda = (\gamma_1, \ldots, \gamma_l, \gamma_0^0, \ldots, \gamma_0^r, \ldots, \gamma_N^0, \ldots, \gamma_N^r, \gamma_1', \ldots, \gamma_m'),$$

implies $\lambda = 0$. This condition is clearly equivalent to condition (P):

(P) The equality

$$\gamma_1 \bar{\nabla} h_1(\hat{\mathbf{x}}_0) + \ldots + \gamma_l \bar{\nabla} h_l(\hat{\mathbf{x}}_l) = [\gamma'_1 \bar{\nabla} g_1(\hat{\mathbf{x}}_N) \\ \ldots + \gamma'_m \bar{\nabla} g_m(\hat{\mathbf{x}}_N)] \mathbf{J}_{N-1}(\hat{\mathbf{x}}_{N-1}, \hat{\mathbf{u}}_{N-1}) \ldots \mathbf{J}_0(\hat{\mathbf{x}}_0, \hat{\mathbf{u}}_0) \tag{7}$$

holds only for

$$\gamma_1 = \ldots = \gamma_l = \gamma'_1 = \ldots = \gamma'_m = 0 .$$

Condition (P) is by no means necessary for the validity of the optimum principle. However, it is fulfilled in a number of important cases, e.g. when no restrictions are imposed upon the final point of the trajectory, or when no restrictions are imposed upon the initial point and $|J_0(\hat{x}_0, \hat{u}_0)| \neq 0, \ldots, |J_{N-1}(\hat{x}_{N-1}, \hat{u}_{N-1})| \neq 0$.

Assume (P) holds. Hence the vectors

$$\bar{\nabla} h_1, \ldots, \bar{\nabla} h_l, \bar{\nabla} g_1 \mathbf{J}_{N-1} \ldots \mathbf{J}_0, \ldots, \bar{\nabla} g_m \mathbf{J}_{N-1} \ldots \mathbf{J}_0 , \tag{8}$$

are linearly independent. Then, as is well known, there exist indices $0 \leqslant i_1 < i_2 < \cdots < i_{l+m} \leqslant r + 1$ such that the $(l + m)$-dimensional vectors formed by the i_1th, i_2th, $\ldots$, i_{l+m}th coordinates of the vectors (8) are linearly independent. Assume, without loss of generality, that $i_1 = 0, i_2 = 1, \ldots , i_{l+m} = l + m - 1$. Then

$$|\hat{\mathbf{J}}_*| = \left. \frac{\partial(h_1, \ldots, h_l, y_1^0 - \Phi_0^0, \ldots, y_N^r - \Phi_{N-1}^r, g_1, \ldots, g_m)}{\partial(y_0^0, \ldots, y_0^{l+m-1}, y_1^0, \ldots, y_1^r, \ldots, y_N^0, \ldots, y_N^r)} \right|_{\mathbf{y}_0 = \hat{\mathbf{x}}_0, \ldots, \mathbf{y}_N = \hat{\mathbf{x}}_N, a_0 = 0, \ldots, a_{N-1} = 0} \neq 0 , \tag{9}$$

since the matrix $\hat{J}_*$, arising from $\hat{J}$ when the $(l + m + 1)$th through $(r + 1)$th columns are left out, has linearly independent rows. By the implicit function theorem, there exist functions

$$Y_0^0(y_0^{l+m}, \ldots, y_0^r, a_0, \ldots, a_{N-1}), \ldots, Y_0^{l+m-1}(y_0^{l+m}, \ldots, a_{N-1}), \\ Y_1^0(y_0^{l+m}, \ldots, a_{N-1}), \ldots, Y_N^r(y_0^{l+m}, \ldots, a_{N-1}),$$

defined in a neighborhood of $(\hat{x}_0^{l+m}, \ldots , \hat{x}_0^r, 0, \ldots , 0)$, such that in this neighborhood (4) and (5) hold if and only if

$$y_0^0 = Y_0^0, \ldots, y_0^{l+m-1} = Y_0^{l+m-1}, y_1^0 = Y_1^0, \ldots, y_N^r = Y_N^r .$$

(Restriction to nonnegative $a_0, \ldots , a_{N-1}$ is not necessary.)

Introduce Lagrange multipliers $\hat{\lambda}_n = (\hat{\lambda}_n^0, \ldots , \hat{\lambda}_n^r)$, $n = 1, \ldots , N$, for conditions (4) and multipliers $\hat{\alpha}_1, \ldots , \hat{\alpha}_l$ and $\hat{\beta}_1, \ldots , \hat{\beta}_m$ for conditions (5). Form the Lagrange function

$$L(\mathbf{y}_0, \mathbf{y}_1, \ldots, \mathbf{y}_N, a_0, \ldots, a_{N-1}) \\ = g_0(\mathbf{y}_N) - \sum_{n=1}^{N} \hat{\boldsymbol{\lambda}}_n \cdot (\mathbf{y}_n - \boldsymbol{\Phi}_{n-1}(\mathbf{y}_{n-1}, a_{n-1})) + \sum_{i=1}^{l} \hat{\alpha}_i h_i(\mathbf{y}_0) + \sum_{i=1}^{m} \hat{\beta}_i g_i(\mathbf{y}_N) .$$

Choose $\hat{\lambda}_1, \ldots , \hat{\lambda}_N, \hat{\alpha}_1, \ldots , \hat{\alpha}_l$ and $\hat{\beta}_1, \ldots , \hat{\beta}_m$ such that

(10)
$$\frac{\partial}{\partial y_j^i} L(\hat{\mathbf{x}}_0, \ldots, \hat{\mathbf{x}}_N, 0, \ldots, 0) = 0\,,$$
$$i = 0\,, \quad j = 0, \ldots, l + m - 1\,, \quad i = 1, \ldots, N\,, \quad j = 0, \ldots, r\,.$$

By (9), relations (10) have a unique solution. It is easy to check that (10) can be written as

(11)
$$0 = \overline{\nabla} g_0(\hat{\mathbf{x}}_N) - \hat{\boldsymbol{\lambda}}_N + \sum_{i=1}^{m} \hat{\beta}_i \overline{\nabla} g_i(\hat{\mathbf{x}}_N)\,,$$

(12)
$$0 = \hat{\boldsymbol{\lambda}}_{n+1} \boldsymbol{J}(\hat{\mathbf{x}}_n, \hat{\boldsymbol{u}}_n) - \hat{\boldsymbol{\lambda}}_n\,, \quad n = 0, \ldots, N - 1\,,$$

(13)
$$0 = \hat{\lambda}_0^j - \sum_{i=1}^{l} \hat{\alpha}_i \frac{\partial}{\partial x^j} h_i(\hat{\mathbf{x}}_0)\,, \quad j = 0, \ldots, l + m - 1\,.$$

For $n = 0$, (12) defines $\hat{\lambda}_0$. By (11)–(13), the Lagrange multipliers do not depend on the control $\{u_0, \ldots, u_{N-1}\}$ chosen arbitrarily at the beginning. Furthermore $\hat{\lambda}_N \neq 0$ by (11), because the $\overline{\nabla} g_i(\hat{x}_N)$, $i = 0, \ldots, m$, are linearly independent according to the assumption.

Finally, consider the function

$$\tilde{g}_0(y_0^{l+m}, \ldots, y_0^r, a_0, \ldots, a_{n-1}) = g_0(Y_N(y_0^{l+m}, \ldots, y_0^r, a_0, \ldots, a_{n-1}))\,.$$

As $\tilde{g}_0(y_0^{l+m}, \ldots, y_0^r, 0, \ldots, 0)$ attains its maximum at $y_0^{l+m} = \hat{x}_0^{l+m}, \ldots, y_0^r = \hat{x}_0^r$, we have

$$\begin{aligned}
0 &= \frac{\partial}{\partial y_0^j} \tilde{g}(\hat{x}_0^{l+m}, \ldots, \hat{x}_0^r\, 0, \ldots, 0) \\
&= \frac{\partial}{\partial y_0^j} L(Y_0^0, \ldots, Y_0^{l+m-1}, y_0^{l+m}, \ldots, y_0^r, Y_1^0, \ldots, Y_N^r, a_0, \ldots, a_{N-1}) \\
&= \frac{\partial L}{\partial y_0^j} + \sum_{k=0}^{l+m-1} \frac{\partial L}{\partial y_0^k} \frac{\partial Y_0^k}{\partial y_0^j} + \sum_{i=1}^{N} \sum_{k=0}^{r} \frac{\partial L}{\partial y_i^k} \frac{\partial Y_i^k}{\partial y_0^j}\bigg|_{y_0 = \hat{x}_0, \ldots, y_N = \hat{x}_N, a_0 = 0, \ldots, a_{N-1} = 0}\,,
\end{aligned}$$
$$j = l + m, \ldots, r\,.$$

By (10) it follows that

(14)
$$0 = \frac{\partial}{\partial y_0^j} L(\hat{\mathbf{x}}_0, \ldots, \hat{\mathbf{x}}_N, 0, \ldots, 0)\,, \quad j = l + m, \ldots, r\,.$$

(14) is equivalent to

(15)
$$0 = \hat{\lambda}_0^j - \sum_{i=1}^{l} \hat{\alpha}_i \frac{\partial}{\partial x^j} h_i(\hat{\mathbf{x}}_0)\,, \quad j = l + m, \ldots, r\,.$$

Furthermore,

$$0 \geqq \frac{\partial}{\partial a_j} \tilde{g}_0 = \frac{\partial}{\partial a_j} L(\hat{\mathbf{x}}_0, \ldots, \hat{\mathbf{x}}_N, 0, \ldots, 0)\,, \quad j = 0, \ldots, N - 1\,.$$

for sufficiently small $a_0 \geqslant 0, \ldots, a_{N-1} \geqslant 0$. Consequently

$$0 \geqq \frac{\partial}{\partial a_j} \tilde{g}_0 = \frac{\partial}{\partial a_j} L(\hat{\mathbf{x}}_0, \ldots, \hat{\mathbf{x}}_N, 0, \ldots, 0), \quad j = 0, \ldots, N-1.$$

That is,

$$0 \geqq \hat{\boldsymbol{\lambda}}_{n+1} \cdot (\mathbf{F}_n(\hat{\mathbf{x}}_n, \mathbf{u}_n) - \mathbf{F}_n(\hat{\mathbf{x}}_n, \hat{\mathbf{u}}_n)), \quad n = 0, \ldots, N-1. \tag{16}$$

Since u_n was chosen arbitrarily, inequality (16) expresses an optimum principle. We formulate it in the following theorem.

THEOREM 1. *Let* $\{\hat{x}_0, \ldots, x_N\}, \{\hat{u}_0, \ldots, \hat{u}_{N-1}\}$ *be a solution of the optimization problem formulated at the beginning of this section, satisfying condition* (P). *Then there exists a nonzero sequence* $\hat{\lambda}_0, \ldots, \hat{\lambda}_N$ *such that*

$$\hat{\boldsymbol{\lambda}}_n = \hat{\boldsymbol{\lambda}}_{n+1} \mathbf{J}(\hat{\mathbf{x}}_n, \hat{\mathbf{u}}_n), \quad n = 0, \ldots, N-1, \tag{17}$$

and

$$\mathscr{H}_n(\hat{\boldsymbol{\lambda}}_{n+1}, \hat{\mathbf{x}}_n, \hat{\mathbf{u}}_n) = \operatorname*{opt}_{\mathbf{u} \in U} \mathscr{H}_n(\hat{\boldsymbol{\lambda}}_{n+1}, \hat{\mathbf{x}}_n, \mathbf{u}), \quad n = 0, \ldots, N-1, \tag{18}$$

$$\hat{\boldsymbol{\lambda}}_0 = \sum_{i=1}^{l} \hat{\alpha}_i \nabla h_i(\hat{\mathbf{x}}_0), \quad \hat{\boldsymbol{\lambda}}_N = \sum_{i=0}^{m} \hat{\beta}_i \nabla g_i(\hat{\mathbf{x}}_N) \tag{19}$$

for suitable $\hat{\alpha}_1, \ldots, \hat{\alpha}_l$ *and* $\hat{\beta}_0, \ldots, \hat{\beta}_m, \hat{\beta}_0 > 0$.

A control $\{\hat{u}_0, \ldots, \hat{u}_{N-1}\}$ and the corresponding trajectory $\{\hat{x}_0, \ldots, \hat{x}_N\}$, $\hat{x}_0 \in I$, $\hat{x}_N \in F$, is called *extremal* when (17)–(19) hold for a suitable nonzero sequence $\hat{\lambda}_0, \ldots, \hat{\lambda}_N$. Relations (19) are the *transversality conditions.*

§3. Examples

EXAMPLE 1. In this example the optimum principle does not hold:

$$U = [-2, -1] \cup [1, 2], \quad x \in R^1,$$

$$F_0(x, u) = x + u, \quad F_1(x, u) = x^2 + u^2, \quad x_0 = 0.$$

It is required to minimize $x_2 = u_0^2 + u_1^2$. The choice $\hat{u}_0 = \hat{u}_1 = 1$ is an optimal control, and

$$\hat{x}_1 = 1, \quad \hat{\lambda}_2 = 1, \quad \hat{\lambda}_1 = \hat{\lambda}_2(2\hat{x}_1) = 2;$$

$$\hat{\lambda}_1 F_0(\hat{x}_0, \hat{u}_0) = 2 > \min_{u \in U} \hat{\lambda}_1 F_0(\hat{x}_0, u) = -4.$$

EXAMPLE 2. Consider a numerical solution of Example 1 of Chapter I. The state vector transformation

$$x_n^0 = x_{n-1}^0 + 2(x_{n-1}^1 + u_{n-1}x_{n-1}^3) + x_{n-1}^2 + (1 - u_{n-1})\, x_{n-1}^3 .$$

$$x_n^1 = x_{n-1}^1 + u_{n-1}x_{n-1}^3,$$

$$x_n^2 = x_{n-1}^2 + (1 - u_{n-1})\, x_{n-1}^3,$$

$$x_n^3 = 0.5\,(x_{n-1}^1 + u_{n-1}x_{n-1}^3) + 0.9\,(x_{n-1}^2 + (1 - u_{n-1})\, x_{n-1}^3)$$

defines the Hamiltonian

$$\mathscr{H}_n(\lambda_{n+1}, \mathbf{x}_n, u) = \lambda^0_{n+1}[x^0_n + 2(x^1_n + ux^3_n) + x^2_n + (1-u)\, x^3_n]$$
$$(20) \qquad + \lambda^1_{n+1}(x^1_n + ux^3_n) + \lambda^2_{n+1}(x^2_n + (1-u)\, x^3_n) + \lambda^3_{n+1}[0.5\,(x^1_n + ux^3_n)$$
$$+ 0.9\,(x^2_n + (1-u)\, x^3_n)]\,.$$

We have to maximize x^0_5 when the initial state is $x_0 = (0, 0, 0, y)$. From (20) we can see that the optimal control $\{\hat{u}_0, \dots, \hat{u}_4\}$ must satisfy

$$(21) \qquad \begin{aligned} \hat{u}_n &= 0 \text{ when } \hat{\lambda}^0_{n+1} + \hat{\lambda}^1_{n+1} - \hat{\lambda}^2_{n+1} - 0{,}4\hat{\lambda}^3_{n+1} < 0\,, \\ \hat{u}_n &= 1 \text{ when } \hat{\lambda}^0_{n+1} + \hat{\lambda}^1_{n+1} - \hat{\lambda}^2_{n+1} - 0{,}4\hat{\lambda}^3_{n+1} > 0\,. \end{aligned}$$

We need to know the Jacobi matrix for $u = 0$ and $u = 1$. We have

$$J_n(\mathbf{x}, 0) = \begin{pmatrix} 1, & 2, & 1, & 1 \\ 0, & 1, & 0, & 0 \\ 0, & 0, & 1, & 1 \\ 0, & 0{,}5, & 0{,}9, & 0{,}9 \end{pmatrix}, \quad J_n(\mathbf{x}, 1) = \begin{pmatrix} 1, & 2, & 1, & 2 \\ 0, & 1, & 0, & 1 \\ 0, & 0, & 1, & 0 \\ 0, & 0{,}5, & 0{,}9, & 0{,}5 \end{pmatrix}.$$

TABLE 4.

n	$\hat{\lambda}_n$	$\hat{\lambda}^0_n + \hat{\lambda}^1_n - \hat{\lambda}^2_n - 0{,}4\hat{\lambda}^3_n$	$\hat{u}_{n-1}$
5	(1. 0. 0. 0)	1	1
4	(1. 2. 1. 2)	1.2	1
3	(1. 5. 3.8. 5)	0.2	1
2	(1. 9.5. 9.3. 9.5)	—2.6	0
1	(1. 16.25. 18.85. 18.85)	—9.14	0

The solution (see Table 4) starts with $\hat{\lambda}_N = (1, 0, 0, 0)$. Then we successively decide, for $n = N, N-1, \dots, 1$, whether $\hat{u}_{n-1} = 0$ or $\hat{u}_{n-1} = 1$, according to the sign of the expression $\hat{\lambda}^0_n + \hat{\lambda}^1_n - \hat{\lambda}^2_n - 0.4\hat{\lambda}^3_n$, and compute $\boldsymbol{\lambda}_{n-1}$ from $\hat{\boldsymbol{\lambda}}_{n-1} = \hat{\lambda}_n J_{n-1}(\hat{x}_{n-1}, \hat{u}_{n-1})$. The Jacobi matrix does not depend on the state vector. The only extremal control is $\{\hat{u}_0, \dots, \hat{u}_4\}$. It is also an optimal control, because such a control obviously exists. The optimal trajectory for $y = 1$ is described in Table 5.

TABLE 5.

n	$\hat{x}^0_n$	$\hat{x}^1_n$	$\hat{x}^2_n$	$\hat{x}^3_n$
0	0	0	0	1
1	1	0	1	0.9
2	1.9	0	1.9	1.71
3	7.22	1.71	1.9	2.565
4	17.67	4.275	1.9	3.8475
5	35.815	8.1225	1.9	5.77125

EXAMPLE 3 (FAN AND WANG)—*extraction of a solute from a solution.* Consider

a system consisting of N stages, where a certain substance is extracted from a solution by addition of wash water. The system is illustrated in Figure 2.

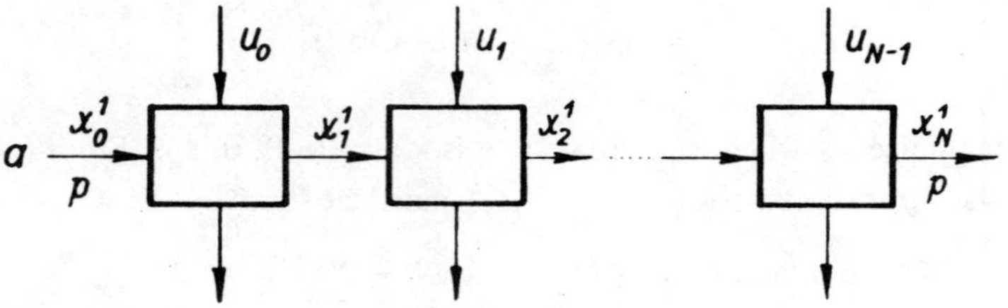

FIGURE 2.

Here p is the flow rate of the solution, w_n the amount of water passing through the $(n+1)$th stage in a time unit, $u_n = w_n/p$ is control parameter, x_n^1 the concentration of the substance in the solution entering the $(n+1)$th stage, and a the initial concentration.

Further, let $z = \varphi(x^1)$ be the concentration of the substance in water when x^1 is its concentration in the solution leaving a stage (φ is an increasing differentiable function of x^1).

Hence

(22) $$x_n^1 = x_{n-1}^1 - u_{n-1}\,\varphi(x_n^1), \quad n = 1, 2, \ldots, N .$$

Relation (22) defines the transformation

$$x_n^1 = F^1(\mathbf{x}_{n-1}, u_{n-1}) .$$

The net profit is equal to the amount of solute obtained (in a time unit), reduced by the cost of the needed water. That is,

$$p(a - x_N^1) - \varrho \sum_{n=0}^{N-1} w_n .$$

Taking into account the initial condition $x_0^1 = a$, we may write

$$p[x_0^1 - x_N^1 - \varrho \sum_{n=0}^{N-1} u_n] .$$

To maximize the profit means to maximize x_N^0, where

$$x_n^0 = x_{n-1}^0 + x_{n-1}^1 - x_n^1 - \varrho u_{n-1}, \quad x_0^0 = 0 ,$$

or, equivalently,

$$x_n^0 = x_{n-1}^0 + u_{n-1}[\varphi(F^1(\mathbf{x}_{n-1}, u_{n-1})) - \varrho] = F^0(\mathbf{x}_{n-1}, u_{n-1}) .$$

To find the Jacobi matrix, we use (22). From $F^1 - x^1 + u\varphi(F^1) = 0$ we get

$$\frac{\partial F^1}{\partial x^1} = \frac{1}{1 + u\,\varphi'(F_1)}, \quad \frac{\partial F^1}{\partial u} = -\frac{\varphi(F^1)}{1 + u\,\varphi'(F_1)} .$$

Hence

$$J(\mathbf{x}_n, u_n) = \begin{pmatrix} 1, & \dfrac{u_n \varphi'(x^1_{n+1})}{1 + u_n \varphi'(x^1_{n+1})} \\ 0, & \dfrac{1}{1 + u_n \varphi'(x^1_{n+1})} \end{pmatrix}.$$

The Lagrange multipliers satisfy

$$\lambda^0_n = \lambda^0_{n+1}, \quad \lambda^1_n = \frac{\lambda^0_{n+1} u_n \varphi'(x^1_{n+1}) + \lambda^1_{n+1}}{1 + u_n \varphi'(x^1_{n+1})}. \tag{23}$$

The Hamiltonian is

$$\mathscr{H}_n(\lambda_{n+1}, \mathbf{x}_n, u) = \lambda^0_{n+1} F^0(\mathbf{x}_n, u) + \lambda^1_{n+1} F^1(\mathbf{x}_n, u).$$

When $\{u_0, \ldots, u_{N-1}\}$ is an optimal control, it follows that

$$0 = \frac{d}{du} \mathscr{H}(\lambda_{n+1}, \mathbf{x}_n, u)\Big|_{u=u_n} = \lambda^0_{n+1}\left[\varphi(x^1_{n+1}) - \frac{u_n \varphi'(x^1_{n+1}) \varphi(x^1_{n+1})}{1 + u_n \varphi'(x^1_{n+1})} - \varrho\right] - \lambda^1_{n+1} \frac{\varphi(x^1_{n+1})}{1 + u_n \varphi'(x^1_{n+1})},$$

or, equivalently,

$$u_n = \frac{\lambda^0_{n+1}[\varphi(x^1_{n+1}) - \varrho] - \lambda^1_{n+1} \varphi(x^1_{n+1})}{\lambda^0_{n+1} \varrho \varphi'(x^1_{n+1})}. \tag{24}$$

Numerical solution may proceed as follows: Put $\boldsymbol{\lambda}_N = (1, 0)$ and choose x^1_N. Compute u_{N-1} from (24) and x^1_{N-1} from (22). Then $\boldsymbol{\lambda}_{N-1}$ can be computed from (23). Then u_{n-2} and x^1_{n-2} are computed in the same way, etc. This computation is repeated with the corrected value of x^1_N until x^1_0 is in a satisfactory agreement with the initial value a.

EXAMPLE 4—*mixing conditions.* Suppose that, in the system in Example 3, a part of the final product is fed back to the first stage, as shown in Figure 3, with flow rate r. The remaining notation is the same as before. The initial concentration satisfies

$$x^1_0 = \frac{pa + rx^1_N}{p + r}. \tag{25}$$

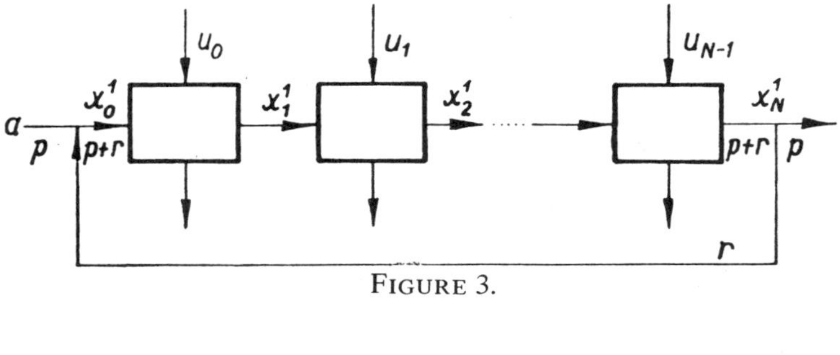

FIGURE 3.

This is the mixing condition, relating the initial and the final trajectory values. We use Theorem 1 to derive restrictions on the Lagrange multipliers following from the mixing conditions. Let the transformations $F_n(x, u)$, $n = 0, 1, \ldots, N-1$, have the properties formulated in §2. The problem is to find a control $\{\hat{u}_0, \ldots, \hat{u}_{N-1}\}$ and the corresponding trajectory $\{\hat{x}_0, \ldots, \hat{x}_N\}$ satisfying

$$m_i(\hat{x}_0, \hat{x}_N) = 0\,, \quad i = 1, \ldots, l\,,$$

such that $g(\hat{x}_N)$ is optimal. The functions g and $m_0, \ldots, m_l$ are assumed to be continuously partially differentiable. In order to derive the transversality condition, extend the state vector to $X = (x, x') = (x^0, \ldots, x^r, x^{r+1}, \ldots, x^{2r+1})$ and define

$$\mathbf{F}_n(\mathbf{X}, u) = (F_n(x, u), x')\,.$$

The Jacobi matrix of this map is

$$\mathbf{J}_n(\mathbf{X}, u) = \begin{pmatrix} J_n(x, u), & 0 \\ 0 & , \mathbf{E} \end{pmatrix}.$$

Let the initial set be $\{X : x = x'\}$, the final set $\{X : m_i(x', x) = 0, i = 1, \ldots, l\}$. Obviously, to every solution $\{\hat{u}_0, \ldots, \hat{u}_{N-1}\}$, $\{\hat{x}_0, \ldots, \hat{x}_N\}$ of the original problem there corresponds the solution

$$\{\hat{u}_0, \ldots, \hat{u}_{N-1}\}, \ \{\hat{\mathbf{X}}_0 = (\hat{x}_0, \hat{x}_0), \ldots, \hat{\mathbf{X}}_N = (\hat{x}_N, \hat{x}_0)\}$$

of the problem with the extended state vector, optimizing $G(\hat{X}_N) = g(\hat{x}_N)$, and conversely. Pick such a solution. Then from

$$\hat{\Lambda}_n = \hat{\Lambda}_{n+1}\, \mathbf{J}_n(\hat{\mathbf{X}}_n, \hat{u}_n)\,, \quad n = 0, \ldots, N-1\,, \quad \hat{\Lambda}_n = (\hat{\lambda}_n, \hat{\lambda}'_n)\,,$$

we obtain $\hat{\lambda}_n = \hat{\lambda}_{n+1} J_n(\hat{x}_n, \hat{u}_n)$ and $\hat{\lambda}'_n = \hat{\lambda}'_{n+1}$. By Theorem 1 (if condition (P) holds), the initial conditions yield $\hat{\lambda}_0 = -\hat{\lambda}'_0 = -\hat{\lambda}'_N$. Further, the final conditions give

$$\hat{\Lambda}_N = \left(\hat{\beta}\frac{\partial g}{\partial x^j} + \sum_{i=1}^{l} \hat{\beta}_i \frac{\partial m_i}{\partial x_N^j}\,, \ j = 0, \ldots, r\,; \ \sum_{i=1}^{l} \hat{\beta}_i \frac{\partial m_i}{\partial x_0^j}\,, \ j = 0, \ldots, r\right)$$
$$= (\hat{\lambda}_N, -\hat{\lambda}_0)\,,$$

where $\hat{\beta} > 0$.

Now relation (25) implies

$$\hat{\lambda}_N^1 = \hat{\beta}_1 r/(p + r)\,, \quad \hat{\lambda}_0^1 = \hat{\beta}_1\,, \quad \text{i.e.} \quad \hat{\lambda}_N^1 = \hat{\lambda}_0^1 r/(p + r)\,.$$

CHAPTER III. THE PONTRJAGIN PRINCIPLE

§1. Introduction

The Pontrjagin principle deals with systems with continuous evolution parameter. We assume that the evolution of the system is described by differential equations, depending on the control parameter. The problem is to find a control maximizing an objective functional, given in integral form.

EXAMPLE 1 (CONNORS AND TEICHROEW)—*advertising costs.* We make the following assumptions: With no advertising, the sales $S(t)$ of a product decrease in proportion to its amount. On the other hand, the sales increase in proportion to the advertising done in the past. Hence

$$\frac{d}{dt} S(t) = -\mu\, S(t) + \gamma \int_{-\infty}^{t} A(\tau)\, e^{\tau - t}\, d\tau ,$$

where $A(t)$ is the advertising level at time t. The right-hand side integral is a discounted value of the advertising activity in the past. Differentiating, we get

$$\frac{d^2}{dt^2} S(t) + (\mu + 1) \frac{d}{dt} S(t) + \mu\, S(t) = \gamma\, A(t) ,$$

with initial conditions $S(0) = s^1$ and $(d/dt)S(0) = s^2$. The aim is to choose $0 \leqslant A(t) \leqslant \bar{A}$ such that the total profit up to time T, given by the integral

$$\int_0^T [S(t) - \alpha\, A(t)]\, dt ,$$

is maximal. Here α, γ and μ are nonnegative constants.

Formulation of the problem. Let $f(x, u) = (f^1(x, u), \ldots, f^r(x, u))$ be r functions of variables $x = (x^1, \ldots, x^r) \in R^r$, $u = (u^1, \ldots, u^s) \in U \subset R^s$, which are differentiable with respect to x^i, $i = 1, \ldots, r$, continuously in (x, u) on $R^r \times U$. Let x be the state vector and u the control parameter of a system satisfying

(1) $$\frac{dx}{dt} = f(x, u) .$$

The control parameter in (1) depends on time, $u = u(t)$. *Admissible controls* will be piecewise continuous and right-continuous functions. Thus (1) has to be understood in the following sense: The trajectory $x(t)$ is a continuous function fulfilling

(2) $$\frac{dx}{dt} = f(x, u(t))$$

at every continuity point of the control $u(t)$. The above assumptions imply that an initial value $x(0)$ and a chosen control $u(t)$ determine the trajectory uniquely. However, sometimes the trajectory exists only in a time interval shorter than that corresponding to the problem's formulation.

Now let us define the *objective functional.* Let $f^0(x, u)$ have the same properties as the functions considered above. The problem is to find a control $u(t)$ and the corresponding trajectory $x(t)$, fulfilling (2), such that

$$\psi = \int_0^T f^0(x(t), u(t))\, dt$$

is minimal. We discuss the minimization of ψ under the following conditions:

1. $T > 0$ is a fixed number. The initial position $x(0) = x_0$ is also fixed.
2. Given an initial set

$$I = \{x : h_i(x) = 0, \quad i = 1, 2, \ldots, l\}$$

and a final set

$$F = \{x : g_i(x) = 0, \quad i = 1, 2, \ldots, m\},$$

where $h_i(x)$, $g_i(x)$ are continuously differentiable, $\nabla h_i(x)$, $i = 1, \ldots, l$, and $\nabla g_i(x)$, $i = 1, \ldots, m$, linearly independent for $x \in I$ or $x \in F$, respectively, the functional ψ is to be minimized under the conditions

$$x(0) \in I, \quad x(T) \in F.$$

No further conditions are imposed upon T. In the important particular case where $f^0(x, u) = 1$, ψ is the time of the transfer from I to F.

We attempt to present the Pontrjagin principle as intelligibly as possible. We therefore begin with a discussion of the optimization problem under conditions 1. Then we use the method of dynamic programming to discuss the problem under conditions 2. The assumptions in this method substantially restrict the validity of the proof. That is why we then prove the Pontrjagin principle once more, in the minimum-time problem, by the method of Boltjanskiĭ.

To simplify the formulation, extend the state vector x to $\mathbf{x} = (x^0, x) = (x^0, \ldots, x^r)$. The trajectory of the system is then $\mathbf{x}(t) = (x^0(t), x(t))$, where

$$\frac{dx^0}{dt} = f^0(x, u).$$

Hence

$$x^0(T) = x^0(0) + \int_0^T f^0(x(t), u(t))\, dt.$$

Thus the zeroth coordinate of the state vector at time T is the objective functional. Denote

$$\mathbf{f}(x, u) = (f^0(x, u), \ldots, f^r(x, u)).$$

Then the system of equations for the extended trajectory may be written as

$$\frac{d\mathbf{x}}{dt} = \mathbf{f}(x, u). \tag{3}$$

§2. Minimization of ψ when T and $x(0)$ are given in advance

We start with some heuristic considerations, looking at a process with continuous evolution parameter as the limit of processes with discrete parameter, when the stage duration Δt tends to zero. We have

$$\mathbf{F}_n(\mathbf{x}, u) = \mathbf{x} + \mathbf{f}(x, u)\,\Delta t.$$

Hence

$$\mathbf{x}(\overline{n + 1}\,\Delta t) = \mathbf{x}(n\,\Delta t) + \mathbf{f}(x(n\,\Delta t), u(n\,\Delta t))\,\Delta t.$$

Passing to the limit as $\Delta t \longrightarrow 0$, $n\Delta t \longrightarrow t$, we get (3). A system of equations for Lagrange multipliers can be derived from (1, II). Namely,

$$\lambda(n\,\Delta t) = \lambda(\overline{n+1}\,\Delta t)\,J(x(n\,\Delta t), u(n\,\Delta t)),$$

where

$$J(x, u) = \left\| \frac{\partial F_n^i(x, u)}{\partial x^j} \right\|_{i,j=0}^r = \|\delta_{ij}\|_{i,j=0}^r + \left\| \frac{\partial f^i(x, u)}{\partial x^j} \right\|_{i,j=0}^r \Delta t\,.$$

Denoting the latter matrix by $j(x, u)$, we obtain

$$\lambda(\overline{n+1}\,\Delta t) - \lambda(n\,\Delta t) = -\lambda(\overline{n+1}\,\Delta t)\,j(x(n\,\Delta t), u(n\,\Delta t))\,\Delta t\,.$$

As $\Delta t \longrightarrow 0$, $n\Delta t \longrightarrow t$ and $N\Delta t \longrightarrow T$, this yields

$$\frac{d}{dt}\lambda = -\lambda j(x(t), u(t))\,, \quad \lambda(T) = (1, 0, \ldots, 0)\,.$$

The Lagrange multipliers have the same meaning as in the discrete case. If

$$*x(t) = x(t) + \varepsilon\delta x + o(\varepsilon)\,,$$

then

$$*x^0(T) = x^0(T) + \varepsilon\,\lambda(t)\,.\,\delta x + o(\varepsilon)\,.$$

Thus the multipliers make up the gradient of the objective function with respect to the state vector.

Now we shall prove our assertions. Let $u(t)$, $0 \leqslant t \leqslant T$, be an admissible control. Denote by $x(t)$ and $*x(t)$ trajectories corresponding to the control $u(t)$; that is, solutions of the system

$$\frac{dx}{dt} = f(x, u(t))\,, \quad t \in [0, T]\,.$$

Assume that the trajectory $x(t)$ exists on the whole interval $[0, T]$.

LEMMA 1. *If*

$$*x(0) = x(0) + \varepsilon\,\delta x(0) + o(\varepsilon)\,, \tag{4}$$

then

$$*x(t) = x(t) + \varepsilon\,\delta x(t) + o(\varepsilon)\,,$$

where $\delta x(t)$ *is continuous and*

$$\frac{d}{dt}\delta x(t) = \delta x(t)\,j(x(t), u(t))' \tag{5}$$

(at every continuity point of $u(t)$*).*

PROOF. It may be proved that for sufficiently small ϵ the trajectory $*x(t)$ also exists on all of $[0, T]$. Clearly $f(x, u(t))$ for $0 \leqslant t \leqslant T$ fulfills the Lipschitz condition

$$\|f(x_1, u(t)) - f(x_2, u(t))\| \leqq k\|x_1 - x_2\|$$

in any bounded region in R^{r+1}. Put $\Delta(t) = \|{}^*x(t) - x(t)\|$ and let $t_1 > 0$ be the first point where $u(t)$ is discontinuous. Then

(6) $$\frac{\mathrm{d}}{\mathrm{d}t}\Delta(t) \leqq \left\|\frac{\mathrm{d}}{\mathrm{d}t}({}^*\mathbf{x}(t) - \mathbf{x}(t))\right\| \leqq k\,\Delta(t)\,, \quad 0 \leqq t \leqq t_1\,,$$

or, equivalently,

$$\frac{\mathrm{d}}{\mathrm{d}t}\,\mathrm{e}^{-kt}\,\Delta(t) \leqq 0\,, \quad \Delta(t) \leqq \mathrm{e}^{kt}\,\Delta(0)\,.$$

Therefore

(7) $$\Delta(t) \leqq \mathrm{e}^{kt}(\varepsilon\|\delta\mathbf{x}(0)\| + o(\varepsilon))\,.$$

Furthermore,

(8) $$\begin{aligned}\frac{\mathrm{d}}{\mathrm{d}t}({}^*\mathbf{x}(t) - \mathbf{x}(t)) &= \mathbf{f}({}^*\mathbf{x}(t), \mathbf{u}(t)) - \mathbf{f}(\mathbf{x}(t), \mathbf{u}(t)) \\ &= ({}^*\mathbf{x}(t) - \mathbf{x}(t))\,\mathbf{j}(\mathbf{x}(t), \mathbf{u}(t))' + o(\Delta(t))\,.\end{aligned}$$

Let $\delta x(t)$ be a solution of (5), with the same $\delta x(0)$ as in (4). Put

$$\vartheta(t) = \|\varepsilon^{-1}({}^*\mathbf{x}(t) - \mathbf{x}(t)) - \delta\mathbf{x}(t)\|\,.$$

From (7) and (8) it follows rather easily that for sufficiently small ϵ we have, by (4),

$$\frac{\mathrm{d}}{\mathrm{d}t}\vartheta(t) \leqq m\,\vartheta(t) + \varepsilon_1\,,$$

where m is a constant and $\epsilon_1 > 0$ is arbitrary. Hence

$$\vartheta(t) \leqq \mathrm{e}^{mt}(\vartheta(0) + \varepsilon_1/m)\,.$$

Since (4) implies that $\vartheta(0) \longrightarrow 0$ as $\epsilon \longrightarrow 0$, and ϵ_1 is arbitrary, we conclude that $\vartheta(t) \longrightarrow 0$. This proves the lemma when $0 \leqslant t \leqslant t_1$. To prove it completely, we just repeat this reasoning for all continuity intervals of $u(t)$.□

LEMMA 2. *If*

(9) $$\frac{\mathrm{d}}{\mathrm{d}t}\boldsymbol{\lambda}(t) = -\boldsymbol{\lambda}(t)\,\mathbf{j}(\mathbf{x}(t), \mathbf{u}(t))\,, \quad t \in [0, T]\,,$$

then

(10) $$\delta\mathbf{x}(t)\,.\,\boldsymbol{\lambda}(t) \equiv \text{const}\,, \quad t \in [0, T]\,.$$

PROOF. At every continuity point of $u(t)$ we have

$$\begin{aligned}\frac{\mathrm{d}}{\mathrm{d}t}(\delta\mathbf{x}(t)\,.\,\boldsymbol{\lambda}(t)) &= \left(\frac{\mathrm{d}}{\mathrm{d}t}\delta\mathbf{x}(t)\right).\,\boldsymbol{\lambda}(t) + \delta\mathbf{x}(t)\,.\,\frac{\mathrm{d}}{\mathrm{d}t}\boldsymbol{\lambda}(t) \\ &= \delta\mathbf{x}(t)\,\mathbf{j}(\mathbf{x}(t), \mathbf{u}(t))'\,\boldsymbol{\lambda}(t)' - \delta\mathbf{x}(t)\,(\boldsymbol{\lambda}(t)\,\mathbf{j}(\mathbf{x}(t), \mathbf{u}(t)))' = 0\,.\end{aligned}$$

This, in view of the continuity of $\delta x(t) \cdot \lambda(t)$, yields (10).□

COROLLARY 1. *If* $\lambda(T) = (1, 0, \ldots, 0)$, *then, under the assumptions of Lemma* 1,

$$\delta \mathbf{x}(t) \cdot \lambda(t) = \delta x^0(T).$$

Let x_0 be a given initial state, and let $u(s)$, $0 \leqslant s \leqslant T$, be an admissible control. Choose a $t \in (0, T)$ and a $v \in U$. Let t be a continuity point of $u(s)$. Consider a variation of the control $u(s)$:

$$^*u(s) = u(s) \quad \text{for} \quad s \notin [t - \varepsilon, t), \quad ^*u(s) = v \quad \text{for} \quad s \in [t - \varepsilon, t).$$

If $x(s)$ and $^*x(s)$ denote the trajectories corresponding to the controls $u(s)$ and $^*u(s)$, respectively, with $x(0) = x_0 = {}^*x(0)$, then

$$^*\mathbf{x}(t) - \mathbf{x}(t) = [\mathbf{f}(\mathbf{x}(t), v) - \mathbf{f}(\mathbf{x}(t), u(t))]\, \varepsilon + o(\varepsilon).$$

Therefore, by Corollary 1,

(11) $$^*x^0(T) - x^0(T) = \lambda(t) \cdot [\mathbf{f}(\mathbf{x}(t), v) - \mathbf{f}(\mathbf{x}(t), u(t))]\, \varepsilon + o(\varepsilon),$$

where $\lambda(t)$ is the solution of (9) satisfying $\lambda(T) = (1, 0, \ldots, 0)$. When a control $u(s)$ minimizes $x^0(T)$, the coefficient of ϵ on the right-hand side of (11) is nonnegative. This proves the following theorem.

THEOREM 1. *A control $u(t)$, $t \in [0, T]$, minimizes the objective functional only if*

(12) $$\mathscr{H}(\lambda(t), \mathbf{x}(t), u(t)) = \min_{v \in U} \mathscr{H}(\lambda(t), \mathbf{x}(t), v), \quad t \in [0, T),$$

where $H(\lambda, x, u) = \lambda \cdot f(x, u)$.

The validity of (12) for those t where $u(t)$ is only right-continuous follows by passing to the limit.

REMARK 1. Choosing $\lambda(T) = (-1, 0, \ldots, 0)$ instead of $\lambda(T) = (1, 0, \ldots, 0)$, we get the necessary condition in the form

$$\mathscr{H}(\lambda(t), \mathbf{x}(t), u(t)) = \max_{v \in U} \mathscr{H}(\lambda(t), \mathbf{x}(t), v), \quad t \in [0, T).$$

This agrees with the name "maximum principle".

REMARK 2. In (3), the right-hand side does not depend on the evolution parameter t. That means that the system is autonomous (time-homogeneous). The reasoning in this section remains valid for nonautonomous systems:

(13) $$\frac{d\mathbf{x}}{dt} = \mathbf{f}(t, \mathbf{x}, u).$$

Condition (12) then takes the form

(12′) $$\mathscr{H}(t, \lambda(t), \mathbf{x}(t), u(t)) = \min_{v \in U} \mathscr{H}(t, \lambda(t), \mathbf{x}(t), v), \quad t \in [0, T),$$

where $H(t, \lambda, x, u) = \lambda \cdot f(t, x, u)$. System (13) becomes autonomous by adding a new variable $x^{r+1} = t$; that is,

$$\frac{dx^{r+1}}{dt} \equiv 1, \quad x^{r+1}(0) = t_0.$$

Using the *Hamiltonian function* $\mathcal{H}(\boldsymbol{\lambda}, \boldsymbol{x}, \boldsymbol{u})$, we may write (3) and (9) as

$$\frac{\mathrm{d}x^j}{\mathrm{d}t} = \frac{\partial \mathcal{H}}{\partial \lambda^j}, \quad \frac{\mathrm{d}\lambda^j}{\mathrm{d}t} = -\frac{\partial \mathcal{H}}{\partial x^j}.$$

The Hamiltonian (at time t) will be $\mathcal{H}(\boldsymbol{\lambda}(t), \boldsymbol{x}(t), \boldsymbol{u})$, as a function of $\boldsymbol{u}$. The following lemma states that $\mathcal{H}(\boldsymbol{\lambda}(t), \boldsymbol{x}(t), \boldsymbol{u}(t))$ is constant when (12) holds.

LEMMA 3. *Let*

$$\mathcal{H}(\boldsymbol{\lambda}(t), \boldsymbol{x}(t), \boldsymbol{u}(t)) = \min_{\boldsymbol{v} \in U} \mathcal{H}(\boldsymbol{\lambda}(t), \boldsymbol{x}(t), \boldsymbol{v}), \quad 0 \leqq t < T,$$

where $\boldsymbol{x}(t)$ *fulfills* (3) *and* $\boldsymbol{\lambda}(t)$ *fulfills* (9). *Then*

$$\mathcal{M}(t) = \mathcal{H}(\boldsymbol{\lambda}(t), \boldsymbol{x}(t), \boldsymbol{u}(t)) \equiv \text{const}, \quad 0 \leqq t < T.$$

PROOF. Let $\tau_1, \tau_2 \in [0, T]$. We have

$$\begin{aligned}\mathcal{M}(\tau_2) - \mathcal{M}(\tau_1) &\leqq \mathcal{H}(\boldsymbol{\lambda}(\tau_2), \boldsymbol{x}(\tau_2), \boldsymbol{u}(\tau_1)) - \mathcal{H}(\boldsymbol{\lambda}(\tau_1), \boldsymbol{x}(\tau_1), \boldsymbol{u}(\tau_1)) \\ &= (\tau_2 - \tau_1)\frac{\mathrm{d}}{\mathrm{d}t}\mathcal{H}(\boldsymbol{\lambda}(t), \boldsymbol{x}(t), \boldsymbol{u}(\tau_1))\Big|_{t=\xi},\end{aligned}$$

with a ξ lying between τ_1 and τ_2. Further,

$$\begin{aligned}&\frac{\mathrm{d}}{\mathrm{d}t}\mathcal{H}(\boldsymbol{\lambda}(t), \boldsymbol{x}(t), \boldsymbol{u}(\tau_1))\Big|_{t=\xi} = \sum_{j=0}^{r} \frac{\partial \mathcal{H}}{\partial \lambda^j}(\boldsymbol{\lambda}(\xi), \boldsymbol{x}(\xi), \boldsymbol{u}(\tau_1))\frac{\mathrm{d}\lambda^j}{\mathrm{d}t}(\xi) \\ &+ \sum_{j=0}^{r} \frac{\partial \mathcal{H}}{\partial x^j}(\boldsymbol{\lambda}(\xi), \boldsymbol{x}(\xi), \boldsymbol{u}(\tau_1))\frac{\mathrm{d}x^j}{\mathrm{d}t}(\xi) = -\sum_{j=0}^{r} \frac{\partial \mathcal{H}}{\partial \lambda^j}(\boldsymbol{\lambda}(\xi), \boldsymbol{x}(\xi), \boldsymbol{u}(\tau_1)) \\ &\cdot \frac{\partial \mathcal{H}}{\partial x^j}(\boldsymbol{\lambda}(\xi), \boldsymbol{x}(\xi), \boldsymbol{u}(\xi)) + \sum_{j=0}^{r} \frac{\partial \mathcal{H}}{\partial x^j}(\boldsymbol{\lambda}(\xi), \boldsymbol{x}(\xi), \boldsymbol{u}(\tau_1))\frac{\partial \mathcal{H}}{\partial \lambda^j}(\boldsymbol{\lambda}(\xi), \boldsymbol{x}(\xi), \boldsymbol{u}(\xi)).\end{aligned}$$

When τ_1 is a continuity point of $\boldsymbol{u}(t)$, this gives

$$\lim_{\tau_2 \to \tau_1} \frac{\mathrm{d}}{\mathrm{d}t}\mathcal{H}(\boldsymbol{\lambda}(t), \boldsymbol{x}(t), \boldsymbol{u}(\tau_1))\Big|_{t=\xi} = 0.$$

Similarly

$$\mathcal{M}(\tau_2) - \mathcal{M}(\tau_1) \geqq (\tau_2 - \tau_1)\frac{\mathrm{d}}{\mathrm{d}t}\mathcal{H}(\boldsymbol{\lambda}(t), \boldsymbol{x}(t), \boldsymbol{u}(\tau_2))\Big|_{t=\xi'}.$$

As $(\mathrm{d}/\mathrm{d}t)\mathcal{H}(\boldsymbol{\lambda}(t), \boldsymbol{x}(t), \boldsymbol{u}(\tau_i))$, $i = 1, 2$, are bounded, we conclude that $\mathcal{M}(t)$ is continuous on $[0, T]$. Moreover, $(\mathrm{d}/\mathrm{d}t)\mathcal{M}(t) = 0$ at every continuity point of $\boldsymbol{u}(t)$. Hence $\mathcal{M}(t) \equiv \text{const}$.□

EXAMPLE 1 CONCLUDED. Put

$$S^0(t) = \int_0^t [S(s) - \alpha A(s)]\,\mathrm{d}s, \quad S^1(t) = S(t), \quad S^2(t) = \frac{\mathrm{d}}{\mathrm{d}t}S(t).$$

Thus

$$\frac{d}{dt}S^0 = S^1 - \alpha A, \quad S^0(0) = 0, \quad \frac{d}{dt}S^1 = S^2,$$

$$\frac{d}{dt}S^2 = -(1+\mu)S^2 - \mu S^1 + \gamma A.$$

Therefore the Hamiltonian may be written as

$$\mathscr{H} = \lambda^0(S^1 - \alpha A) + \lambda^1 S^2 + \lambda^2[-(1+\mu)S^2 - \mu S^1 + \gamma A].$$

Furthermore,

$$\lambda^0 \equiv 1, \quad \frac{d}{dt}\lambda^1 = -1 + \mu\lambda^2,$$

$$\frac{d}{dt}\lambda^2 = -\lambda^1 + (1+\mu)\lambda^2.$$

This gives the following equation for λ^2:

$$\frac{d^2}{dt^2}\lambda^2 - (1+\mu)\frac{d}{dt}\lambda^2 + \mu\lambda^2 - 1 = 0, \quad \lambda^2(T) = 0 = \frac{d}{dt}\lambda^2(T).$$

The solution is

$$\lambda^2(t) = \mu^{-1} + \frac{1}{\mu(\mu-1)}e^{\mu(t-T)} - \frac{1}{\mu-1}e^{t-T} \quad \text{for} \quad \mu \neq 1,$$

$$\lambda^2(t) = 1 + te^{t-T} - (1+T)e^{t-T} \quad \text{for} \quad \mu = 1.$$

Note that $\lambda^2(t)$ is decreasing on $[0, T]$.

We have $H = (\gamma\lambda^2 - \alpha)A + H'$ with H' independent of A. If a control $\{\hat{A}(t), 0 \leqslant t \leqslant T\}$ maximizes the Hamiltonian, then

$$\hat{A}(t) = \bar{A} \quad \text{for} \quad \gamma\lambda^2(t) - \alpha > 0, \quad \hat{A}(t) = 0 \quad \text{for} \quad \gamma\lambda^2(t) - \alpha < 0.$$

If there exists a $\tau \in (0, T]$ such that $\gamma\lambda^2(\tau) = \alpha$, then the optimal result is achieved by advertising until time τ with the maximal intensity. Otherwise the most profitable policy is not to advertise at all.

§3. Linear systems with quadratic objective functional

Let the system of equations describing the trajectory be

$$\frac{d}{dt}\boldsymbol{x} = \boldsymbol{x}\,\boldsymbol{A}(t) + \boldsymbol{u}\,\boldsymbol{B}(t), \tag{14}$$

where $\boldsymbol{A}(t)$ is an $r \times r$ matrix, $\boldsymbol{B}(t)$ is an $s \times r$ matrix and $u \in R^s$. Let the objective functional be given by

$$\psi = \frac{1}{2}\int_0^T [\boldsymbol{x}(t)\,\boldsymbol{Q}(t)\,\boldsymbol{x}(t)' + \boldsymbol{u}(t)\,\boldsymbol{R}(t)\,\boldsymbol{u}(t)']\,dt, \tag{15}$$

where $\boldsymbol{Q}(t)$ is a symmetric positive semidefinite $r \times r$ matrix and $\boldsymbol{R}(t)$ is a symmetric positive definite $s \times s$ matrix. All the matrices are assumed to be continuous functions of t.

The Hamiltonian function is then

$$\mathscr{H}(t, \lambda, x, u) = \tfrac{1}{2}\lambda^0[x\, Q(t)\, x' + u\, R(t)\, u'] + \lambda\,.\,(x\, A(t) + u\, B(t))\,.$$

Here we write $\lambda = (\lambda^1, \ldots, \lambda^r)$. Hence $\lambda^0 \equiv 1$, $\lambda(T) = 0$ and

$$\frac{d\lambda}{dt} = -x\, Q(t) - \lambda\, A(t)'\,. \tag{16}$$

To find the u minimizing the Hamiltonian function, put

$$\nabla\mathscr{H} = u\, R(t) + \lambda\, B(t)' = 0\,.$$

This system has a solution

$$\hat{u} = -\lambda\, B(t)'\, R(t)^{-1}\,. \tag{17}$$

The inverse $R(t)^{-1}$ exists because $R(t)$ is positive definite. The matrix of the second derivatives $\|\partial^2 H/\partial u_i \partial u_j\|_{i,j=1}^s = R(t)$ is also positive definite, and therefore the Hamiltonian function assumes a minimum at $\hat{u}$. For the optimal trajectory, (14) yields

$$\frac{d\hat{x}}{dt} = \hat{x}\, A(t) - \hat{\lambda}\, B(t)'\, R(t)^{-1}\, B(t)\,. \tag{18}$$

We propose to satisfy (16) and (18) using $\hat{\lambda}(t) = \hat{x}(t)K(t)$, with a suitable matrix $K(t)$. Substituting in (16) and using (18), we get

$$\frac{d}{dt}(\hat{x}K) = \hat{x}\, A(t)\, K - \hat{x}K\, B(t)'\, R(t)^{-1}\, B(t)\, K + \hat{x}\frac{dK}{dt} = -\hat{x}\, Q(t) - \hat{x}K\, A(t)'\,.$$

If K fulfills the matrix Riccati equation

$$\frac{dK}{dt} = -A(t)\, K - K\, A(t)' + K\, B(t)'\, R(t)^{-1}\, B(t)\, K - Q(t) \tag{19}$$

with the condition $K(T) = 0$, then $\hat{\lambda}(t) = \hat{x}(t)K(t)$ is a Lagrange multiplier. From (17) we obtain the optimal control

$$\hat{u} = -\hat{x}\, K(t)\, B(t)'\, R(t)^{-1}\,. \tag{20}$$

Relation (20) expresses the optimal value of the control parameter directly as a function of the state vector (and the evolution parameter t). In Chapter I this description of the intended control of an object was called a policy. Now we mention the terms which are more often used in the present context. The formulation of the problem at the beginning of this chapter was based on the following idea: From the information about the initial state of the controlled object we find an optimal control $\hat{u}(t)$ on the interval $0 \leqslant t \leqslant T$. This determination must take into account all the possible trajectories of the object in the time interval $[0, T]$. During the actual control, the instantaneous state of the object is not considered any more–the control parameter is a function of time. This is called *programmed control* (open-loop control) (see Figure 4).

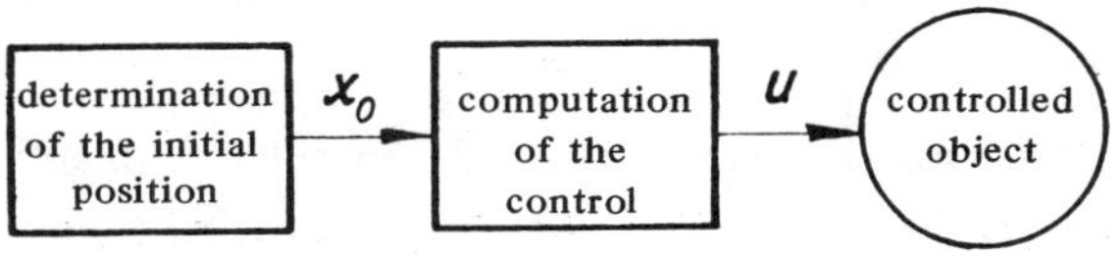

FIGURE 4.

When we know how the optimal control depends on the state of the system, we can determine the control parameter during the control process according to the actual trajectory of the system. This is the so-called *feedback control* (closed-loop control) (Figure 5). The determination of dependence of the control parameter on the state of the system is called the *control synthesis.*

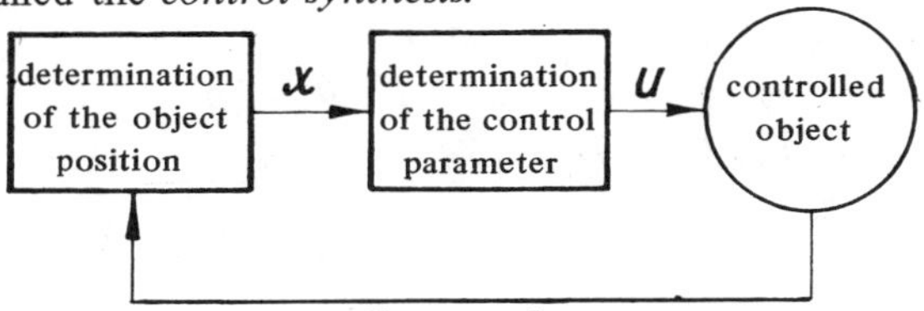

FIGURE 5.

EXAMPLE 2. The problem is to find $u(t)$, $0 \leqslant t \leqslant T$, minimizing the objective functional

$$\tfrac{1}{2}\int_0^T [q\, x(t)^2 + r\, u(t)^2]\, dt,$$

where $q \geqslant 0$ and $r > 0$. Assume that

$$\frac{d}{dt} x(t) = a\, x(t) - b\, u(t), \quad a, b > 0. \tag{21}$$

By (20), we have $\hat{u} = \hat{x}k(t)b/r$, where

$$\frac{d}{dt} k = -2ak + \frac{b^2}{r} k^2 - q, \quad k(T) = 0.$$

The solution is

$$k(t) = (A + B)(1 - e^{2B(t-T)}) \left[1 + \frac{B + A}{B - A} e^{2B(t-T)} \right]^{-1},$$

with

$$A = arb^{-2}, \quad B = b^{-2}\sqrt{(a^2r^2 + b^2qr)}.$$

The optimal trajectory is found by substituting in (21).

§4. Minimization of ψ under the conditions $x(0) \in I$, $x(T) \in F$

Let us proceed to the second problem formulated at the beginning of the chapter. We assume that $\psi = x^0(T)$ (see the end of §1). Denote by $\hat{E}$ the set of $x = (x^0, x) \in R^{r+1}$ such that there exists an optimal trajectory from x to F. That is, there exists a control $\hat{u}(t)$, $0 \leqslant t \leqslant \hat{T}$, and the corresponding trajectory $\hat{x}(t)$, $0 \leqslant t \leqslant \hat{T}$, $\hat{x}(0) = x$, $\hat{x}(\hat{T}) \in F$, such that $\hat{x}^0(\hat{T}) \leqslant x^0(T)$ for all the trajectories $x(t)$, $0 \leqslant t \leqslant T$, satisfying $x(0) = x$ and $x(T) \in F$. For $x \in \hat{E}$ put $\hat{\psi}(x) = \hat{x}^0(\hat{T})$. The equation

$$\hat{\psi}(\mathbf{x}) = C$$

defines a one-parameter family $\{\Sigma\}$ of surfaces, the so-called *boundary surfaces.*

The *Bellman principle,* formulated above for processes with discrete evolution parameter, takes the following form in the continuous case:

LEMMA 4. *If $\hat{x}(t)$, $t_0 \leqslant t \leqslant \hat{T}$, is an optimal trajectory from $\hat{x}(t_0)$ to F, then, for an arbitrary $t_1 \in [t_0, \hat{T}]$, $\hat{x}(t)$, $t_1 \leqslant t \leqslant T$, is an optimal trajectory from $\hat{x}(t_1)$ to F.*

PROOF. Assume that for some $u(t)$, $t_1 \leqslant t \leqslant T$, the trajectory $x(t)$ fulfills $x(t_1) = \hat{x}(t_1)$, $x(T) \in F$ and $x^0(T) < \hat{x}^0(\hat{T})$. Then the trajectory $\bar{x}(t)$, $t_0 \leqslant t \leqslant T$, $\bar{x}(t_0) = \hat{x}(t_0)$, corresponding to the control $\bar{u}(t) = \hat{u}(t)$ for $t_0 \leqslant t < t_1$, $\bar{u}(t) = u(t)$ for $t_1 \leqslant t \leqslant T$, is a trajectory from $\hat{x}(t_0)$ to F, better than $\hat{x}(t)$ (see Figure 6).□

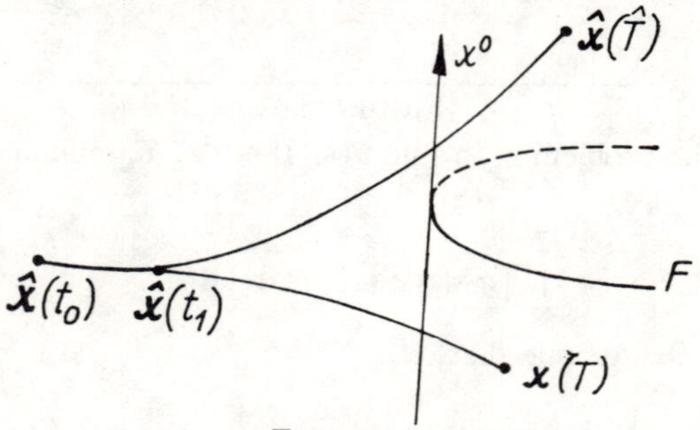

FIGURE 6.

Briefly, the principle says that "*a part of an optimal trajectory is again an optimal trajectory.*"

LEMMA 5. *Let $x(t)$, $0 \leqslant t \leqslant T$, be a trajectory, and let $x(0) \in \hat{E}$ and $\hat{\psi}(x(0)) = C$. Then*

$$\hat{\psi}(x(t)) < C \tag{22}$$

holds for no $t \in [0, T]$. If $x(t)$ is an optimal trajectory from $x(0)$ to F, then $\hat{\psi}(x(t)) = C$ for $t \in [0, T]$.

PROOF. The first assertion follows from the Bellman principle. For if (22) holds, then there exists a trajectory $y(s)$, $t \leqslant s \leqslant T'$, from $x(t)$ to F such that $y^0(T') = \hat{\psi}(x(t)) < C$. Defining $y(s) = x(s)$ for $0 \leqslant s \leqslant t$, we get a trajectory from $x(0)$ to F. Therefore $y^0(T') \geqslant C$, a contradiction. The second assertion follows from the first one.□

A point $x \in \hat{E}$ is called *regular* when x is an interior point of $\hat{E}$ and $\overline{\nabla}\hat{\psi}$ is continuous in a neighborhood of x. Since $\partial\hat{\psi}/\partial x^0 = 1$, we have $\overline{\nabla}\hat{\psi} \neq 0$. A trajectory $x(t)$, $0 \leqslant t \leqslant T$, is called *regular* if all the points $x(t)$, $0 \leqslant t < T$, are regular.

Let $\hat{x}(t)$, $0 \leqslant t \leqslant \hat{T}$, be a regular optimal trajectory from $\hat{x}(0)$ to F corresponding to a control $\hat{u}(t)$. Let $\hat{\psi}(\hat{x}(0)) = C$. The boundary surface Σ, defined by $\hat{\psi}(x) = C$, divides $\hat{E}$ in two parts

$$\hat{E}_- = \{\mathbf{x} \in \hat{E} : \hat{\psi}(\mathbf{x}) < C\}, \quad \hat{E}_+ = \{\mathbf{x} \in \hat{E} : \hat{\psi}(\mathbf{x}) \geqq C\}.$$

By Lemma 5, the trajectory $\hat{x}(t)$ is entirely contained in Σ. Look at the behavior of the tangent hyperplane of Σ during motion along the trajectory. The tangent hyperplane at $\hat{x}(0)$ may be expressed as

$$\pi_0 = \{\mathbf{x} : \mathbf{x} = \hat{\mathbf{x}}(0) + \mathbf{z},\ \hat{\lambda}(0) \,.\, \mathbf{z} = 0\},$$

where $\lambda(0) = -\overline{\nabla}\hat{\psi}(\hat{x}(0))$ is the vector normal to Σ, directed into $\hat{E}_-$. Denote $L_0 = \{z\colon \hat{\lambda}(0) \cdot z = 0\}$. For any $z \in L_0$ there exists a curve $x_\epsilon \in \Sigma$, depending on a non-negative parameter ϵ, such that

$$\mathbf{x}_\varepsilon = \hat{\mathbf{x}}(0) + \varepsilon\mathbf{z} + o(\varepsilon) \quad \text{for} \quad \varepsilon \to 0. \tag{23}$$

Consider the trajectory $x_\epsilon(t)$ corresponding to the control $\hat{u}(t)$ and beginning at x_ϵ. By Lemma 1,

$$\mathbf{x}_\varepsilon(s) = \hat{\mathbf{x}}(s) + \varepsilon\, \mathbf{z}(s) + o(\varepsilon), \tag{24}$$

where

$$\frac{\mathrm{d}}{\mathrm{d}t}\mathbf{z}(t) = \mathbf{z}(t)\,\mathbf{j}(\hat{\mathbf{x}}(t), \hat{\mathbf{u}}(t))', \quad 0 \leqq t \leqq s, \quad \mathbf{z}(0) = \mathbf{z}. \tag{25}$$

Denote $L_s = \{z(s)\colon z \in L_0\}$. It follows from properties of the system (25) that L_s is an r-dimensional space. We claim that

$$\pi_s = \{\mathbf{x} : \mathbf{x} = \hat{\mathbf{x}}(s) + \mathbf{z},\ \mathbf{z} \in L_s\}$$

is the tangent hyperplane of Σ at $\hat{x}(s)$. This is proved when we verify that in a sufficiently small neighborhood of $\hat{x}(s)$ the hyperplane π_s does not intersect Σ. Assume that π_s does intersect Σ. Then there exists a $z' \in L_s$, directed into $\hat{E}_-$ (see Figure 7). That is,

$$\hat{\mathbf{x}}(s) + \varepsilon\mathbf{z}' \in \hat{E}_-$$

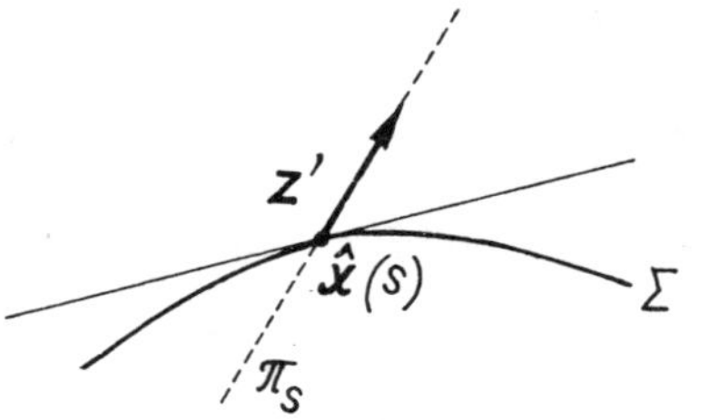

FIGURE 7.

for all sufficiently small ϵ. At the same time, the distance of $\hat{x}(s) + \epsilon z'$ from Σ is of order ϵ. Choose the initial value $z \in L_0$ in (25) such that $z(s) = z'$. When a curve on Σ satisfies (23), by (24) the corresponding trajectory $x_\epsilon(t)$ fulfills

$$\mathbf{x}_\varepsilon(s) = \hat{\mathbf{x}}(s) + \varepsilon\mathbf{z}' + o(\varepsilon).$$

Therefore $x_\epsilon(s) \in E_-$ for all sufficiently small ϵ. This contradicts Lemma 5.

Consider a solution of the system

$$\text{(26)} \qquad \frac{d}{dt}\hat{\lambda}(t) = -\hat{\lambda}(t)\,j(\hat{x}(t), \hat{u}(t)), \quad t \in [0, T],$$

with the initial condition

$$\text{(27)} \qquad \hat{\lambda}(0) = -\overline{\nabla}\hat{\psi}(\hat{x}(0)).$$

Since $\hat{\lambda}(0) \cdot z = 0$ for $z \in L_0$, we have $\hat{\lambda}(s) \cdot z = 0$ for $z \in L_s$ by (25) and Lemma 2. Thus $\hat{\lambda}(s)$ is a normal vector to Σ at $\hat{x}(s)$. Furthermore, $\hat{\lambda}^0 \equiv -1$ by (26), and therefore

$$\text{(28)} \qquad \hat{\lambda}(s) = -\overline{\nabla}\hat{\psi}(\hat{x}(s)).$$

Now we use these results to prove the following theorem. It is assumed that the zeroth coordinate of the initial position $x^0(0)$ is given in advance.

THEOREM 2. *Let $\hat{x}(t)$, $0 \leqslant t \leqslant \hat{T}$, be a regular optimal trajectory from I to F, corresponding to a control $\hat{u}(t)$. Let $\hat{\lambda}(t)$ be a solution of* (26) *with the initial condition* (27). *Then*

$$\text{(29)} \qquad 0 \equiv \mathscr{H}(\hat{\lambda}(t), \hat{x}(t), \hat{u}(t)) = \max_{u \in U} \mathscr{H}(\hat{\lambda}(t), \hat{x}(t), u), \quad 0 \leqq t < \hat{T},$$

where

$$\text{(30)} \qquad \mathscr{H}(\lambda, x, u) = \lambda \,.\, f(x, u).$$

Furthermore,

$$\text{(31)} \qquad (\lambda^1(0), \ldots, \lambda^n(0)) = \sum_{i=1}^{l} \alpha_i \overline{\nabla} h_i(\hat{x}(0)).$$

PROOF. Let $\hat{\psi}(\hat{x}(0)) = C$. By Lemma 5, $\hat{\psi}(\hat{x}(t)) = C$ for $t \in [0, \hat{T}]$. Define a variation of the control $\hat{u}(s)$ by

$$u(s) = \hat{u}(s) \quad \text{for} \quad s \notin [t, t+\varepsilon), \quad u(s) = u \quad \text{for} \quad s \in [t, t+\varepsilon),$$

and denote by $x(s)$ the corresponding trajectory, beginning at $\hat{x}(0)$. Then

$$x(t+\varepsilon) = \hat{x}(t) + f(\hat{x}(t), u)\,\varepsilon + o(\varepsilon).$$

Hence, by (28) and (30),

$$\begin{aligned} \hat{\psi}(x(t+\varepsilon)) &= \hat{\psi}(\hat{x}(t)) + \overline{\nabla}\hat{\psi}(\hat{x}(t)) \,.\, f(\hat{x}(t), u)\,\varepsilon + o(\varepsilon) \\ &= C - \hat{\lambda}(t) \,.\, f(\hat{x}(t), u)\,\varepsilon + o(\varepsilon) = C - \mathscr{H}(\hat{\lambda}(t), \hat{x}(t), u)\,\varepsilon + o(\varepsilon). \end{aligned}$$

If $H(\hat{\lambda}(t), \hat{x}(t), u) > 0$, then $\hat{\psi}(x(t+\epsilon)) < C$ for all sufficiently small ϵ. Since $\hat{\psi}(x(0)) = C$, this contradicts Lemma 5. Therefore $H(\hat{\lambda}(t), \hat{x}(t), u) \leqslant 0$.

Furthermore,

$$C = \hat{\psi}(\hat{\boldsymbol{x}}(t+\varepsilon)) = C - \mathscr{H}(\hat{\lambda}(t), \hat{\boldsymbol{x}}(t), \hat{\boldsymbol{u}}(t))\,\varepsilon + o(\varepsilon),$$

and hence $H(\hat{\lambda}(t), \hat{x}(t), \hat{\boldsymbol{u}}(t)) = 0$. This proves (29).

It remains to verify the *transversality condition* (31). As $\hat{x}(t)$ is an optimal trajectory from I to F, $\hat{\psi}(x^0(0), \boldsymbol{x})$ attains its minimum over I at $\hat{\boldsymbol{x}}(0)$. Therefore there exist Lagrange multipliers $\alpha_1, \ldots, \alpha_l$ such that

(32) $$\overline{\nabla}[\hat{\psi}(\hat{\boldsymbol{x}}(0)) + \alpha_1 h_1(\hat{\boldsymbol{x}}(0)) + \ldots + \alpha_l h_l(\hat{\boldsymbol{x}}(0))] = 0\,.$$

Here $\overline{\nabla}$ is the gradient with respect to the variables $x^1, \ldots, x^r$. By (28), (32) coincides with (31).□

REMARK 1. Let us consider, without insisting on precision, the question of whether the system of conditions in Theorem 2 is large enough to determine the solution of the optimization problem. Assume first that F contains a single point $^{+}\boldsymbol{x}$. Relation (29) determines $\hat{\boldsymbol{u}}(t)$ as depending on $\hat{\lambda}(t)$ and $\hat{x}(t)$. A general solution of the systems

$$\frac{d\hat{\boldsymbol{x}}}{dt} = \boldsymbol{f}(\hat{\boldsymbol{x}}, \hat{\boldsymbol{u}}(t)), \quad \frac{d\hat{\lambda}}{dt} = -\hat{\lambda}\boldsymbol{j}(\hat{\boldsymbol{x}}(t), \hat{\boldsymbol{u}}(t))$$

contains $2r+2$ independent constants. Another undetermined quantity is $\hat{T}$, and there are l constants in (31). This adds up to $2r+l+3$ unknown quantities. To determine them, we have r conditions (31), r conditions given by $\hat{x}(\hat{T}) = {}^{+}\boldsymbol{x}$ and l conditions

$$h_1(\hat{\boldsymbol{x}}(0)) = 0, \ldots, h_l(\hat{\boldsymbol{x}}(0)) = 0\,.$$

Moreover, the initial value $x^0(0)$ is prescribed according to the assumption, and the Hamiltonian must be zero by (29). This is $2r+l+2$ conditions altogether. The number is sufficient, since in order to solve the problem we have to know the function $\hat{\lambda}(t)$ only up to a positive multiple.

When F is defined by

(33) $$g_1(\boldsymbol{x}) = 0, \ldots, g_m(\boldsymbol{x}) = 0\,,$$

for $m < r$ this gives us $r+l+m+2$ conditions, and that is not enough. However, it can be proved that the transversality condition at the final point holds; that is,

(34) $$(\lambda^1(\hat{T}), \ldots, \lambda^r(\hat{T})) = \sum_{i=1} \beta_i \overline{\nabla} g_i(\hat{\boldsymbol{x}}(\hat{T}))\,.$$

Equations (34) represent r relations with m new constants. Thus we have $2r+l+m+2$ equations for determining $2r+l+m+3$ constants. This is just the needed number.

EXAMPLE 3. As in Example 2, let the system trajectory be described by

$$\frac{d}{dt} x^1(t) = a\, x^1(t) - b\, u(t)\,, \quad a, b > 0\,.$$

The task is to transfer the system from a position $x_0^1 > 0$ to the position 1, keeping the quantity

$$\frac{1}{2}\int_0^T [q\, x^1(t)^2 + r\, u(t)^2]\, dt\,, \quad q \geqq 0\,, \quad r > 0\,,$$

minimal.

Relation (29) takes the form

$$\begin{aligned} 0 &\equiv -\tfrac{1}{2}[q\,\hat{x}^1(t)^2 + r\,\hat{u}(t)^2] + \hat{\lambda}^1(t)\,[a\,\hat{x}^1(t) - b\,\hat{u}(t)] \\ &= \max_u \{-\tfrac{1}{2}[q\,\hat{x}^1(t)^2 + ru^2] + \hat{\lambda}^1(t)\,[a\,\hat{x}^1(t) - bu]\} \\ &= \max_u \{-\tfrac{1}{2}q\,\hat{x}^1(t)^2 + \hat{\lambda}^1(t)\,a\,\hat{x}^1(t) + b^2\,\hat{\lambda}^1(t)^2/(2r) \\ &\quad - \tfrac{1}{2}r(u + b\hat{\lambda}^1(t)/r)^2\}\,. \end{aligned}$$

Hence

$$\begin{aligned} \hat{\lambda}^1(t) &= b^{-2}[-ar \pm \sqrt{(a^2r^2 + b^2qr)}]\,\hat{x}^1(t)\,, \\ \hat{u}(t) &= b^{-1}[a \mp \sqrt{(a^2 + b^2q/r)}]\,\hat{x}^1(t)\,. \end{aligned}$$

Therefore

$$\frac{d}{dt}\,\hat{x}^1(t) = \pm\,\sqrt{(a^2 + b^2q/r)}\,\hat{x}^1(t)\,,$$

or, with $C = \sqrt{a^2 + b^2q/r}$,

$$\begin{aligned} \hat{x}^1(t) &= x_0^1 e^{Ct} \quad \text{for} \quad x_0^1 < 1\,, \\ \hat{x}^1(t) &= x_0^1 e^{-Ct} \quad \text{for} \quad x_0^1 > 1\,. \end{aligned}$$

The time needed to reach the position 1 is $\hat{T} = |\ln x_0^1|/C$. Computation gives

$$\hat{\psi}(\mathbf{x}_0) = x_0^0 \pm [q + rb^{-2}(a \mp C)^2]\,[(x_0^1)^{-1} - x_0^1]/(4C)\,.$$

Let us investigate *the transversality condition at the final point.* Let $\hat{\mathbf{x}}(t)$, $0 \leqslant t \leqslant \hat{T}$, be the optimal trajectory from ${}_+\mathbf{x}$ to F corresponding to a control $\hat{\mathbf{u}}(t)$. Then $\hat{\mathbf{y}}(t) = \hat{\mathbf{x}}(\hat{T} - t)$ is the trajectory from F to ${}_+\mathbf{x}$ for the control $\hat{\mathbf{v}}(t) = \hat{\mathbf{u}}(\hat{T} - t)$ (at the continuity points), corresponding to the system

$$\frac{d\mathbf{y}}{dt} = -\mathbf{f}(\mathbf{y}, \mathbf{v})\,,$$

and $\hat{x}^0(\hat{T})$ is, for given $\hat{x}^0(0)$, the smallest possible value of the zeroth coordinate for the transfer from ${}_+\mathbf{x}$ to F. Therefore $\hat{y}^0(\hat{T}) = \hat{x}^0(0)$ is in turn the largest possible value for the transfer from F to ${}_+\mathbf{x}$ with $\hat{y}^0(0) = \hat{x}^0(\hat{T})$. Thus $\hat{\mathbf{y}}(t)$ is an optimal (in the maximization sense) trajectory from F to ${}_+\mathbf{x}$. By Theorem 2,

$$0 \equiv -\mathscr{H}(\hat{\mathbf{v}}(t), \hat{\mathbf{y}}(t), \hat{\mathbf{v}}(t)) = \min_{\mathbf{v} \in U}(-\mathscr{H}(\hat{\mathbf{v}}(t), \hat{\mathbf{y}}(t), \mathbf{v}))\,, \quad 0 \leqq t \leqq \hat{T}\,,$$

$$(\hat{v}^1(0), \ldots, \hat{v}^r(0)) = \sum_{i=1}^m \beta_i\,\nabla g_i(\hat{y}(0))\,,$$

where $\hat{\mathbf{v}}(t)$ satisfies

$$\frac{d}{dt}\,\hat{\mathbf{v}}(t) = \hat{\mathbf{v}}(t)\,\mathbf{j}(\hat{\mathbf{y}}(t), \hat{\mathbf{v}}(t))\,.$$

Substituting back $\hat{\mathbf{x}}(t) = \hat{\mathbf{y}}(\hat{T} - t)$, $\hat{\mathbf{u}}(t) = \hat{\mathbf{v}}(\hat{T} - t)$ and $\hat{\boldsymbol{\lambda}}(t) = \hat{\mathbf{v}}(\hat{T} - t)$, we get (26), (29) and (34). Note that the validity of both transversality conditions in the general problem of transfer from I to F cannot be immediately deduced from our considerations.

§5. The minimum-time problem

In this section we mention another method of proof of the Pontrjagin principle in the problem of transfering a controlled object from a state ${}_{+}x$ to a state ${}^{+}x \neq {}_{+}x$ in a minimum time. Let $\hat{x}(t)$, $0 \leqslant t \leqslant \hat{T}$, be a trajectory in R^r satisfying

$$\frac{\mathrm{d}\hat{x}}{\mathrm{d}t} = f(\hat{x}, \hat{u}(t)), \quad \hat{x}(0) = {}_{+}x, \quad \hat{x}(\hat{T}) = {}^{+}x,$$

and such that $\hat{T} \leqslant T$ for each trajectory $x(t)$, $0 \leqslant t \leqslant T$, $x(0) = {}_{+}x$, $x(T) = {}^{+}x$. Let $\hat{T}$ be a continuity point of the control $\hat{u}(t)$. Consider a variation ${}^{*}u(t)$ of the control $\hat{u}(t)$:

$${}^{*}u(t) = \hat{u}(t), \quad t \notin [\tau - l\varepsilon, \tau), \quad u(t) = v, \quad t \in [\tau - l\varepsilon, \tau),$$

where $l > 0$, $v \in U$ and $\tau < \hat{T}$ is a continuity point of $\hat{u}(t)$. As $\epsilon \longrightarrow 0$, $\epsilon > 0$, we have

$$\text{(35)} \qquad {}^{*}x(\tau) = \hat{x}(\tau) + l[f(\hat{x}(\tau), v) - f(\hat{x}(\tau), \hat{u}(\tau))]\,\varepsilon + o(\varepsilon).$$

Hence, by Lemma 1,

$${}^{*}x(\hat{T}) = {}^{+}x + \delta\hat{x}(\hat{T})\,\varepsilon + o(\varepsilon),$$

where $\delta\hat{x}(t)$ is a solution of

$$\text{(36)} \qquad \frac{\mathrm{d}}{\mathrm{d}t}\,\delta\hat{x}(t) = \delta\hat{x}(t)\, j(\hat{x}(t), \hat{u}(t))',$$

satisfying

$$\delta\hat{x}(\tau) = l[f(\hat{x}(\tau), v) - f(\hat{x}(\tau), \hat{u}(\tau))]$$

and

$$j(x, u) = \left\| \frac{\partial f^i(x, u)}{\partial x^j} \right\|_{i,j=1}^{r}.$$

Consider also a change of the time $\hat{T}$ and the corresponding state change, given by

$$\hat{x}(\hat{T} - l\varepsilon) = {}^{+}x - lf({}^{+}x, \hat{u}(\hat{T}))\,\varepsilon + o(\varepsilon), \quad l > 0.$$

Let K_0 be the set of $y \in R^r$ such that either $y = {}^{+}x + \delta\hat{x}(\hat{T})$ for a certain variation of the control $\hat{u}(t)$ or $y = {}^{+}x - lf({}^{+}x, \hat{u}(\hat{T}))$, $l \geqslant 0$. Then K_0 is a cone with the apex at ${}^{+}x$ (if ${}^{+}x + z \in K_0$, then ${}^{+}x + \alpha z \in K_0$, $\alpha \geqslant 0$). The variation of the end of the trajectory corresponding to a simultaneous variation of the control at several points is the sum of single variations. Therefore put

$$K = \{y \in R^r : y = {}^{+}x + z_1 + \ldots + z_n,\ {}^{+}x + z_i \in K_0,$$
$$i = 1, \ldots, n,\ n = 1, 2, \ldots\}.$$

Again K is a cone with the apex ${}^+x$; and, besides, it is convex [we have

$$y_i = {}^+x + z_i \in K\,, \quad i = 1, 2\,, \quad 0 \leqq p \leqq 1\,, \quad \Rightarrow {}^+x + pz_1 \in K\,,$$
$${}^+x + (1-p)\, z_2 \in K \Rightarrow {}^+x + pz_1 + (1-p)\, z_2 = py_1 + (1-p)\, y_2 \in K]\,.$$

We claim that the optimality of $\hat{x}(t)$ implies $K \neq R^r$. Assume, on the contrary, that $K = R^r$. In particular,

$${}^+x + f({}^+x, \hat{u}(\hat{T})) \in K\,.$$

Here $f({}^+x, \hat{u}(\hat{T}))$ is the direction of the tangent to the trajectory at ${}^+x$. It follows that $\delta\hat{x}(\hat{T}) = f({}^+x, \hat{u}(\hat{T}))$ holds for a suitable (multiple) variation of the control $\hat{u}(t)$. Hence

$${}^*x(\hat{T} - \varepsilon) = {}^*x(\hat{T}) - f({}^*x(\hat{T}), \hat{u}(\hat{T}))\, \varepsilon + o(\varepsilon)$$
$$= {}^+x + f({}^+x, \hat{u}(\hat{T}))\, \varepsilon + o(\varepsilon) - f({}^+x, \hat{u}(\hat{T}))\, \varepsilon + o(\varepsilon) = {}^+x + o(\varepsilon)\,.$$

This means that at time $\hat{T} - \epsilon$ the new trajectory lies within a distance of order $o(\epsilon)$ from ${}^+x$. Thus, by means of a small variation of the trajectory, ${}^+x$ may be reached before time $\hat{T}$.

The exact proof is as follows: Let C be the sphere in R^r with radius 1 and center at 0. By assumption, ${}^+x + C \subset K$. Therefore for each $c \in C$ there exists a variation of the control $\hat{u}(t)$ such that the corresponding trajectory satisfies

$${}^*x_c(\hat{T} - l_c\varepsilon) = {}^+x + \varepsilon c + o(\varepsilon)\,.$$

We can achieve that ${}^*x_c(t)$ is a continuous function of (c, t), l_c is a continuous function of c and $o(\epsilon)$ holds uniformly with respect to $c \in C$. Hence, for ϵ sufficiently small (fixed in the sequel), ${}^+x$ is inside the region bounded by the surface

$$L = \{{}^*x_c(\hat{T} - l_c\varepsilon) : c \in C\}\,.$$

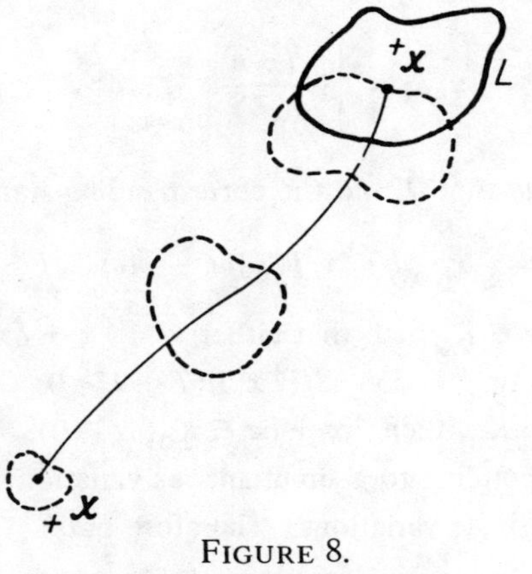

FIGURE 8.

Introduce an auxiliary parameter σ, $0 \leqslant \sigma \leqslant 1$, and denote

$$L_\sigma = \{{}^*x_c(\sigma(\hat{T} - l_c\varepsilon)) : c \in C\}\,.$$

In Figure 8, L_σ is marked by a dashed line; L_σ depends continuously on σ and $L_0 = \{_+x\}$. Therefore $^+x \in L_{\sigma'}$ for a certain σ', $0 < \sigma' < 1$; that is, $^*x_c(T) = {}^+x$ for a certain $c \in C$, where $T = \sigma'(\hat{T} - l_c\epsilon) < \hat{T}$. This contradicts the minimality of $\hat{T}$. This contradiction proves that $K \neq R^r$.

Now K is a convex cone with the apex ^+x, distinct from the whole space. Consequently, there exists a hyperplane Π in R^r, which contains ^+x, and such that K lies entirely in one of the half-spaces defined by Π. Denote by ν the normal direction to Π; that is,

$$\Pi = \{x \in R^r : x = {}^+x + y\,, \quad \nu \,.\, y = 0\}\,.$$

Let ν be pointed into the half-space not containing K. Then $^+x + y \in K$ implies $\nu \cdot y \leqslant 0$. In particular,

$$\nu \,.\, \delta\hat{x}(\hat{T}) \leqq 0 \tag{37}$$

for every variation according to (35), and

$$-\nu \,.\, f({}^+x, \hat{u}(\hat{T})) \leqq 0\,. \tag{38}$$

Consider a solution of the system

$$\frac{\mathrm{d}}{\mathrm{d}t}\hat{\lambda} = -\hat{\lambda}j(\hat{x}(t), \hat{u}(t)) \tag{39}$$

satisfying $\hat{\lambda}(\hat{T}) = \nu$. Consider further the Hamiltonian $H(\hat{\lambda}(t), \hat{x}(t), u)$, where $H(\lambda, x, u) = \lambda \cdot f(x, u)$. It follows from Lemma 2, taking (36) and (37) into account, that for every variation according to (35) we have

$$\hat{\lambda}(t) \,.\, \delta\hat{x}(t) \equiv \text{const} = \hat{\lambda}(\hat{T}) \,.\, \delta\hat{x}(\hat{T}) \leqq 0\,.$$

Hence

$$\hat{\lambda}(\tau) \,.\, [f(\hat{x}(\tau), v) - f(\hat{x}(\tau), \hat{u}(\tau))] \leqq 0\,,$$

or, equivalently,

$$H(\hat{\lambda}(\tau), \hat{x}(\tau), v) \leqq H(\hat{\lambda}(\tau), \hat{x}(\tau), \hat{u}(\tau)) \tag{40}$$

at every continuity point of the control $\hat{u}(t)$. The validity of (40) can be easily extended to all $t \in [0, \hat{T}]$.

We have proved the following assertion: *If* $\hat{x}(t)$, $0 \leqslant t \leqslant \hat{T}$, *is an optimal trajectory, then there exists a nonzero solution of* (39) *such that*

$$H(\hat{\lambda}(t), \hat{x}(t)\, \hat{u}(t)) = \max_{v \in U} H(\hat{\lambda}(t), \hat{x}(t), v)\,, \quad 0 \leqq t \leqq \hat{T}\,, \tag{41}$$

and

$$H(\hat{\lambda}(\hat{T}), \hat{x}(\hat{T}), \hat{u}(\hat{T})) \geqq 0\,. \tag{42}$$

Inequality (42) follows from (38). Note that, by Lemma 3,

$$H(\hat{\lambda}(t), \hat{x}(t), \hat{u}(t)) \equiv \text{const.}$$

§6. The Pontrjagin principle

Now we sum up the *Pontrjagin principle* in a theorem. As in §4, we consider a trajectory in the space R^{r+1}.

THEOREM 3. *Let $\hat{x}(t)$, $0 \leqslant t \leqslant \hat{T}$, be the optimal trajectory from I to F, corresponding to a control $\hat{u}(t)$. Denote $H(\lambda, x, u) = \lambda \cdot f(x, u)$. Then there exists a nonzero solution $\hat{\lambda}(t)$ of the system*

$$\frac{d}{dt}\hat{\lambda} = -\hat{\lambda}j(\hat{x}(t), \hat{u}(t)) \tag{43}$$

such that

$$0 \equiv \mathcal{H}(\hat{\lambda}(t), \hat{x}(t), \hat{u}(t)) = \max_{u \in U} \mathcal{H}(\hat{\lambda}(t), \hat{x}(t), u), \quad 0 \leqq t < T, \tag{44}$$

$$\lambda^0(t) \equiv \text{const} \leqq 0, \tag{45}$$

$$(\hat{\lambda}^1(0), \ldots, \hat{\lambda}^r(0)) = \sum_{i=1}^{l} \alpha_i \overline{\nabla} h_i(\hat{x}(0)), \tag{46}$$

$$(\hat{\lambda}^1(\hat{T}), \ldots, \hat{\lambda}^r(\hat{T})) = \sum_{i=1}^{m} \beta_i \overline{\nabla} g_i(\hat{x}(\hat{T})). \tag{47}$$

The controls and trajectories satisfying the necessary conditions in Theorem 3 will be called *extremal.* When using the Pontrjagin principle, we find all the extremal controls and then determine which of them are optimal.

Now we turn to *nonautonomous systems.* As we noted in Remark 2, the maximum principle for a nonautonomous system

$$\frac{dx}{dt} = f(t, x, u) \tag{48}$$

follows from the results for autonomous systems. We consider time as an additional state coordinate $t = x^{r+1}$, i.e. $dx^{r+1}/dt \equiv 1$ and $x^{r+1}(0) = t_0$. Let $\hat{x}(t)$, $0 \leqslant t \leqslant \hat{T}$, be an optimal trajectory from I to F satisfying (48) with a control $\hat{u}(t)$. Denote

$$\mathcal{H}(t, \lambda, x, u) = \lambda \,.\, f(t, x, u), \quad j(t, x, u) = \left\| \frac{\partial f^i(t, x, u)}{\partial x^j} \right\|_{i,j=0}^{r}.$$

From Theorem 3 it follows that there exists a nonzero solution $(\hat{\lambda}(t), \hat{\lambda}^{r+1}(t))$ of the system

$$\frac{d}{dt}\hat{\lambda} = -\hat{\lambda}j(t, \hat{x}(t), \hat{u}(t)), \quad \frac{d}{dt}\hat{\lambda}^{r+1} = -\sum_{k=0}^{r} \hat{\lambda}^k \frac{\partial}{\partial t} f^k(t, \hat{x}(t), \hat{u}(t)),$$

such that

$$0 \equiv \mathcal{H}(t, \hat{\lambda}(t), \hat{x}(t), \hat{u}(t)) + \hat{\lambda}^{r+1}(t) = \max_{u \in U} \mathcal{H}(t, \hat{\lambda}(t), \hat{x}(t), u) + \hat{\lambda}^{r+1}(t).$$

Relations (45)–(47) continue to hold. Since $x^{r+1}(\hat{T}) = \hat{T} + t_0$ is undetermined, we have an additional transversality condition $\hat{\lambda}^{r+1}(\hat{T}) = 0$. Hence

$$\hat{\lambda}^{r+1}(t) = \int_t^{\hat{T}} \sum_{k=0}^{r} \hat{\lambda}^k(s) \frac{\partial}{\partial s} f^k(s, \hat{\boldsymbol{x}}(s), \hat{\boldsymbol{u}}(s))\, ds ,$$

and therefore $(\hat{\lambda}(t), \hat{\lambda}^{r+1}(t))$ is nonzero only if $\hat{\lambda}(t)$ is nonzero. We conclude that, for a nonautonomous system, (44) ought to be replaced by the relation

$$\begin{aligned} \mathscr{H}(t, \hat{\lambda}(t), \hat{\boldsymbol{u}}(t)) &= \max_{\boldsymbol{u} \in U} \mathscr{H}(t, \hat{\lambda}(t), \hat{\boldsymbol{x}}(t), \boldsymbol{u}) \\ &= -\int_t^{\hat{T}} \sum_{k=0}^{r} \hat{\lambda}^k(s) \frac{\partial}{\partial s} f^k(s, \hat{\boldsymbol{x}}(s), \hat{\boldsymbol{u}}(s))\, ds , \quad 0 \leqq t < \hat{T} . \end{aligned} \tag{49}$$

§7. Examples

EXAMPLE 4 (ZERMELO). A boat is carried by a stream, whose direction will be identified with the positive x^1 direction, with velocity s. The boat can move with respect to the stream with velocity not greater that one in any direction. The problem is to transfer the boat from the initial point, identified with the origin of coordinates ${}_{+}\boldsymbol{x} = (0, 0)$, to a point ${}^{+}\boldsymbol{x} = ({}^{+}x^1, {}^{+}x^2)$ in minimum time. The control parameter is the vector $\boldsymbol{u}$ of the relative boat velocity with respect to the stream. Thus

$$(u^1)^2 + (u^2)^2 \leqq 1 .$$

Trajectory equations are

$$\frac{dx^1}{dt} = s + u^1 , \quad \frac{dx^2}{dt} = u^2 .$$

Invoking the results in §5, we have

$$H(\lambda, x, u) = \lambda^1(s + u^1) + \lambda^2 u^2 ,$$

whence

$$\frac{d\hat{\lambda}^i}{dt} = -\frac{\partial H}{\partial x^i} = 0 , \quad i = 1, 2 .$$

Therefore $\hat{\lambda}^i \neq \text{const}, i = 1, 2$. The control maximizing the Hamiltonian is $\hat{\boldsymbol{u}} \equiv \hat{\lambda}/\|\hat{\lambda}\|$. Note that the extremal trajectories are rectilinear. When $s < 1$, the boat can reach any point ${}^{+}\boldsymbol{x} \in R^2$. When $s \geqslant 1$, it can reach only the ${}^{+}\boldsymbol{x} \in R^2$ satisfying

$${}^{+}x^1 > 0 , \quad -(s^2 - 1)^{-1/2} \leqq {}^{+}x^2/{}^{+}x^1 \leqq (s^2 - 1)^{-1/2} .$$

EXAMPLE 5 —*system moving along a line under the influence of a limited force.* The trajectory equations are

$$\frac{dx^1}{dt} = x^2 , \quad \frac{dx^2}{dt} = u , \quad |u| \leqq 1 .$$

The problem is to transfer the system from an initial state ${}_{+}\boldsymbol{x}$ to a state ${}^{+}\boldsymbol{x} = (0, 0)$ in a minimum time. We have

$$H(\lambda, x, u) = \lambda^1 x^2 + \lambda^2 u .$$

The system (39) now takes the form

$$\frac{d}{dt}\hat{\lambda}^1 \equiv 0 , \quad \frac{d}{dt}\hat{\lambda}^2 = -\hat{\lambda}^1 ;$$

hence $\hat{\lambda}^1(t) \equiv c_1$ and $\hat{\lambda}^2(t) \equiv c_2 - c_1 t$. We can see that $\hat{u}(t) = 1$ for $\hat{\lambda}^2(t) > 0$ and $\hat{u}(t) = -1$ for $\hat{\lambda}^2(t) < 0$. In view of the linearity of $\hat{\lambda}^2(t)$, the switch appears at most once.

Consider the system of equations

$$\text{(50)} \qquad \frac{dx^1}{dt} = x^2 , \quad \frac{dx^2}{dt} = 1 ; \quad \frac{dx^1}{dt} = x^2 , \quad \frac{dx^2}{dt} = -1 .$$

The solutions are

$$x^2(t) = \pm(t - t_0) + x^2(t_0) ,$$
$$x^1(t) = \pm\tfrac{1}{2}(t - t_0)^2 + (t - t_0)\, x^2(t_0) + x^1(t_0) .$$

Therefore

$$x^1(t) = \pm\tfrac{1}{2}x^2(t)^2 \mp \tfrac{1}{2}x^2(t_0)^2 + x^1(t_0) .$$

Thus the trajectories are parabolas. If the trajectory corresponding to one of systems (50) passes through the origin, then

$$\text{(51)} \qquad x^1(t_0) = \tfrac{1}{2}x^2(t_0)^2 , \quad x^2(t_0) \leqq 0 , \quad \text{or} \quad x^1(t_0) = -\tfrac{1}{2}x^2(t_0)^2 , \quad x^2(t_0) \geqq 0 .$$

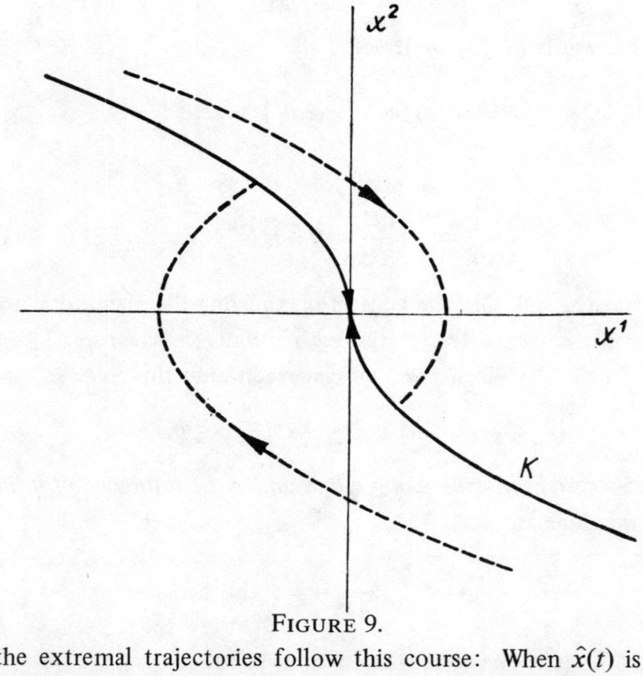

FIGURE 9.

We see that the extremal trajectories follow this course: When $\hat{x}(t)$ is above the curve K defined by (51), we have $\hat{u}(t) = -1$; when $\hat{x}(t)$ is under K, we have $\hat{u}(t)$

$= 1$. When the trajectory reaches K, the control parameter is switched and the trajectory follows K to the point $(0, 0)$ (see Figure 9). This describes the dependence of the extremal control on the position of the system (feedback).

EXAMPLE 6 (LEITMANN)—*maximum range of a thrust-limited rocket, with aerodynamic effects neglected.* Let x^0 and x^1 be the horizontal and vertical coordinates of the rocket position. (The flight is assumed to take place in a vertical plane.) Further, let x^2 and x^3 be the velocity components, and x^4 the mass of the rocket. Control parameters are u^1 and u^2 (the direction cosine and sine of the thrust vector), and u^3 (the fuel flow rate). We have

$$(u^1)^2 + (u^2)^2 = 1\,, \quad 0 \leqq u^3 \leqq \bar{u}^3\,.$$

Denote by g the gravitational acceleration, and by c the effective exhaust speed. We wish to transfer the rocket from the initial state ${}_{+}x$ into the state set

$$F = \{x : x^1 = {}^{+}x^1,\ x^4 = {}^{+}x^4\} \tag{52}$$

in such a way that $x^0(T)$ is maximal. In (52), the value ${}^{+}x^1$ is the altitude where the maximum range is to be attained, and ${}^{+}x^4$ is the mass of the rocket without fuel.

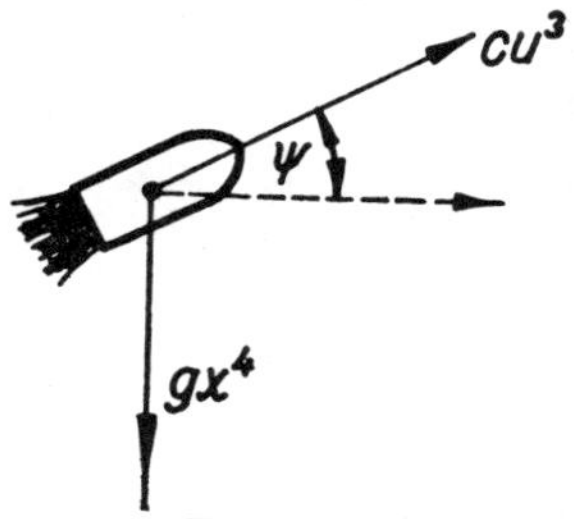

FIGURE 10.

The system of trajectory equations is (see Figure 10)

$$\begin{aligned} &\frac{dx^0}{dt} = x^2\,, \quad \frac{dx^1}{dt} = x^3\,, \\ &\frac{dx^2}{dt} = \frac{c}{x^4} u^1 u^3\,, \quad \frac{dx^3}{dt} = \frac{c}{x^4} u^2 u^3 - g\,, \quad \frac{dx^4}{dt} = -u^3\,. \end{aligned} \tag{53}$$

Further,

$$\begin{aligned} \mathscr{H}(\boldsymbol{\lambda}, \boldsymbol{x}, \boldsymbol{u}) &= \lambda^0 x^2 + \lambda^1 x^3 + \lambda^2 \frac{c}{x^4} u^1 u^3 \\ &+ \lambda^3 \left(\frac{c}{x^4} u^2 u^3 - g\right) - \lambda^4 u^3 = \left[\frac{c}{x^4}(\lambda^2 u^1 + \lambda^3 u^2) - \lambda^4\right] u^3 \\ &+ \lambda^0 x^2 + \lambda^1 x^3 - \lambda^3 g\,. \end{aligned}$$

As the aim is to maximize $x^0(T)$, extremal controls are those minimizing the Hamiltonian. Equations for the Lagrange multipliers are

$$\frac{d\hat{\lambda}^0}{dt} = 0 = \frac{d\hat{\lambda}^1}{dt}, \quad \frac{d\hat{\lambda}^2}{dt} = -\hat{\lambda}^0,$$

(54)

$$\frac{d\hat{\lambda}^3}{dt} = -\hat{\lambda}^1, \quad \frac{d\hat{\lambda}^4}{dt} = \frac{c}{(\hat{x}^4)^2}(\hat{\lambda}^2\hat{u}^1 + \hat{\lambda}^3\hat{u}^2)\,\hat{u}^3.$$

The transversality conditions are

$$\hat{\lambda}^2(\hat{T}) = 0 = \hat{\lambda}^3(\hat{T}).$$

Therefore

$$\hat{\lambda}^0(t) \equiv \text{const}, \quad \hat{\lambda}^1(t) \equiv \text{const},$$
$$\hat{\lambda}^2(t) = (\hat{T} - t)\,\hat{\lambda}^0, \; \hat{\lambda}^3(t) = (\hat{T} - t)\,\hat{\lambda}^1.$$

It is not difficult to check that $\hat{\lambda}^0 \equiv 0 \equiv \hat{\lambda}^1$ implies $\hat{\lambda}(t) = 0$, whereas $\hat{\lambda}(t)$ has to be nonzero according to the assumption. The Hamiltonian is minimal when

$$\hat{u}^1(t) = \frac{-\hat{\lambda}^2(t)}{\sqrt{[\hat{\lambda}^2(t)^2 + \hat{\lambda}^3(t)^2]}} \equiv \frac{-\hat{\lambda}^0}{\sqrt{[(\hat{\lambda}^0)^2 + (\hat{\lambda}^1)^2]}},$$

$$\hat{u}^2(t) = \frac{-\hat{\lambda}^3(t)}{\sqrt{[\hat{\lambda}^2(t)^2 + \hat{\lambda}^3(t)^2]}} \equiv \frac{-\hat{\lambda}^1}{\sqrt{(\hat{\lambda}^0)^2 + (\hat{\lambda}^1)^2]}}.$$

Thus the extremal thrust direction is constant. Consider $\hat{u}^3(t)$. Set

$$\sigma(t) = \frac{c}{\hat{x}^4(t)}(\hat{\lambda}^2(t)\,\hat{u}^1 + \hat{\lambda}^3(t)\,\hat{u}^2) - \hat{\lambda}^4(t)$$

$$= \frac{c}{\hat{x}^4(t)}(-\sqrt{[(\hat{\lambda}^0)^2 + (\hat{\lambda}^1)^2]}\,(\hat{T} - t)) - \hat{\lambda}^4(t).$$

Then $\hat{u}^3(t) = \bar{u}^3$ when $\sigma(t) < 0$, and $\hat{u}^3(t) = 0$ when $\sigma(t) > 0$. By means of (54) we obtain

$$\frac{d\sigma(t)}{dt} = \frac{c}{\hat{x}^4(t)}\sqrt{[(\hat{\lambda}^0)^2 + (\hat{\lambda}^1)^2]}.$$

Hence $\sigma(t)$ is increasing. Therefore

$$\hat{u}^3(t) = \bar{u}^3 \quad \text{for} \quad 0 \leqq t < T', \quad \hat{u}^3(t) = 0 \quad \text{for} \quad T' \leqq t \leqq \hat{T}.$$

The switching time T' is determined by the obvious relation

$$\bar{u}^3 T' = {}_+x^4 - {}^+x^4.$$

It remains to find the extremal thrust direction. Certainly $\hat{\lambda}^0 \neq 0$ (if $\hat{\lambda}^0 = 0$, then $\hat{u}^1 = 0$; consequently $\hat{x}^0(t) \equiv {}_+x^0$, and the latter is clearly not optimal). From the relation

$$0 = \mathscr{H}(\hat{\lambda}(\hat{T}), \hat{\boldsymbol{x}}(\hat{T}), \hat{\boldsymbol{u}}(\hat{T})) = \hat{\lambda}^0\hat{x}^2(\hat{T}) + \hat{\lambda}^1\hat{x}^3(\hat{T})$$

we get for the angle $\hat{\psi}$ of the extremal thrust

$$\operatorname{tg} \hat{\psi} = -\hat{x}^2(\hat{T})/\hat{x}^3(\hat{T}) .$$

Integrating the third and fourth equations (53) over the interval $[T', \hat{T}]$, we get

$$\hat{x}^2(\hat{T}) = \hat{x}^2(T'), \quad \hat{x}^3(\hat{T}) = \hat{x}^3(T') - g(\hat{T} - T') .$$

From the second equation it follows, by means of the relation

$$\hat{x}^3(T')(\hat{T} - T') - \tfrac{1}{2}g(\hat{T} - T')^2 = {}^+x^1 - x^1(T'),$$

that

$$\hat{x}^3(\hat{T}) = -[\hat{x}^3(T')^2 + 2g(x^1(T') - {}^+x^1)]^{1/2} .$$

The corresponding equations in (53) yield

$$\hat{x}^2(T') = {}_+x^2 + c(\cos \hat{\psi}) \ln \frac{{}_+x^4}{{}^+x^4},$$

$$\hat{x}^3(T') = {}_+x^3 + c(\sin \hat{\psi}) \ln \frac{{}_+x^4}{{}^+x^4} - gT',$$

$$\hat{x}^1(T') = {}_+x^1 + c(\sin \hat{\psi}) \int_0^{T'} \left(\ln \frac{{}_+x^4}{{}_+x^4 - \bar{u}^3 t} \right) dt - \tfrac{1}{2} g T'^2 .$$

This system of equations, together with the previous ones, may be used for the computation of $\hat{\psi}$.

EXAMPLE 7 (MEDITCH)—*rocket landing.* Let $y(t)$ be the vertical coordinate of a rocket at time t, $m(t)$ its mass, and $g = \text{const}$ the gravitational acceleration. Let the initial position of the rocket be $y(0) = {}_+x^1 > 0$ and its initial velocity $(d/dt)y(0) = {}_+x^2 < 0$. The difference $m(0) - m(t)$ represents the fuel consumption up to time t. The propulsion system is capable of creating a force in the positive y direction. The only other force acting on the rocket is its own weight mg (see Figure 11). The thrust force is given by $-k\, dm/dt$, where $k > 0$ is the velocity of the exhaust gases with respect to the rocket. Hence the motion of the rocket is governed by the differential equation

$$(55) \qquad \frac{d^2 y}{dt^2} = -\frac{k}{m}\frac{dm}{dt} - g .$$

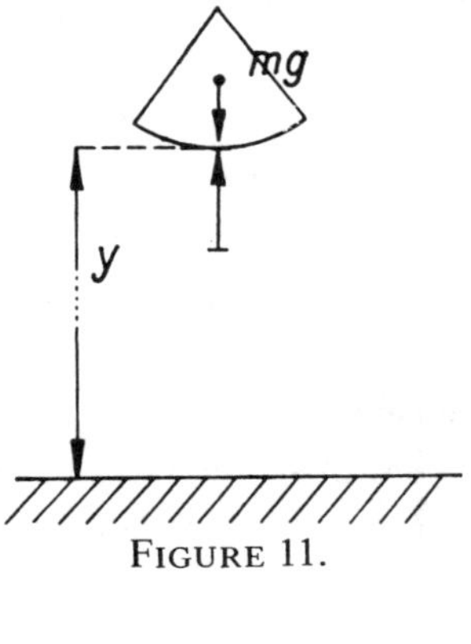

FIGURE 11.

Here dm/dt assumes values in an interval $[-\omega, 0]$ with $\omega > 0$. The problem is to transfer the rocket to the state $y(T) = 0$, $dy(T)/dt = 0$ with a minimum expenditure of fuel $m(0) - m(T)$.

By integrating (55) we get

$$-{}_+x^2 = \frac{d}{dt}y(T) - \frac{d}{dt}y(0) = -k\ln(m(T)/m(0)) - gT.$$

Therefore

$$m(0) - m(T) = m(0)\left[1 - \exp\left\{\frac{{}_+x^2 - gT}{k}\right\}\right].$$

Thus the fuel consumption is minimal when the landing time is minimal.

The rocket is a system whose state is characterized by coordinates (x^1, x^2, x^3), where $x^1(t) = y(t)$, $x^2(t) = dy(t)/dt$ and $x^3(t) = m(t)$. The equations of motion are

$$(56)\qquad \frac{d}{dt}x^1 = x^2, \quad \frac{d}{dt}x^2 = -\frac{k}{x^3}u - g, \quad \frac{d}{dt}x^3 = u.$$

Here $u = dm/dt$ is the control parameter, which takes values in $U = [-\omega, 0]$. The problem is to transfer the system from the initial state ${}_+x$ to the set $F = \{x\colon x^1 = x^2 = 0\}$ in the shortest time. We have

$$H(\lambda, x, u) = \lambda^1 x^2 - \lambda^2\left(\frac{k}{x^3}u + g\right) + \lambda^3 u = \lambda^1 x^2 - \lambda^2 g + h(\lambda, x)\,u,$$

where

$$h(\lambda, x) = \lambda^3 - \lambda^2 k/x^3.$$

The extremal trajectory $\hat{x}(t)$ and the corresponding control must satisfy

$$H(\lambda(t), \hat{x}(t), \hat{u}(t)) = \max_{u\in U} H(\lambda(t), \hat{x}(t), u) \geqq 0,$$

i.e. $\hat{u}(t) = 0$ for $\hat{h}(t) = h(\hat{\lambda}(t), \hat{x}(t)) > 0$, and $\hat{u}(t) = -\omega$ for $\hat{h}(t) < 0$. At the same time,

$$(57)\qquad \frac{d}{dt}\hat{\lambda}^1 = 0, \quad \frac{d}{dt}\hat{\lambda}^2 = -\hat{\lambda}^1, \quad \frac{d}{dt}\hat{\lambda}^3 = -\hat{\lambda}^2\frac{k\hat{u}(t)}{(\hat{x}^3(t))^2}.$$

At the final point, the transversality condition $\hat{\lambda}^3(\hat{T}) = 0$ must hold.

Using (56) and (57), we obtain $(d/dt)\hat{h}(t) = \hat{\lambda}^1 k/\hat{x}^3(t)$. Hence $\hat{h}(t)$ is a monotone function of t. Therefore, the extremal control consists of a period of free fall followed by full thrust until the rocket stops. (It is not excluded that the full thrust takes place all the time.) The controls working in the opposite way do not agree with the physical meaning of the problem.

We try to implement the control synthesis. The engine will be switched on in the state $\tilde{x}$ with the property that when the rocket starts at $\tilde{x}$ and uses full thrust, it has velocity 0 at altitude 0. The full thrust satisfies

$$\frac{d}{dt}x^1 = x^2, \quad \frac{d}{dt}x^2 = \frac{k\omega}{x^3} - g, \quad \frac{d}{dt}x^3 = -\omega.$$

Hence

$$x^3(t) = \tilde{x}^3 - \omega t, \quad x^2(t) = \tilde{x}^2 - k\ln\left(1 - \frac{\omega t}{\tilde{x}^3}\right) - gt,$$

$$\begin{aligned} x^1(t) &= \tilde{x}^1 + \tilde{x}^2 t + \frac{k\tilde{x}^3}{\omega}\left(1 - \frac{\omega t}{\tilde{x}^3}\right)\ln\left(1 - \frac{\omega t}{\tilde{x}^3}\right) + kt - \tfrac{1}{2}gt^2 \\ &= \tfrac{1}{2}gt^2 + \left(k - \frac{\tilde{x}^3 g}{\omega}\right)t + \tilde{x}^1 + \frac{\tilde{x}^2\tilde{x}^3}{\omega} + \left(t - \frac{\tilde{x}^3}{\omega}\right)x^2(t). \end{aligned}$$

Denoting by T the time when altitude 0 is attained, we get

$$0 = \tilde{x}^2 - k\ln\left(1 - \frac{\omega T}{\tilde{x}^3}\right) - gT, \tag{58}$$

$$0 = \tfrac{1}{2}gT^2 + \left(k - \frac{\tilde{x}^3 g}{\omega}\right)T + \tilde{x}^1 + \frac{\tilde{x}^2\tilde{x}^3}{\omega}. \tag{59}$$

We compute T from (59) and substitute into (58) to get a relation of the form $B(\tilde{x}^1, \tilde{x}^2, \tilde{x}^3) = 0$ for the region of full thrust.

BIBLIOGRAPHY

1. R. Aris, *The optimal design of chemical reactors,* Academic Press, New York, 1961. MR **23** #B473.

2. R. Bellman, *Dynamic programming,* Princeton Univ. Press, Princeton, N. J., 1957. MR **19**, 820.

3. R. Bellman and S. Dreyfus, *Applied dynamic programming,* Princeton Univ. Press, Princeton, N. J., 1962. MR **25** #3791.

4. V. G. Boltjanskiĭ, *Mathematical methods of optimal control,* "Nauka", Moscow, 1966; English transl., Holt, Reinhart and Winston, New York, 1971. MR **34** #7230; **50** #5567.

5. M. M. Connors and D. Teichroew, *Optimal control of dynamic operations research models,* Internat. Textbook Co., Scranton, Pa., 1967. Reviewed in Management Sci. 14, B128.

6. Liang-tseng Fan and Chiu-sen Wang, *The discrete maximum principle,* Wiley, New York, 1964.

7. H. Halkin, *A maximum principle of the Pontryagin type for systems described by nonlinear difference equations,* SIAM J. Control **4** (1966), 90–111. MR **33** #7155.

8. G. Leitmann, *An introduction to optimal control,* McGraw-Hill, New York, 1966.

9. P. Mandl, *Controlled Markov chains,* Kybernetika (Prague) 5 (1969), příloha, 1–74. (Czech)

10. J. Meditch, *The Pontryagin maximum principle and some of its applications,* Advances in Control Systems, Vol. 1, Academic Press, New York, 1964, pp. 55–74. MR **32** #2241.

11. L. S. Pontrjagin et al., *The mathematical theory of optimal processes,* Fizmatgiz, Moscow, 1961; English transl., Wiley, New York, 1962; Macmillan, New York, 1964. MR **29** #3316a, b; **32** #3896.

12. D. J. Wilde and C. S. Beightler, *Foundations of optimization,* Prentice-Hall, Englewood Cliffs, N. J., 1967.

Translated by JAN K. PACHL

Amer. Math. Soc. Transl.
(2) Vol. 111, 1978

ON CONTROL TACTICS FOR COMPLEX SYSTEMS AND THEIR RELATION TO PHYSIOLOGY*

I. M. GEL'FAND, V. S. GURFINKEL' AND M. L. CETLIN

The present paper concerns certain mathematical models which arose in the course of our studies in physiology. In recent years physiologists have had some success in applying ideas borrowed from modern mathematics and engineering. Especially fruitful has been the mathematical concept of feedback as developed in the famous works of Norbert Wiener and proposed still earlier in the studies of N. A. Bernšteĭn and P. K. Anohin. Other useful notions include the Van der Pol relaxation model.

However, despite the significant results achieved by the use of mathematical methods, these have proved inadequate for describing basic physiological phenomena, i.e. behavior and its relationship to the structures of the central nervous system. We use the term "behavior" in a very broad sense, including in it such notions as the structuring of movements, the role and interaction of the various levels of the central nervous system in the regulation of particular functions, etc.

1. Once embarked on our physiological studies, we saw that the mathematical language and the mathematical way of thinking evolved largely for the purpose of dealing with physical, mechanical, and engineering problems were too remote from the specific requirements of physiology, that they were inadequate for dealing with basic physiological phenomena and concepts. Hence, while investigating specific physiological problems, we tried, as mathematicians, to maintain a keen awareness of the possibility of discovering new mathematical models and methods which might later shed light on specific problems of physiology and assist the development of an appropriate mathematical language.

Our purpose here is to discuss some of the models which occurred to us during our physiological research. Two of these models relate to the problems of behavior in the broad sense; one has to do with the structure of the nervous system.

2. The first model is related to the problem of attaining a goal in a varying medium. This problem is in many respects similar to the mathematical problem of finding the minimum of a nonstationary function of several variables. Since the function varies, the search for the minimum must yield a result in a strictly limited time, i.e. before the function has had time to change substantially. For example, in attempting to forecast tomorrow's weather on the basis of today's data we must see to it that the calculations require not more than 24 hours. Thus, even though mathematics has at its disposal exact classical algorithms for finding the minimum of a function of several variables, these algorithms are practically useless for our purposes. The goal-attainment problems which arise in

---

AMS (MOS) subject classifications (1970). Primary 92A05, 68A35.

*Translation of *Biological Aspects of Cybernetics*, Izdat. Akad. Nauk SSSR, Moscow, 1962, pp. 66–73.

physiology are usually very complex, and attempts to solve them by means of exact algorithms must fail.

Problems of this type require a *search* which ensures progress towards the goal and also yields the information required for continued searching. The search process can be made successful only through *organization*, i.e. through the exploitation of those properties of the problem which assist attainment of the goal in an acceptable time. These properties are not known in advance; they are merely more or less plausible. Organization is effected by setting out hypotheses (plausible assumptions about the structure of the problem) and constructing search tactics based on these hypotheses. Search is accompanied by hypothesis verification: that hypothesis is valid which ensures successful search. The use of hypotheses precludes consideration of all possibilities; this narrowing of choice is a prerequisite for successful search even in minimally complex situations.

In the minimum search problem we make use of the hypothesis that the variables determining the value of the function can be divided into two groups. The first group (which includes almost all the variables) consists of those variables whose variation results in substantial variations of the function being minimized. The values of these parameters (called *nonessential* parameters) can be chosen relatively simply and rapidly.

The second group of parameters includes a small number of variables (or functions of a small number of variables) whose variation results in relatively small variations in the value of the function being minimized. These parameters are called *essential*. (The essential variables can be functions of a large number of parameters; all that matters is that these functions themselves be well-organized.)

The essential and nonessential parameters are not identifiable in advance. We merely assume that such a breakdown is possible and that the choice of essential parameters is effected automatically during search. Some highly illustrative examples of breaking down parameters into the essential and nonessential categories will be found in N. A. Bernšteĭn's paper [1].

The minimum search tactic which exploits these properties of function organization has been called the "gully tactic". It consists in an alternation of local refinement of the result (choice of values of the nonessential variables) with rapid nonlocal progression (extrapolation) in the essential variables. An important property of this tactic is *adaptation*, i.e. considerable acceleration of search during the search process itself, which occurs as a result of increasingly precise isolation of the essential variables. The description of the tactic itself necessarily includes the various *control levels* which interact in a special way (see [2]). We have already described this tactic in a separate paper; we merely note here that it turned out to be effective in several computational problems (e.g. in problems of phase analysis of elementary particle scattering).

One of the interesting problems of physiology is that of determining which search tactics (search being inevitable here) arise in the structuring of movements of man and the higher animals. Detailed studies of specific motor problems, especially the thorough investigations of the human standing mechanism carried out by V. S. Gurfinkel', indicate that the structuring of movements involves such elements as search, local refinement, extrapolation, adaptation, tactic realization levels, etc.; i.e. that the language which has arisen on the basis of the gully tactic is a convenient means of attacking this problem. The depth of the analogies we have noted will be the subject of a future experimental study.

G. V. Geršuni has noted the usefulness of this language for the study of phenomena having to do with the operation of the auditory analyzer, where the problem of search and analysis also arises. In his paper in the present collection G. M. Frank [3] has interesting things to say about the role of search tactics in the self-regulation of cell processes. M. M. Bongard has used algorithms (tactics) somewhat like those described here in his studies of the recognition problem. This is quite natural, as the recognition problem is a complex one and requires the discernment of organization.

3. Consideration of animal behavior naturally gives rise to the question as to the nature of the purposefulness of behavior and the constituent elements of this purposefulness. The multilevel highly complex and ramified nervous systems of the higher animals pose yet another problem: in view of the fact that all the nerve centers operate jointly, what are the principles governing the operation of each individual center, in what measure is it autonomous, and in what way is it controlled (does it interact)?

We became interested in seeing whether there might not be some common principle (e.g. something similar to the variational principles of mechanics) which might help us to deal with these questions.

Such a principle became evident in our studies of even the simplest models of continuous control systems (see [4]). However, we shall formulate it here in connection with models of the behavior of automata in random media.

Let us imagine an object capable of carrying out a certain number of operations. Let this object be immersed in some medium which reacts to these operations in some way. Let us assume, moreover, that these reactions of the medium are perceived by the object as either favorable or unfavorable. The unfavorable and favorable reactions of the medium can be called "penalties" and "nonpenalties", respectively. Taking account of these actions, the object chooses its own actions in such a way as to score a maximum number of favorable reactions and a minimum of unfavorable ones.

Let us assume that the object just described is a finite automaton. Finite automata are well known to the mathematician. They are devices with a finite memory which are capable of performing a finite number of actions and of receiving a finite number of signals (actions) in each operating cycle. To be specific we also assume that the medium "penalizes" the automaton with a specific probability for each of the actions of the latter.

If the automaton does not in any way relate its actions to the reactions of the medium, we naturally assume that it "pays a penalty" with a probability equal to the average probability of a "penalty" in the given medium. Let us agree to say that such an automaton does not behave expediently. If, on the other hand, the automaton "pays a penalty" with a less than average probability, then this penalty reduction can serve as a measure of the expedience of its behavior.

Let us cite the simplest example of an automaton with expedient behavior. The tactic of its behavior is extremely simple: having performed a certain action, the automaton repeats it if it has not been "penalized" and goes on to perform another action in the case of a "penalty". This automaton has a very small memory, so that the expedience of its behavior is relatively small.

It is possible, however, to indicate an automaton structure (an automaton with a linear tactic) which, given a sufficiently large memory, performs only those actions for which the "penalty" is minimal, i.e. whose behavior is optimal. We note that the simplest automaton described above also employs a linear tactic.

In an earlier paper [5], the third author has investigated the behavior of an automaton in a medium whose properties vary with time. On that occasion he demonstrated the existence of an optimal automaton complexity and an optimal memory size which ensure a minimal "penalty". This optimal memory volume depends, of course, on the properties of the medium and on the rate at which these properties vary.

The existence of optimal memory size can be illustrated by means of a simple everyday example. Let us be given the task of finding a scientist to direct the solution of an important scientific problem. If we base our decision on the candidates' achievements over the past 50 years our choice will hardly be the right one, i.e. our memory would be too large. On the other hand, if we consider only their achievements over the past six months, our decision is again likely to be wrong, i.e. our memory would be too short. Clearly, some optimal memory is required.

This model of behavior was used by M. A. Alekseev, M. S. Zalkind and V. M. Kušnarev in experiments on the solution of a choice problem by humans with probabilistic reinforcement of motor reactions. The results of this study are described in their paper [6].

The formation of the structure of an automaton with purposeful behavior constitutes a very interesting problem. This problem can be understood as follows. Let us take an automaton with a random tactic and let us vary this tactic in accordance with its degree of expediency (see [7]). The automaton then employs the most expedient tactic and pays the minimum penalty.

We have been giving much attention lately to the problem of the behavior of several automata. The "penalties" are then paid by each automaton of the group on the basis not only of its own actions but of the actions of the other automata as well. A certain natural mathematical formalization of this problem yields a system of functions analogous to the so-called payoff function of game theory.

It is interesting to note that such a system of automata is ergodic, i.e. that the behavior states which become established in it do not depend on the initial states of the automata. These steady states can be determined by a suitable choice of payoff functions. In other words, by suitable choice of a penalty system it is possible either to make the automata perform prescribed combinations of actions, or (and this is particularly interesting) prescribed sequences of such combinations. Thus, a system of very simple automata has a common purpose and pursues the same end, exhibiting fairly rich and complex behavior. The interacting automata need not know the payoff matrices. This distinguishes the above situation from that considered by von Neumann.

We hope that this game-theoretical model of the behavior of automata will help to explain certain interesting features of nerve center interaction. It is natural to suppose, for example, that automata are analogous to nerve centers of a certain level, that the actions of the medium correspond to ascending afference, and that the role of higher-level centers consists in the specification of the payoff functions which define the game.

The central nervous system includes a large number of individual nerve centers which control a relatively small number of effectors, so that each act of behavior (e.g. movement) is the result of their joint action. A change in the functioning of one of the centers necessarily alters the activity of the others.

In our view, the mathematical model just proposed makes it possible to some extent to describe the interaction of nerve centers without resorting to a complex system of links and coordinations of their activity. Each center solves its own "particular" problem (as does the individual automaton in our model) and thus takes part in the solution of the common problem.

Of great importance in this connection is the question of what it is in a physiological object which corresponds to the "penalty" organizing the functioning of the mathematical model. What, then, is the physiological nature of this factor which ensures the coordinated functioning of individual nerve centers? In other words, what "particular" or "private" problem does the individual nerve center solve?

In our view the role of the "penalty" for each nerve center is played by the incoming afferent signals. The "particular" expediency of an individual nerve center then consists in the fact that the outgoing signals are selected in such a way as to reduce the afferent signals arriving at this center. It is our opinion that this role of afference is universal, i.e. that it operates in one way or another for all nerve centers.

Since the centers are interconnected and since the afferent signals emitted by each of them are received by other centers (or by effectors), it follows that as the centers interact ("play the game") they strive to realize a situation of minimal afference. We call the principle of nerve center operation just described the *least-interaction principle.* (While the automata in the mathematical model minimize the "penalty" paid, nerve centers minimize the incoming afferent signals and therefore the interaction of nerve centers.)

A control system constructed on the basis of this principle is characterized by the fact that whatever the problem solved by the system as a whole, each center solves the same "particular", "private" problem, i.e. that of minimizing the incoming afferent signals. For every set of external conditions the task of the system is to bring itself into a state of minimal afference. Such a system could function in such a way that adjustments at higher afference levels are more frequent than at lower afference levels. For this reason each center is a steady state spends most of its time in states of minimal incoming afferent signals.

It is, of course, true that the behavior of each center is in the final analysis determined by the medium external to the system (to the organism). However, the center need not concern itself with this fact; it need only solve its own "particular" task of minimizing afference. It is "informed" about a change in medium due to a change in afferent activity either directly or through other centers which have altered their operating states ultimately as a result of changes in the external medium.

In solving their own "particular" problems, the centers endow the system as a whole with purpose. An important feature of the model just described is the fact that control over nerve centers of a given level is considerably simplified: once the payoff functions have been specified there is no need to "enter into the details of the private behavior" of these centers; they themselves will choose sequences of actions which will render the "penalty" minimal.

This method of control is, moreover, highly flexible. The structure of the system of payoff functions can be corrected by several nerve centers (having at least an indirect relationship to the task and not coordinated in their activity). The latter are thus able to refine the "fine structure" of behavior.

We were able recently to construct several examples of such behavior, some of which bear considerable resemblence to physiological concepts. Thus, the behavior of a pair of automata in the situation known as the "zero-sum game" in game theory is very similar to the so-called reciprocal relations in physiology. We also found it possible to model synergistic and other types of relationships.

The least-interaction principle introduces organization into a system of nerve centers, rendering its operation purposeful and economical. We note too that stable minimal-afference situations do not imply very little or no afferent activity. The entire system must be active, and this necessarily implies a certain level of afference. In coarse terms, while we try to expend as little energy as possible on the solution of each individual problem, our total productiveness depends on the over-all level of our activity, which must, of course, be high.

4. Much experimental (computer) work on models of nerve center structure which we have called "continuous control systems" has been done by N. N. Rikko and I. P. Lukaševič. We ourselves hope in the near future to devise a model of myocardium and respiratory center functioning. (We note in this connection the interesting experimental study of the earthworm axon by M. B. Berkinblit, Ju. I. Aršavskiĭ and S. A. Kovalev [8].) The cerebellum makes a very interesting object for such studies. The uniform structure of its cortex and the relative simplicity of its connections suggest that its functioning can be described by means of various models of continuous control systems. We are presently conducting an experimental study of this organ.

5. Materials on certain medical instruments based on ideas closely related to those which we have discussed are to be presented in Leningrad at a conference of the A. S. Popov Scientific and Engineering Society of Radio Engineering and Electrical Communications, whose theme will be the medico-biological applications of electronics.

BIBLIOGRAPHY

1. N. A. Bernšteĭn, *Trends in the development of physiology, and associated problems in cybernetics*, Biological Aspects of Cybernetics, Izdat. Akad. Nauk SSSR, Moscow, 1962, pp. 52–65. (Russian)

2. I. M. Gel'fand and M. L. Cetlin, *Some control methods for complex systems,* Uspehi Mat. Nauk **17** (1962), no. 1 (103), 3–25 = Russian Math. Surveys **17** (1962), no. 1, 95–117. MR **25** #1084.

3. G. M. Frank, *Self-regulation of cell processes,* Biological Aspects of Cybernetics, Izdat. Akad. Nauk SSSR, Moscow, 1962, pp. 33–44. (Russian)

4. I. M. Gel'fand and M. L. Cetlin, *Continuous models of control systems*, Dokl. Akad. Nauk SSSR **131** (1960), 1242–1245 = Soviet Math. Dokl. **1** (1960), 409–412. MR **22** #9316.

5. M. L. Cetlin, *On the behavior of finite automata in random media*, Automat. i Telemeh. **22** (1961), 1345–1354 = Automat. Remote Control **22** (1961), 1210–1219. MR **25** #4973.

6. M. A. Alekseev, M. S. Zalkind and V. M. Kušnarev, *Solution of a choice problem by humans with probabilistic reinforcement of motor reactions*, Biological Aspects of Cybernetics, Izdat. Akad. Nauk SSSR, Moscow, 1962, pp. 198-209. (Russian)

7. V. I. Varšavskiĭ, I. I. Voroncova and M. L. Cetlin, *Teaching stochastic automata*, Biological Aspects of Cybernetics, Izdat. Akad. Nauk SSSR, Moscow, 1962, pp. 192–197. (Russian)

8. Ju. I. Aršavskiĭ, M. B. Berkinblit and S. A. Kovalev, a) *Periodic transformation of rhythm in individual nerve fibers*, Biofizika **7** (1962), 449–459. (Russian)*

b) *The site where transformation of rhythm originates in a nerve fiber with artificially created heterogeneity,* ibid. 7 (1962), 619–623. (Russian)*

Translated by A. YABLONSKY

**Editor's note.* The cover-to-cover translation of this journal, Biophysics, was not published for this one volume.